Clean Energy Production Technologies

Series Editors
Neha Srivastava, Department of Chemical Engineering and Technology
IIT (BHU) Varanasi, Varanasi, Uttar Pradesh, India

P. K. Mishra, Department of Chemical Engineering and Technology
IIT (BHU) Varanasi, Varanasi, Uttar Pradesh, India

The consumption of fossil fuels has been continuously increasing around the globe and simultaneously becoming the primary cause of global warming as well as environmental pollution. Due to limited life span of fossil fuels and limited alternate energy options, energy crises is important concern faced by the world. Amidst these complex environmental and economic scenarios, renewable energy alternates such as biodiesel, hydrogen, wind, solar and bioenergy sources, which can produce energy with zero carbon residue are emerging as excellent clean energy source. For maximizing the efficiency and productivity of clean fuels via green & renewable methods, it's crucial to understand the configuration, sustainability and techno-economic feasibility of these promising energy alternates. The book series presents a comprehensive coverage combining the domains of exploring clean sources of energy and ensuring its production in an economical as well as ecologically feasible fashion. Series involves renowned experts and academicians as volume-editors and authors, from all the regions of the world. Series brings forth latest research, approaches and perspectives on clean energy production from both developed and developing parts of world under one umbrella. It is curated and developed by authoritative institutions and experts to serves global readership on this theme.

More information about this series at http://www.springer.com/series/16486

Neha Srivastava • Manish Srivastava •
P. K. Mishra • Vijai Kumar Gupta
Editors

Biofuel Production Technologies: Critical Analysis for Sustainability

Springer

Editors
Neha Srivastava
Department of Chemical Engineering & Technology
Indian Institute of Technology (Banaras Hindu University)
Varanasi, Uttar Pradesh, India

Manish Srivastava
Department of Physics & Astrophysics
University of Delhi
Delhi, India

Department of Chemical Engineering & Technology
Indian Institute of Technology (Banaras Hindu University)
Varanasi, Uttar Pradesh, India

P. K. Mishra
Department of Chemical Engineering & Technology
Indian Institute of Technology (Banaras Hindu University)
Varanasi, Uttar Pradesh, India

Vijai Kumar Gupta
School of Science, Department of Chemistry and Biotechnology
Tallinn University of Technology
Tallinn, Estonia

ISSN 2662-6861 ISSN 2662-687X (electronic)
Clean Energy Production Technologies
ISBN 978-981-13-8636-7 ISBN 978-981-13-8637-4 (eBook)
https://doi.org/10.1007/978-981-13-8637-4

This Springer imprint is published by the registered company Springer Nature Singapore Pte Ltd.
The registered company address is: 152 Beach Road, #21-01/04 Gateway East, Singapore 189721, Singapore

Foreword

Rising human population and growing emphasis on improving the quality of human life have increased the demand of food and energy. Efforts to provide food for all have yielded results through the use of agrochemicals and fertilizers and increase in the area of cultivable land. The current agricultural practices have become energy intensive. Direct energy needs of people in all countries have also increased. The energy availability and security has thus become an important priority for all countries and more so for developing countries. The currently used conventional energy sources are finite in nature, and their extensive use causes pollution. Scientists now have come to a unanimous conclusion that global increase in carbon dioxide and other gases and deforestation is responsible for increasing the stress on the global environment. More efficient use of energy through appropriate policies and demand management, implementation of better energy-conservation technologies, and the intensive development of renewable energy sources are being talked about.

Various renewable energy options are biomass, geothermal, hydro, ocean current, solar, tidal, and wind. It is estimated that the developed countries are meeting around 3–4% of their energy needs through renewable energy resources as against 18–19% of developing countries. Out of the various renewable energy options, biomass is a quite versatile resource as it can be used for supporting both energetic and non-energetic requirements of human society. The diversified nature of biomass resources is quite often expressed in terms of 6-F concept, namely, food, feed, fuel, fiber, fertilizer, and feedstock, the essential requirements of modern human society. Thus, it is amply clear that there are several competitive areas that will be controlling the demand for biomass and energy is one of these.

Lignocellulosic biomass can be used for producing biogas, bio-hydrogen, hythane, bio-butanol, and ethanol. Algal biomass and sewage sludge are being considered as suitable feedstocks for producing biodiesel and bio-oil. Out of all the candidate biomasses, lignocellulosic biomass has the greatest potential for use as feedstock for producing chemicals and fuels. This however requires selection of the candidate biomass and a suitable pretreatment technology to make the selected biomass as a suitable substrate for bioconversion to a fuel. Consistent research efforts in the past two decades have resulted in considerable improvement in the pretreatment and bioconversion processes, having high productivity, high titer value, and high yield as well as inexpensive downstream processing. However, consistent efforts are still required to make biomass-based chemicals and biofuel production routes economically comparable to petrochemical-based routes.

The book entitled ***Biofuel Production Technologies: Critical Analysis for Sustainability*** edited by Dr. Neha Srivastava, Dr. Manish Srivastava, Prof. P. K. Mishra, and Dr. Vijai Kumar Gupta is focused on the different biofuels, including biogas, bio-alcohols (butanol, ethanol), biohydrogen, and biodiesel. It makes an attempt to present a summary of the state of the art as it exists today. Various factors that control the nature of biofuel production technologies and their scale are also included as and where required. The 12 intensive chapters comprising this book focus on various available production technologies together with the strategies required for economical production of various biofuels. The current bottlenecks impeding the wide-scale exploitation of available technologies are also discussed together with future perspectives in each case. The book is likely to be a valuable reference for academicians, researchers, students, and professionals interested in the area of biofuel production.

Ex-Director, IT-BHU Varanasi
Varanasi, India

Emeritus Professor, IIT (BHU) Varanasi
Varanasi, India

S. N. Upadhyay

Acknowledgments

The editors are thankful to all the academicians and scientists whose contributions have enriched this volume. We also express our deep sense of gratitude to our parents whose blessings have always prompted us to pursue academic activities deeply. It is quite possible that in a work of this nature, some mistakes might have crept in text inadvertently, and for these, we owe undiluted responsibility. We are grateful to all the authors for their contribution to present this book. We are also thankful to Springer Nature for giving this opportunity to editors and the Department of Chemical Engineering & Technology, IIT (BHU), Varanasi, UP, India, for all their technical support. We thank them from the core of our heart. Editor Manish Srivastava acknowledges the Department of Science and Technology (DST), Government of India, for awarding the DST-INSPIRE Faculty Award [IFA13-MS-02] 2014, and also Science and Engineering Research Board for SERB-Research Scientist Award-2019.

Contents

About the Editors

Dr. Neha Srivastava is currently working as postdoctorate fellow in the Department of Chemical Engineering and Technology, IIT (BHU), Varanasi, India. She has published 25 research articles in the peer-reviewed journals with three patents and one technology transfer. She has completed her PhD from the Department of Molecular and Cellular Engineering, SHIATS, India, in 2016 in the area of bioenergy and has received six Young Scientist Awards. Presently, she is working on biofuel production (cellulase enzymes, production and enhancement, biohydrogen production from waste biomass, bioethanol production). Her **fields of expertise** are on biofuel production, microbial bioprocessing, and enzyme technologies.

Dr. Manish Srivastava has worked as DST INSPIRE faculty in the Department of Physics and Astrophysics, University of Delhi, India, during June 2014 to June 2019. Currently, he is working as SERB-Research Scientist in the Department of Chemical Engineering and Technology, IIT (BHU), Varanasi, India. He has published 45 research articles in the peer-reviewed journals, authored several book chapters, and filed one patent. He worked as a postdoctorate fellow in the Department of BIN Fusion Technology, Chonbuk National University from August 2012 to August 2013. He was an assistant professor in the Department of Physics, DIT School of Engineering, Greater Noida, from July 2011 to July 2012. He received his PhD in Physics from the Motilal Nehru National Institute of Technology, Allahabad, India, in 2011. Presently, he is working on the synthesis of graphene-based metal oxide hybrids and their applications as catalysts. His areas of interest are synthesis of nanostsructured materials and their applications as catalyst for the development of electrode materials in energy storage, biosensors, and biofuel production. His **fields of expertise** are synthesis of nanomaterials and their application as catalyst for the development of electrode materials in energy storage, biosensors, and biofuel productions.

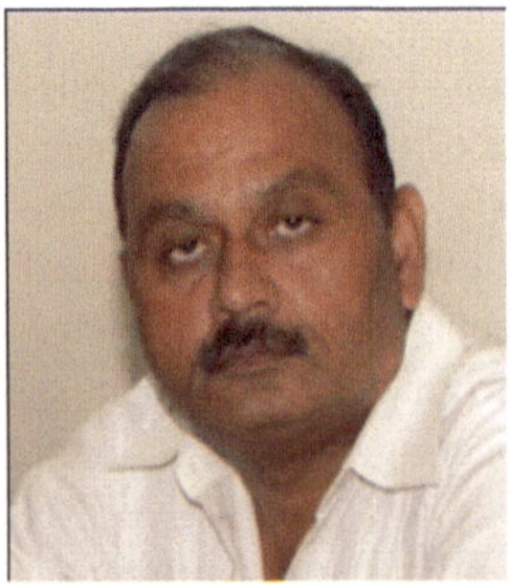

Professor P. K. Mishra is currently professor and head in the Department of Chemical Engineering and Technology, Indian Institute of Technology, Banaras Hindu University, Varanasi, India, where he obtained his PhD degree in Chemical Engineering in 1995. He has authored and coauthored over 60 technical papers published in reputed national and international journals and supervised more than 20 doctoral students. He has received several awards at national and international levels and has five patents with one technology transfer. He has also made significant contribution toward the development of entrepreneurship ecosystem in Eastern part of the country. He is fellow of the Institution of Engineers (India), coordinator in Technology Business Incubator at the Indian Institute of Technology, Banaras Hindu University, and member of the executive committee NISBUD, Ministry of Skill Development, Government of India. His **fields of expertise are** biofuel production, microbial bioprocessing, and enzyme technologies.

Dr. Vijai Kumar Gupta, ERA chair of Green Chemistry, Department of Chemistry and Biotechnology, School of Science, Tallinn University of Technology, Tallinn, Estonia, is one of the leading experts in the area of microbial biology and biotechnology. He is the member of the International Subcommission on Trichoderma and Hypocrea, Austria, and International Society for Fungal Conservation, UK, and secretary of the European Mycological Association. He is the fellow of the prestigious The Linnaean Society of London, UK, of Indian Mycological Association, and of Mycological Society of India. He has been honored with several awards in his career, including Indian Young Scientist

Award, for his advanced research achievements in the field of fungal biology and biotechnology. He is the editor of few leading scientific journals of high repute, having many publications in his hands with h-index 21. He has edited many books for publishers of international renown such as CRC Press; Taylor & Francis, USA; Springer, USA; Elsevier Press, the Netherlands; Nova Science Publishers, USA; Walter de Gruyter, Germany; and CABI, UK.

Biofuels: Types and Process Overview

1

Pietro Bartocci, Roman Tschentscher, Yunjun Yan, Haiping Yang, Gianni Bidini, and Francesco Fantozzi

Abstract

The term biofuels refer mainly to fuels derived from biomass, which can be considered as plants and organic residues. In this chapter attention will be focused on liquid biofuels that can be used mainly for transportation. As reported in the IEA Technology Road Map for biofuels, presented in 2011, they can be divided in two main categories, based on the type of technologies used: conventional biofuels (sugar- and starch-based ethanol, conventional biodiesel, biogas) and advanced biofuels (cellulosic ethanol, hydrotreated vegetable oil, biomass-to-liquids, biosynthetic syngas, etc.). The production of these biofuels is object of big research efforts directed through process intensification and increase of the efficiency of biomass conversion into an energy vector. For this reason this chapter takes into account the production of first-generation biodiesel, first-generation bioethanol, second-generation biodiesel, second-generation bioethanol, and hydrotreated vegetable oils focusing on their market and the most importantly production techniques.

P. Bartocci (✉) · G. Bidini · F. Fantozzi
Department of Engineering, University of Perugia, Perugia, Italy
e-mail: bartocci@crbnet.it

R. Tschentscher
SINTEF Energy Research, Oslo, Norway

Y. Yan
Key Lab of Molecular Biophysics of Ministry of Education, College of Life Science and Technology, Huazhong University of Science and Technology, Wuhan, China

H. Yang
State Key Laboratory of Coal Combustion, Huazhong University of Science and Technology, Wuhan, Hubei, China

China-EU Institute for Clean and Renewable Energy, Huazhong University of Science and Technology, Wuhan, Hubei, China

N. Srivastava et al. (eds.), *Biofuel Production Technologies: Critical Analysis for Sustainability*, Clean Energy Production Technologies,
https://doi.org/10.1007/978-981-13-8637-4_1

Keywords
Biofuels · Biodiesel · Bioethanol · Vegetable oils · BTL · Fischer-Tropsch

1.1 Introduction to Biofuels

Compared to gasoline, diesel, and natural gas, alternative liquid biofuels derived from biomasses have one main selling point: they are renewable. While there are significant differences among liquid biofuels with regard to production, all are argued to have a lower environmental impact at both the extraction and consumption stages (Renewable Fuels Association 2015; Skutsch et al. 2011; Slade and Bauen 2013). Thus, while biofuels are economically marginal in the marketplace, they are socially and politically useful (Solomon et al. 2007).

The major benefits of biofuels by an economic, environmental, and energetic point of view are shown in Table 1.1.

Biomasses, based on their composition, can be divided into three main categories: sugar/starch crops, lignocellulosic biomass, and oil plants. The composition and main characteristics of these feedstocks will be better explained in the next paragraph. Figure 1.1 shows the most important conversion processes to produce biofuels from biomass. From sugar/starch crops, bioethanol can be produced through milling, hydrolysis, fermentation, and refining. Bioethanol can be also produced with similar processes from lignocellulosic materials. Lignocellulosic biomass can

Table 1.1 Benefits linked with the use of biofuels (Demirbas 2009a)

Economic impacts	Sustainability
	Fuel diversity
	Increased number of rural manufacturing jobs
	Increased income taxes
	Increased investments in plant and equipment
	Agricultural development
	International competitiveness
	Reducing the dependency on imported petroleum
Environmental impacts	Greenhouse gas reductions
	Reducing of air pollution
	Biodegradability
	Higher combustion efficiency
	Improved land and water use
	Carbon sequestration
Energy security	Domestic targets
	Supply reliability
	Reducing use of fossil fuels
	Ready availability
	Domestic distribution
	Renewability

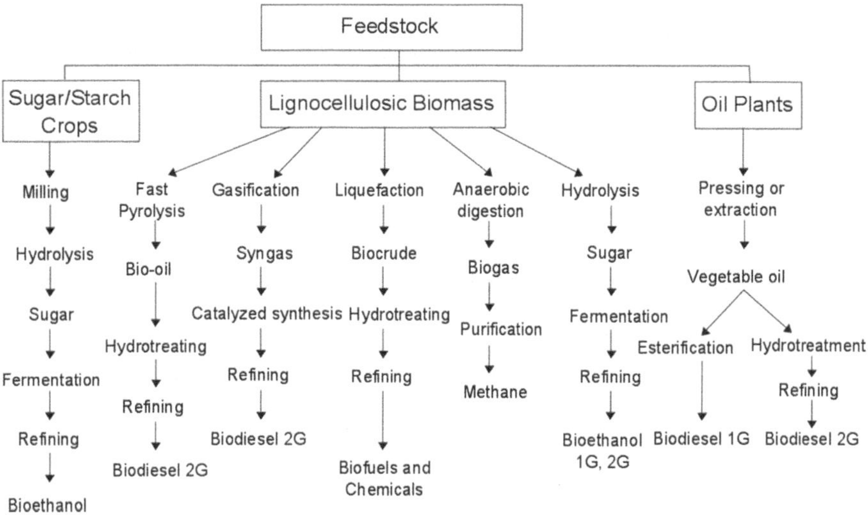

Fig. 1.1 Most important conversion processes to produce biofuels from biomass (Demirbas 2009b)

be converted also through pyrolysis, gasification, hydrothermal liquefaction, and anaerobic digestion. Vegetable oils can be converted through transesterification but also through hydrotreating, producing hydrotreated vegetable oils. This chapter will take into consideration the following biofuels:

- Biodiesel
- Bioethanol
- BTL (Biomass to Liquids)
- HVO (Hydrotreated Vegetable Oils)

These have been selected among the existing ones because they are believed to have higher market potential.

As it is reported in the World Energy Outlook 2018 of the International Energy Agency (IEA), transport accounts for a fifth of global energy demand and is responsible for a quarter of energy-related CO_2 emissions. More than 95% of today's transport sector emissions are from oil (IEA 2018) and the demand for the transport of people and of goods is projected to increase significantly through to 2040.

Global transport biofuel consumption has increased by more than 5% in 2017 and has reached 150 billion liters, of which three-quarters is ethanol. In energy terms, biofuel consumption is about 86 Mtoe, of which two-thirds is ethanol. Biofuel promotion policies are now in place in 68 countries. While large volumes of advanced biofuels could be produced sustainably, their development has been slowed by their costs. In fact, producing a barrel of second-generation biodiesel can cost around \$140/barrel today (IEA 2018). Assuming that advanced biofuels are not responsible of net CO_2 emissions, a carbon tax above \$150 per ton of CO_2 would

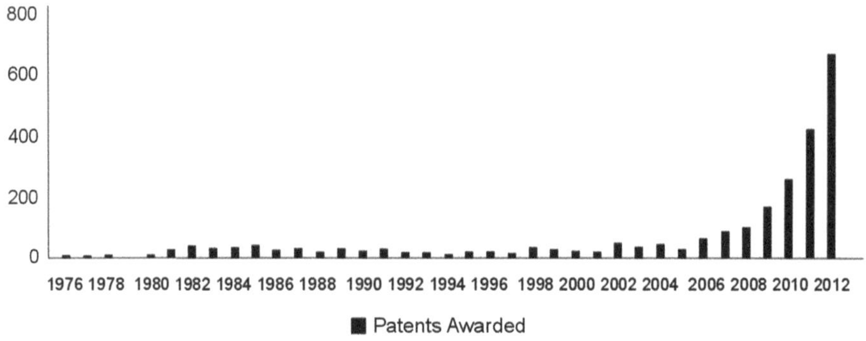

Fig. 1.2 Patents awarded for all biofuels technologies, N = 2587 (Arnold et al. 2019)

be required for them, to be cost-competitive with fossil ones (IEA 2018). Production costs have to be reduced through technological innovation. Continuous innovation must provide constant or increasing returns to innovative efforts, but complexity can increase the costs of those efforts (Costantini et al. 2013, 2015a). In the analysis of (Arnold et al. 2019), an innovation is considered to be a technical novelty that earns a patent. Using the data provided by the United States Patent and Trademark Office (USPTO), a database of liquid biofuel technologies patented since 1976 has been constructed (Arnold et al. 2019).

Figure 1.2 shows that biofuel technologies of all generations show a low level of innovation from the start of the data series through 2005, while from 2006 the level of innovation has risen and continues to rise consistently. This rise in patenting parallels the results seen in other international studies. In fact, biofuel patents rose first in Japan, in the period 1994–2002, and then increased in Europe in 2004 and in the United States in 2005 (Albers et al. 2016). The reasons for this increase can be found in market forces, concerns over supply, price, and air quality.

From 1976 through 2012, the number of patents per author has declined from 0.64 patents per author in 1976 to 0.33 patents per author in 2012. Thus it can be concluded that the productivity of innovation in biofuels has also declined. Besides, the increase in patenting from 2006 on did not affect the trends in patents productivity. The decline of productivity of innovation is evident in both newer technological areas (such as information technology, biotechnology, and nanotechnology) and in older sectors (Strumsky et al. 2010; Tainter et al. 2018). This appears to be the result of increasing complexity in the research process (Strumsky et al. 2010; Tainter et al. 2018). As Kessler and Sperling (Kessler and Sperling 2016) noted, second-generation biofuel technologies are surely more complex than those first-generation ones (Himmel et al. 2007). This implies that nowadays biofuel innovation requires increasing diversity of technical knowledge and a multidisciplinary approach. It is now difficult for a single researcher to master all of the technologies that make up a biofuel patent (Costantini et al. 2015b). As a result, it requires the collaboration of increasing numbers of researchers to develop a patent, who work in interdisciplinary teams (Albers et al. 2016).

1.2 Feedstock

1.2.1 Vegetable Oils

The main feedstock for first-generation biodiesel production worldwide includes oils from energy crops (such as soybean, rapeseed, canola, sunflower, corn, palm kernels, animal fats) and recycled oil. *Jatropha curcas L.* oil has also been used for biodiesel production in tropical areas such as India and Africa.

Vegetable oils from energy crops (which are also often commercialized in the food sector) are considered high-quality materials for biodiesel production because they have a high triglyceride content (92–99%) and low FFA content (<2%). Vegetable oils produced from soybean and rapeseed are the most commonly used feedstock for biodiesel production in the United States and Europe, respectively.

Waste oil is recycled cooking oil used in restaurants, food industry, and households. It contains usually more free fatty acids and water and less triglycerides than fresh vegetable oils. The typical composition of waste oil includes linoleic acid (53%), oleic acid (28%), and palmitic acid (11.73%) (Shah et al. 2007). Because of its high FFA concentration and water content, waste oil usually needs a pretreatment to remove water and transform FFAs to esters. The production of waste oils worldwide is significant; in 2007, in fact more than 15 million tons have been generated in the world (Gui et al. 2008). Vegetable oils composition and main characteristics are proposed in Table 1.2 (Leung et al. 2010).

1.2.2 Starch and Sugars

Ethanol derived from biomass has the potential to be a sustainable transportation fuel that can replace gasoline (Wang 2000; Kim and Dale 2004). Ethanol can be produced from sugar- or starch-containing crops (see Table 1.3) and lignocellulosic biomass (such as agricultural residues, herbaceous crops, forestry wastes, wastepaper, and other wastes) (Wyman 1996). The production of bioethanol from lignocellulosic biomasses is still under development. The composition of lignocellulosic biomasses is presented in the next paragraph.

1.2.3 Lignocellulosic Biomass

Lignocellulosic biomass composition derives directly from the composition of the plant cell wall (Caffall and Mohnen 2009). The lignocellulosic feedstock is represented by the agricultural and forest residuals, which are mainly composed by the cell wall tissue, which remains after the plants have died. Plant cell wall biomass contains mainly cellulose, hemicellulose, and lignin. Different species of plants have significant differences in the proportions of the main components and important differences in the types of hemicellulose which are contained and the ratios of

Table 1.2 Feedstock for biodiesel production (Leung et al. 2010)

Type of oil	Species	Fatty acids composition (wt. %)	Kinematic viscosity (cst. at 40 °C)	Acid value (mg KOH/g)
Edible oil	Soybean	C16:0, C18:1, C18:2	32.9	0.2
	Rapeseed	C16:0, C18:0, C18:1, C18:2	35.1	2.92
	Sunflower	C16:0, C18:0, C18:1, C18:2	32.6	–
	Palm	C16:0, C18:0, C18:1, C18:2	39.6[a]	0.1
	Corn	C16:0, C18:0, C18:1, C18:2, C18:3	34.9[a]	–
	Canola	C16:0, C18:0, C18:1, C18:2, C18:3	38.2	0.4
Nonedible oil	Jatropha curcas	C16:0, C16:1, C18:0, C18:1, C18:2	29.4	28
Other	Used cooking oil	Depends on fresh cooking oil	44.7	2.5

[a]Kinematic viscosity at 38 °C, mm^2/s

Table 1.3 Starch content in energy crops used for 1st generation bioethanol (Zabed et al. 2017)

Crop	Scientific name	Starch content (%)[a]
Corn	*Zea mays*	70–72
Sorghum	*Sorghum bicolor*	68–70.7
Wheat	*Triticum aestivum*	65.3–76
Rice	*Oryza sativa*	87.5
Oat	*Avena sativa*	65.6
Potato	*Solanum tuberosum*	73

[a]Dry weight

monomers in lignin (Pauly and Keegstra 2010). The composition in terms of main components of the most important lignocellulosic feedstocks is shown in Table 1.4.

Woody biomass contains more cellulose and lignin, whereas grass biomass has higher content of hemicellulose (mainly xylan), extractives, and ashes.

Cellulose is a polysaccharide consisting of a linear chain of D-glucose units.

Hemicellulose has a backbone composed of 1, 4-linked β-D-hexosyl residues and may contain pentoses, hexoses, and/or uronic acids. Other sugars, such as rhamnose and fucose, may also be present, and the hydroxyl groups of sugars can be partially substituted with acetyl groups (Gírio et al. 2010). Unlike cellulose, hemicellulose composition varies depending on cell tissue and plant species (Chundawat et al. 2011). In fact it can be noted that:

- The principal hemicellulose of hardwoods is an O-acetyl-4-O-methylglucuronoxylans.
- The main hemicellulose of soft woods is an O-acetylgalactoglucomannan.

Table 1.4 Main chemical composition of some lignocellulosic feedstocks (Zhao et al. 2012)

Feedstock	Cellulose (%)	Xylan (%)	Galactan (%)	Araban (%)	Lignin (%)	Mannan (%)	Extractives (%)	Ash (%)
Hybrid poplar	48.6	14.6	0.3	0.3	21.8	0.5	NA*	0.7
White oak	43.6	18.0	0.4	2.4	23.2	2.9	NA	0.6
Red oak	43.4	18.9	NA*	1.9	25.8	2.7	NA	NA
Walnut	46.2	16.5	NA	1.8	21.9	2.6	NA	NA
Maple	44.9	17.3	NA	2.8	20.7	2.9	NA	0.4
Corn stover	40.9	18.0	1.0	3.0	16.6	0.6	7.3	9.7
Wheat straw	38.2	21.2	0.7	2.5	23.4	0.3	13.0	10.3
Rice straw	34.2	24.5	NA	NA	11.9	NA	17.9	16.1
Switchgrass	31.0	20.4	0.9	2.8	17.6	0.3	17.0	5.8

*NA: data not available

- The main hemicellulose in Gramineae (such as cereal straws) is arabinoxylans, which are similar to harwoods xylan, but the amount of L-arabinose is higher (Peng et al. 2011).
- Lignin is the organic substance which is responsible of binding the cells (Sticklen 2008). The three basic monomeric units constitute lignin: p-Hydroxyphenyls (H); Guaicyls (G); Syringyls (S).

Hardwood lignins are predominantly G and S monolignols with trace amounts of H units. Soft wood lignins are composed of mostly G units. Herbaceous plants contain all three units in significant amounts (Chundawat et al. 2011; Buranov and Mazza 2008).

1.3 First-Generation Biodiesel

1.3.1 Transesterification Reaction

The transesterification reaction with alcohol is represented by the general equation shown in Fig. 1.3a which consists of a number of consecutive, reversible reactions. These are shown in Fig. 1.3b. The first step is the conversion of triglycerides into diglycerides, and then diglycerides are converted into monoglycerides and monoglycerides into FFAs and glycerol. Each step yields one methyl ester molecule (Freedman et al. 1986; Noureddini and Zhu 1997). The transesterification can be both catalyzed by acid and alkali.

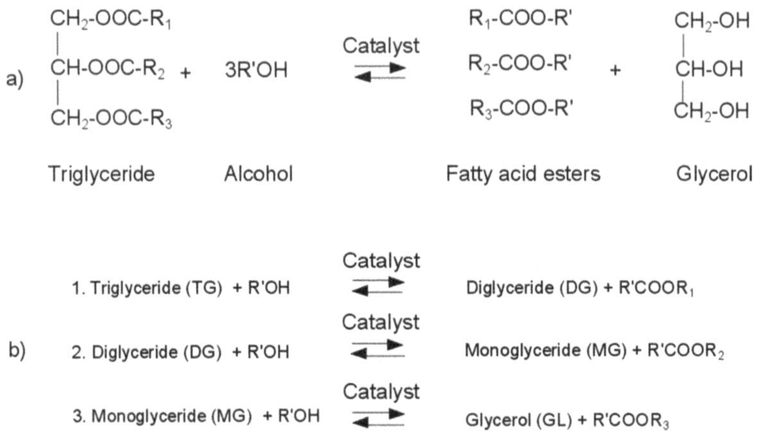

Fig. 1.3 Transesterification reaction (Eckey 1956)

Different types of catalyzed transesterification can be adopted (Lam et al. 2010):

1. Homogeneous base catalyst
2. Heterogeneous base catalyst
3. Homogeneous acid catalyst
4. Heterogeneous acid catalyst
5. Enzymes

State-of-the-art biodiesel production usually is done using base catalyst (e.g., KOH or NaOH). In that case reaction time can vary between 0.33 and 2 hours and yields are between 88% and 98% (Lam et al. 2010; Liu et al. 2010). Catalyst loading is between 1 and 6%wt, while methanol excess is between 7:1 and 9:1, expressed in molar ratio. Temperature can range between 60 and 87 °C (Lam et al. 2010). Also two-step catalysis can be a solution: first acid catalyst, then followed by basic catalysis.

Homogeneous acid catalyst usually can use H_2SO_4 or HCl. Heterogeneous basic catalysis can be performed using basic zeolites, alkaline earth metal oxides (e.g., CaO), and hydrotalcites. Heterogeneous acid catalyst can be zirconium oxide (ZrO_2), titanium oxide (TiO_2), tin oxide (SnO2), zeolites, sulfonic ion-exchange resin, sulfonated carbon-based catalyst, and heteropoly acids (HPAs) (Lam et al. 2010).

Enzyme catalysts lipases can be produced from several microorganisms, such as *Mucor miehei* (Lipozyme IM 60), *Pseudomonas cepacia* (PS 30), *C. antarctica* (Novozyme 435), *Bacillus subtilis*, *Rhizopus oryzae*, and *Penicillium expansum* (Lam et al. 2010; Liu et al. 2010, 2011a, b; Yan et al. 2014; Fan et al. 2016, 2017; Su et al. 2016; Li et al. 2017).

1.3.2 Biodiesel Production Process Diagram

Figure 1.4 shows the flow sheet of the production processes used for biodiesel.

Based on Fig. 1.4 scheme, it is assumed that alcohol, catalyst, and oil are inserted in the reactor and agitated for approximately 1 h at 60°C. Small plants often use batch reactors (Stidham et al. 2000), while larger plants (higher than 4 million liters/year) use continuous stirred-tank reactors (CSTR) or plug flow reactors operated in continuous mode (Assman et al. 1996).

Once methyl ester has been produced, it must be separated from the glycerol (through phase separation because glycerol is much heavier). Methyl esters undergo a neutralization step and then pass through a methanol stripper. A vacuum flash process or a falling film evaporator can be used for this purpose. Before washing with water, acid is added to the biodiesel to neutralize any residual catalyst and to split any soap that may have formed during the reaction. Soaps react with the acid to form water-soluble salts and FFAs. The salts are removed during the water washing step. The water washing step removes traces of catalyst, soap, salts, methanol, or free glycerol remained in the biodiesel. After the wash process, remaining water is removed from the biodiesel by flash-vacuum distillation.

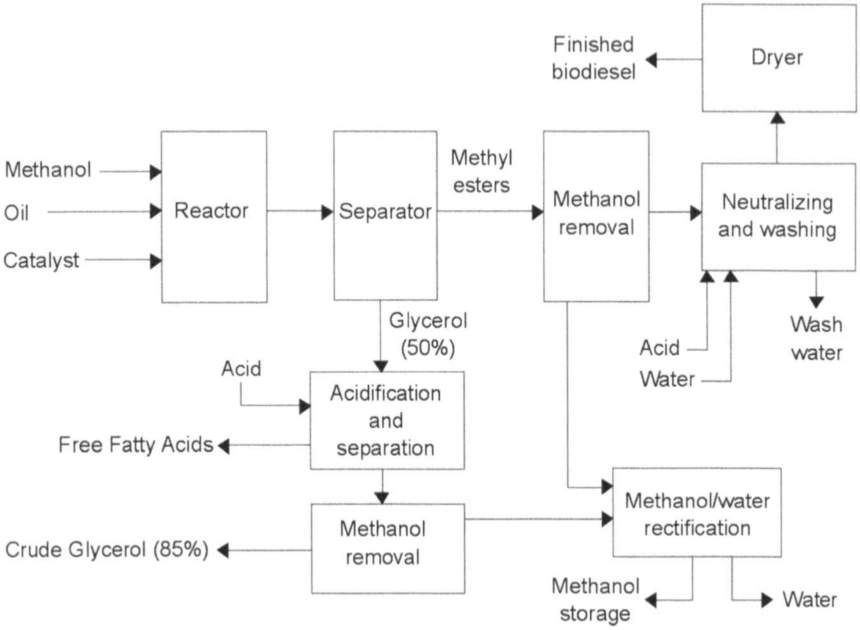

Fig. 1.4 First-generation biodiesel production flow sheet (Van Gerpen 2005)

The glycerol stream leaving the separator is composed only by about 50%wt glycerol. It contains also methanol, catalyst, and soap. The first step in glycerol refining is to add acid to split the soaps into FFAs and salts. The free fatty acids are not soluble in the glycerol and will rise to the top, so that they can be removed and recycled. The salts generally remain with the glycerol. The methanol mixed with the glycerol is removed by flash-vacuum distillation or conventional distillation. A purity of approximately 85% is reached; this allows to sell the glycerol to a refiner. Glycerol refining can be performed using vacuum distillation or ion exchange processes. A purity comprised between 99.5%wt and 99.7%wt is achieved.

The methanol that is removed from the methyl ester and from glycerol is mixed with water after separation has been performed. This water should be removed in a distillation column before the methanol is recycled into the process.

1.4 Bioethanol

1.4.1 Production of Ethanol from Sucrose

Sugar cane, sugar beet, and sweet sorghum are crops which contain sugars, which can be used as feedstock for ethanol production. Their main advantages are high yield of sugar per hectare and low conversion costs.

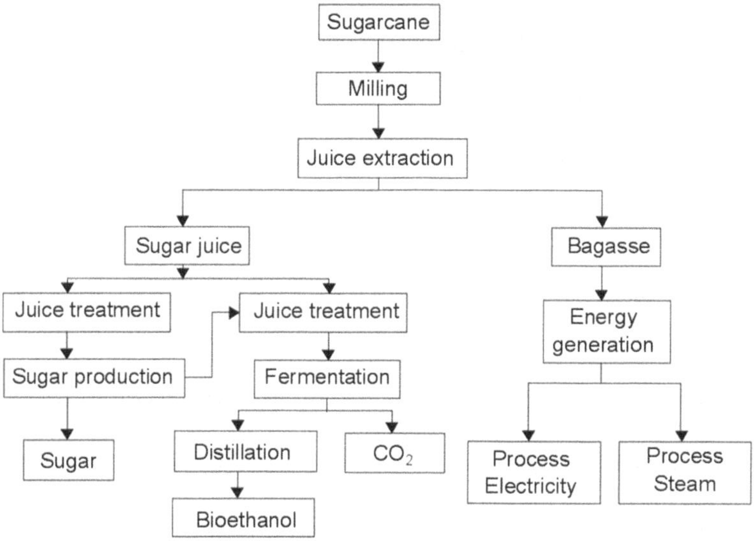

Fig. 1.5 Bioethanol production from sugar juice (Vohra et al. 2014)

Sugarcane has become a very effective source of biofuel, given that:

1. Bagasse can be used to generate process steam and electricity.
2. Vinasse (the liquid effluent) can be used as a fertilizer and irrigation supply to the cane fields (Kojima and Johnson 2005).

Sugar cane must be processed within 24–72 h after being harvested. Sugar is first extracted by crushing the stalks, to release the juice (Fig. 1.5). Calcium hydroxide is then added to precipitate the fiber and the sludge, and the mixture is then filtered. The filtrate solution is evaporated to concentrate and crystallize the sugar. The sugar is removed by centrifugation.

The sugar which is not crystallized and the accompanying salts are concentrated to form a syrup called molasses. These are used to produce ethanol (Kojima and Johnson 2005). To achieve the optimum fermentation efficiency of yeast *Saccharomyces cerevisiae*, the sugar content in the molasses has to be adjusted in the range of 14–18%wt. The typical temperature of the fermentation process is about 33–35 °C, while cell density is about 8–17% (v/v). Cell recycle system can be used to concentrate yeasts and recycle them into the process, obtaining high cell densities, which shorten fermentations to 6–10 h (Wyman 2004). Fermentation is interrupted at concentration of approximately 10% (v/v) ethanol. Fermentation reaction is the following:

$$C6H12O6 \rightarrow 2C2H6O + 2CO2 \quad (1.1)$$

The next step is represented by distillation and rectification. An azeotropic solution of 95% (v/v) ethanol is obtained. Further concentration to absolute ethanol is finally achieved by molecular sieves or azeotropic distillation (using benzene or cyclohexane) (Chiaramonti 2007).

1.4.2 Bioethanol Production from Starch

As reported in Table 1.3, grains (corn, wheat, or barley) mainly provide starch. This is made up of long chains of glucose units. The amylose structure contains 1000 monomeric units, while the amylopectin structure contains 1000–6000 units. Starch is the most utilized for ethanol production in North America and Europe. To produce ethanol it is necessary to hydrolyze the starch into monomers. The hydrolytic reaction is catalyzed by glucoamylase enzyme. D-glucose, which is an isomer of glucose, is obtained as final hydrolysis product. Enzymatic hydrolysis is then followed by fermentation, distillation, and dehydration to yield anhydrous ethanol (Kumar et al. 2010).

There are two distinct methods for processing corn: wet milling and dry milling. Dry mills are usually smaller in size and are built primarily to produce only ethanol. Wet mill facilities also produce a list of high-valued co-products such as high-fructose corn syrup, corn oil, and corn gluten.

Corn dry-milling process is carried out in five steps (Vohra et al. 2014):

(i) Biomass handling (milling)
(ii) Liquefaction
(iii) Hydrolysis (saccharification)
(iv) Fermentation
(v) Distillation and recovery

In dry-grind process, the corns are milled to a powder and heated with water at 85°C (Kojima and Johnson 2005). Then hot water and alpha-amylase enzymes are added and the mixture is heated at 110–150 °C for an hour. This causes the liquefaction of starch. When liquefaction is completed, the mixture is cooled down and glucoamylases are added to produce dextrose. In dry-grind milling plants, often the glucoamylases are directly added into the fermentor. The process is known as "simultaneous saccharification and fermentation" (SSF) (Fig. 1.6).

In the fermentation process, yeasts convert glucose into ethanol and carbon dioxide. The process is completed in about 40–50 h. During fermentation, the mash is continuously mixed and it is cooled down. The beer obtained from fermentation is transferred to the distillation columns where ethanol is separated from the stillage (Singh et al. 2001). The stillage contains protein, oil, and fiber and are dried to obtain dried grains with solubles (DDGS) or just distillers dried grains (DDG). DDGS contain the process syrup combined with the solids, while DDG don't contain it.

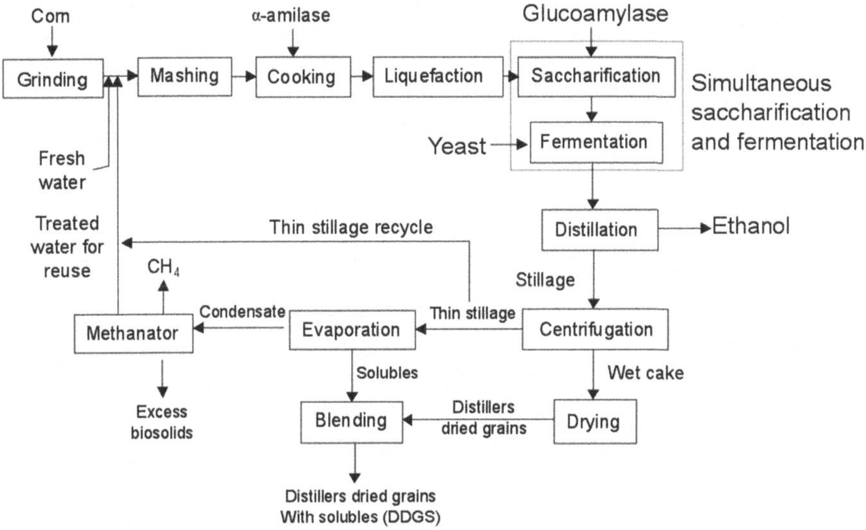

Fig. 1.6 Corn dry milling process flow diagram (Vohra et al. 2014)

In wet milling the shelled corns pass through mechanical cleaners designed to remove unwanted material, such as pieces of cobs, sticks, husks, meal, and stones. The cleaned corns are then fed into "steep" tanks, where they are soaked in dilute sulfuric acid and remain for 24 to 48 h at a temperature of about 52°C. Steeping softens the kernel and helps to break down the protein holding the starch particles. Generally, water drained from the steep tank, called "light steep water" contains about 6% of the original dry weight of the grains and is discharged to multiple-effect evaporators. The solids from steep water are rich in protein and are concentrated to 30–55% solids. The resulting steeping liquor can be sold as animal feed (May 1994) (Fig. 1.7).

The germ is removed from the steeped corn in the degerminating mills, which break the kernel to free the germ, the starch, and the gluten. The germ is separated in liquid cyclones from the mixture of fiber, starch, and gluten. It is then washed, dewatered, dried, and further processed to extract corn oil (Bothast and Schlicher 2005).

The starch and gluten are separated from the fiber by further washing, grinding, and screening operations. The solids and the fiber are used as a feed. The starch is separated from the gluten by centrifugation (May 1994; Bothast and Schlicher 2005). Once the pure starch slurry is obtained, the wet-mill process is similar to that of dry milling. First, the pH of the slurry is adjusted to 5.8–6.2 with calcium hydroxide, and then alpha-amylase is added to convert the starch into soluble short-chain dextrins (liquefaction). Calcium is often added (20–100 ppm) to enhance enzyme stability.

The slurry from the liquefaction stage is mixed with heat-sterilized steep water and sent for saccharification. The steep water provides both the fermentation

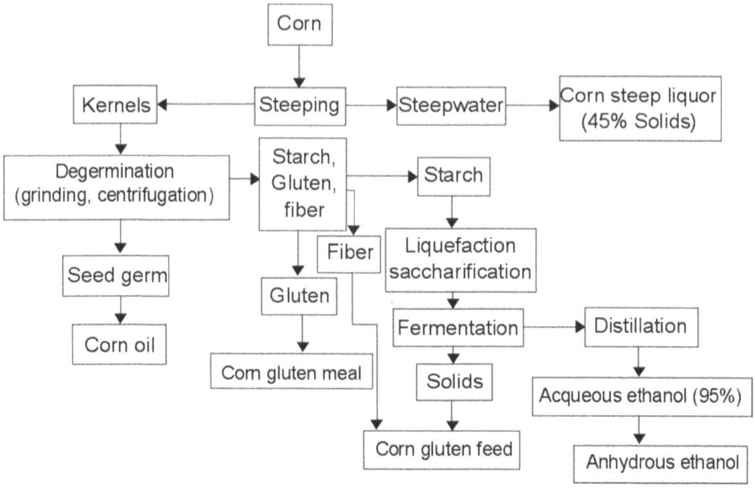

Fig. 1.7 Corn wet milling process flow diagram (Vohra et al. 2014)

nutrients and pH adjustment for saccharification, in which the glucoamylase converts the dextrins to glucose at a pH of 4.5 and a temperature of 65 °C. Then *S. cerevisiae* is added to ferment the sugars to ethanol and CO_2. The total fermentation time varies from 20 to 60 h. The final product from a continuous process will have an ethanol content of 8–10%v (Kojima and Johnson 2005; Bothast and Schlicher 2005).

1.4.3 Bioethanol Production From Lignocellulosic Feedstock

Bioethanol produced from lignocellulosic materials is commonly known as second-generation bioethanol. There have been huge research efforts in developing economically feasible advanced technologies for ethanol production; however, some challenges are still remaining (Cheng and Timilsina 2011). Chemical composition of lignocellulosic materials is the key factor affecting efficiency of biofuel production. Cellulose and hemicellulose are more present in hardwoods (78.8%) than softwoods (70.3%), while lignin is more present in softwoods (29.2%) than hardwoods (21.7%) (Balat 2011). According to (Vohra et al. 2014), the technologies for the conversion of lignocellulosic feedstocks to ethanol can be grouped into two broad macrocategories:

- The sugar platform (biochemical conversion)
- The syngas platform (thermochemical conversion)

The sugar platform uses enzymes to convert lignocellulosic biomass materials into sugars, while the syngas platform gasifies biomass and converts syngas into ethanol (Datta et al. 2011).

The biochemical platform consists of three main processes (Cotana et al. 2015; Barbanera et al. 2018; Buratti et al. 2015, 2018; Cavalaglio et al. 2016):

- Pretreatment
- Enzymatic hydrolysis
- Fermentation

1.4.3.1 Pretreatment

The pretreatment process significantly affects all the downstream processes and ultimately influences the overall biofuel yield and cost.

Pretreatment step can be performed through biological, physical, and chemical processes or a combination of them. Chemical methods use dilute acids (such as sulfuric or hydrochloric acid), alkalis (such as calcium hydroxide), and liquid ammonia (the ammonia fiber explosion pretreatment), while a physical method is represented by steam explosion (Ruane et al. 2010).

Pretreatment with dilute acid and intermediate temperatures is generally considered quite cost-effective. It loosens the cell wall matrix through degradation of hemicelluloses. Lignin is unaffected by this process. Accessibility to cellulose microfibrils is increased to provide a higher yield of sugars for fermentation. Acid treatment will result in other high-value products like furfural, hydroxyl-methyl furfural (HMF), phenolics, aldehydes, and aliphatic compounds. These products have to be removed before using the residues for further biochemical treatments. Acid pretreatment processes have to be followed by neutralization and detoxification (Kurian et al. 2013).

Steam explosion is the physical treatment where the lignocellulosic biomass is subjected to high pressures and temperatures for short duration, followed by the rapid decrease to atmospheric pressure, which will break the polymeric bonds in the substrate. Temperatures can range between 180 and 250 °C, pressures can range between 1 and 5 MPa (Jacquet et al. 2011).

Steam explosion has the following advantages:

- Lower capital investment
- Significantly lower environmental impact
- More potential for energy efficiency
- Less hazardous process conditions
- Complete sugar recovery

To compare steam explosion conditions, the severity factor has to be taken into account, defined as (Li et al. 2005)

$$S_0 = \log\{\exp[(T - 100)/14.75]t\} \qquad (1.2)$$

where T is the temperature (°C) and t is the duration of treatment (min).

Steam explosion is considered the most cost-effective option for hardwood and agriculture residues, while it is less effective for softwood. Acid catalysis can be used also within the steam explosion treatment and is found to reduce the temperature and the retention time. Another advantage is that complete hydrolysis of hemicellulose can be achieved (Mood et al. 2013).

1.4.3.2 Hydrolysis

During the hydrolysis, polysaccharides are broken down to simple sugars. Two examples of hydrolysis methods of cellulose into glucose are (Lynd et al. 2002):

1. Concentrated acid (H_2SO_4 30–70%, 40 °C, a few hours to achieve >90% glucose yields)
2. Enzymatic hydrolysis (cellulase mixture, 50 °C several days to reach 75–95% glucose yields)

The current trend is to use enzymatic hydrolysis to avoid costly recovery and wastewater treatment requirements, resulting from the use of acid hydrolysis. Enzymatic hydrolysis is attractive because it produces better yields than acid-catalyzed hydrolysis and enzyme producers have recently reduced their cost using biotechnology (Ruane et al. 2010). The conversion of cellulose and hemicellulose is catalyzed by cellulase and hemicellulase enzymes, respectively.

1.4.3.3 Fermentation

The ability to use the hemicellulose component in biomass feedstock is critical for any bio-ethanol plant. *Saccharomyces cerevisiae* and *Zymomonas mobilis*, the commonly employed organisms in alcohol fermentation, are not able to ferment hemicellulose-derived pentose (C5) sugars. There are organisms that can ferment C5 sugars (e.g., *Pichia stipitis*, *Pachysolen tannophilus*, *Candida shehatae*), but their efficiencies are low. They also need microaerophilic conditions. This implies that for more than 20 years research activities have focused on the development of improved microorganisms for the fermentation of pentose sugars (Hahn-hägerdal et al. 2007). Besides this, currently there are not known natural organisms that have the ability to convert both these C6 and C5 sugars at high yields. While pentose fermentation has been achieved on ideal substrates, (i.e., laboratory preparations of sugars designed to imitate a perfectly pretreated feedstock), significant work remains to apply this to real lignocellulosic feedstocks (Sims et al. 2008). Lignocellulosic biomass conversion into bioethanol flow diagram is shown in Fig. 1.8.

A typical process for making cellulosic ethanol starts with pretreatment and separation of the insolubles. The insoluble fraction is then hydrolyzed with cellulase and glycosidases to release glucose, which is fermented to produce ethanol. The residual insoluble material, mostly lignin, is burned to generate energy (Ruane et al. 2010). If the fermentation process is performed after the hydrolysis, this is called separate hydrolysis and fermentation (SHF). The fermentation process produces wastewaters which can be used to recover a nutrient-rich microbial cell mass (Kurian et al. 2013). Pentose fermentation, when it is carried out, is accomplished in an independent unit. The advantage of SHF is the ability to carry out each step under

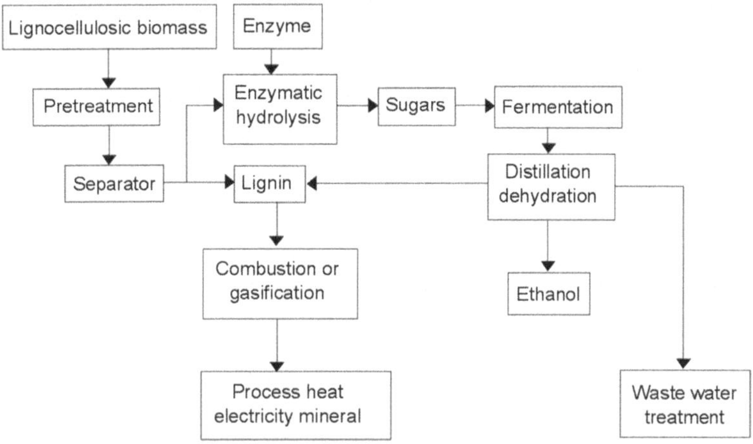

Fig. 1.8 Lignocellulosic biomass conversion into bioethanol process flow diagram (Vohra et al. 2014)

optimal conditions, i.e., enzymatic hydrolysis at 45–50 °C and fermentation at about 35 °C (Cardona and Sanchez 2007; Kurian et al. 2013). Hydrolysis and fermentation can also be performed through integrated techniques, such as simultaneous saccharification and fermentation (SSF), simultaneous saccharification and co-fermentation (SSCF), and consolidated bioprocessing (CBP) (Vohra et al. 2014).

1.4.3.4 Bioethanol Production Through Syngas Fermentation

Syngas conversion using microbial catalysts offers three main advantages:

- It requires significantly lower temperature and pressure conditions (usually atmospheric conditions).
- It is less susceptible to varying feed gas compositions.
- Chemical catalysts are more susceptible to poisoning, compared to microbial processes (Köpke et al. 2011).

After biomass gasification has been performed, cleaned gas is cooled to the normal ambient temperature and stored at a high pressure. The gas is then fed into an ethanol conversion chamber, where microbes ferment it into ethanol and acetic acid. After fermentation is completed, the liquid is distilled to separate ethanol from other products. Then ethanol is dehydrated (Dwivedi et al. 2009); see Fig. 1.9.

A large number of bacterial strains have been isolated that have the ability to ferment producer gas (composed by CO, CO2, and H2) to ethanol, acetic acid, and other useful liquid products; see, for example, Clostridium ljungdahlii (Henstra et al. 2007), *Butyribacterium methylotrophicum*, and *Clostridium autoethanogenum* (Abubackar et al. 2011).

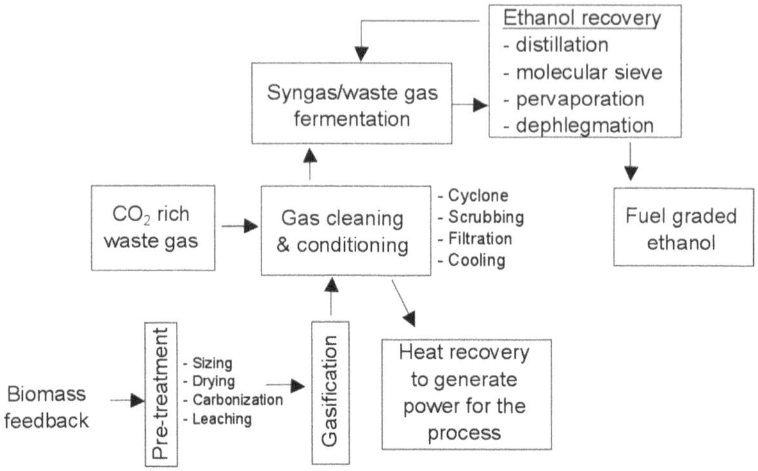

Fig. 1.9 Microbial fermentation of syngas (Vohra et al. 2014)

Producer gas fermentation is a technology which has not yet reached the market, because of low productivity of the bioreactor. This is due to several factors, such as (Ungerman and Heindel 2007):

- Low cell density
- Lack of regulation of metabolic pathways to yield only the desired product
- Inhibition of the biological catalysts by products and substrates
- Low gas–liquid mass transfer

At mild temperatures, CO and H_2 have aqueous solubilities of 60% and 4%, with respect to oxygen, on a mass basis. This results in low concentration gradients and, hence, low mass transfer rates. Higher mass transfer rates can be achieved using:

- An agitator system
- Increasing the operating pressure
- Producing micro-bubble dispersions (bubbles with diameters of about 50–100 mm have been used to provide a large gas transport area at low power consumption (Lewis et al. 2006))

1.5 BTL (FT-Diesel, Methanol and DME)

1.5.1 Introduction

The term "BTL" is applied to a liquid fuel obtained through thermo-chemical processes, such as pyrolysis and gasification, applied to biomass.

While large-scale coal-to-liquid (CTL) and gas-to-liquid (GTL) processes have been commercialized for decades (e.g., Sasol and Shell plants) (Dimitriou et al. 2018), this is not the case of BTL processes. Only a few plants have been built to date on pilot and demonstration scale:

– In the late nineties, Choren started operating a 1MWth BTL plant in Freiberg, Germany, which is not working anymore (Dimitriou et al. 2018).
– NSE Biofuels Oy operated a 12MWth (656 t/yr of fuels) BTL demonstration plant in Finland from 2009 to 2011, based on a circulating fluidized bed (CFB) gasifier designed by Foster Wheeler (Neste Oil Corporation n.d.).
– In 2010, five French partners and Uhde launched BioTfueL with two pilot plants currently on operation in France: a biomass torrefaction unit in Venette and an entrained flow gasification and Fischer-Tropsch (FT) synthesis plant near Dunkirk (Dimitriou et al. 2018).
– The Karlsruhe Institute of Technology (KIT) bioliq pilot plant with a capacity of 1 t/day has been in operation since 2014 using a process similar to the Topsoe TIGAS process.

Biomass-to-liquid (BTL) is a multistep process which consists of the following phases:

1. Reception, storage, handling, and preparation
2. Biomass gasification
3. Gas cleaning and conditioning
4. Fuel synthesis

1.5.2 Reception, Storage, Handling, and Preparation

Biomass, which is mainly transported by road, after storage is conveyed to a magnetic separator (to separate iron parts and impurities) and then screened to keep particle sizes within appropriate limits.

Biomass drying can be performed either by hot air (rotary dryer) or steam (superheated steam dryer). Air rotary dryers are the most common (WA A 1998), while superheated steam dryers (SSD) are less common but are safer with respect to fire hazards. Fuel synthesis processes (such as FT synthesis) generate significant amounts of steam, which can be reused to dry biomass (at the temperature of 200 °C and pressure of 12 bar) (Dimitriou et al. 2018).

A grinder (hammer mill) has to be placed after the dryer in case the fuel will be used in an entrained flow gasifier, to reduce the wood chips size to 1mm (Van der Drift et al. 2004; Swanson et al. 2010). If a circulating fluidized bed gasifier is used, this is capable of handling a wider variety of biomass particle sizes (Bridgwater and Maniatis 2014), so no grinding would be required.

1.5.3 Biomass Gasification

The two gasification technologies best suited for large-scale BTL plants are the circulating fluidized bed (CFB) and the entrained flow (EF) gasification (Swanson et al. 2010; Bridgwater and Maniatis 2014; The Royal Society 2008; Boerrigter 2006; The German Energy Agency 2006). For circulating fluidized bed gasifiers, operating temperature varies between 700 and 1100 °C. EF gasifiers can operate at much higher gasification temperatures (about 1200–1400 °C); this results in higher carbon conversion, very low tar and methane content, and thus lower gas cleaning requirements (Van der Drift et al. 2004; Swanson et al. 2010; Boerrigter 2006). EF gasification has the advantage that extensive experience is available from coal entrained flow gasification plants (e.g., 2000 t/d coal-fired Shell gasifier in Buggenum, Netherlands) (Hofbauer et al. 2009; Dimitriou et al. 2018). For both reactors best operating conditions are oxygen-blown and pressurized (using CO2) (Dimitriou et al. 2018). For example, oxygen at 95% purity and steam can fed into the gasifiers operating at a pressure of 28 bar and temperatures of 870 °C for the CFB and 1400 °C for the EF gasifier, respectively (Swanson et al. 2010; Dimitriou et al. 2018).

Generally the entrained flow reactor produces a syngas with higher concentration of hydrogen and carbon monoxide, as a result of reforming of light hydrocarbons. The CFB gasifier, on the other hand, produces more tar and a significant amount of methane and other light hydrocarbons (Table 1.5).

Table 1.5 Producer gas composition, depending on the reactor (Dimitriou et al. 2018)

	CFB gasifier	EF gasifier
P (bar)	28	28
T (°C)	870	1400
Oxygen (kg/kg fry feed)	0.32	0.6
Steam (kg(kg dry feed)	0.17	0.15
Gas composition (vol% wet basis [dry basis])		
H2O	12.6 [0]	25 [0]
H2	28.3 [32.4]	25.9 [34.5]
CO	26 [29.8]	37.1 [49.5]
CO2	21.2 [24.2]	10.8 [14.4]
CH4	10.5 [12]	0 [0]
C2+	0.52 [0.6]	0 [0]
Ar	0.27 [0.3]	0.42 [0.55]
N2	0.56 [0.62]	0.75 [0.99]
NH3	0.005 [5.8×10^{-3}]	0 [0]
H2S	0.02 [0.024]	0.017 [0.023]
HCl	0.01 [0.013]	0.009 [0.013]
HCN	5×10^{-4} [6×10^{-4}]	0 [0]

1.5.4 Gas Cleaning

Gas cleaning is the biggest challenge to the development of a successful BTL plant. The impurities in syngas need to be reduced to the level demanded by the catalytic fuel synthesis processes.

For CFB gasification, a cyclone can be used for particulates separation, and then syngas should pass through a tar cracker, where tars are destroyed at 875 °C by addition of oxygen and steam. The tar-free syngas is then cooled down to 280 °C using a heat exchanger. The cooled syngas passes through a bag filter (Hofbauer et al. 2009) and then is fed to the Rectisol unit, where CO_2 and sulfur compounds are removed (Hofbauer et al. 2009).

For the EF gasification concept, if the H_2/CO molar ratio of the dust-free syngas produced by the entrained flow gasifier is lower than the required ratio (H2/CO = 2) for FT and methanol synthesis, a water-gas-shift (WGS) reactor should be added before the Rectisol process. In that way the carbon dioxide produced in the WGS unit can be removed soon after in the Rectisol unit. So the dust-free syngas is fed to a direct water quench where it is cooled to the operating temperature of the WGS reactor (200 °C) (Swanson et al. 2010). The cooled syngas then passes through a bag filter to remove particulates and then enters the Rectisol unit.

1.5.5 Fuel Synthesis

After the Rectisol unit, liquid fuels can be produced from syngas using:

- FT synthesis
- Methanol synthesis followed by the MTG process
- The TIGAS process

These three processes are currently the most reliable syngas conversion technologies for transport fuel production available on the market. FT synthesis has already been used in large-scale coal-to-liquid (CTL) and gas-to-liquid (GTL) plants worldwide (Mangena 2012; Fleisch et al. 2002). Both the MTG and the TIGAS technologies have been successfully proven at demonstration scale (Fürnsinn 2007; Topp-Jorgensen 1988).

1.5.5.1 Fischer-Tropsch Synthesis

Fischer-Tropsch synthesis is a process for catalytically converting syngas to mainly hydrocarbon products of different chain lengths (typically from C1 to C100). Among the most widely known fuel synthesis plants in the world are:

- The CTL Fischer-Tropsch plants operated by Sasol in South Africa, which is the world's largest CTL production facility producing 27% of South Africa's total liquid fuel production (Mangena 2012).

- The Pearl GTL is the largest implementation of FT synthesis, located in Qatar and owned by Shell (Fleisch et al. 2002).
- CHOREN's has realized a 1MWth Carbo-V gasifier coupled to a Fischer-Tropsch reactor in 2002 (Dimitriou et al. 2018).

If cobalt-based catalyst is used, the FT synthesis takes place at 230 °C and 25 bar (Fleisch et al. 2002). The product distribution can be estimated using the Anderson-Schulz-Flory (ASF) model with an alpha value of 0.85 which favors the production of middle distillates (Swanson et al. 2010; Fürnsinn 2007; Taschler 2009). A product distribution of 60% diesel, 25% gasoline, and 25% kerosene can be achieved after the hydrocracking unit, as reported for the Shell Middle Distillate Synthesis (SMDS) process (Eilers et al. 1990).

1.5.5.2 Methanol-to-Gasoline (MTG) Synthesis

In the methanol-to-gasoline (MTG) process, the first step is represented by methanol synthesis from syngas at 50 bar and 250 °C (LeBlanc et al. 1994; Lee 1990). The produced methanol is vaporized, before it enters a dehydration reactor, where a mixture of DME, methanol, and water is produced at 404 °C. The effluent from the DME reactor is combined with the recycled gas from the product separator and enters the MTG reactor, where it is converted at 415 °C and 21.2 bar to mainly hydrocarbons and water over zeolite catalysts (ZSM-5) (Maiden 1988). The gasoline fraction in the product stream is usually about 36 wt% of the methanol and DME input (Yurchak 1988). The hot reactor effluent is cooled by heat exchange with the gas recycled from the vapor-liquid separator. It is then further cooled to about 200 °C before it passes to the vapor-liquid separator, where gas, liquid gasoline, and water are the outputs.

The MTG process was developed by Mobil scientists in the 1970s (Keil 1999). A Mobil MTG plant was operated in Motunui, New Zealand, from 1985 to 1997 and produced 14,500 bbl/d of gasoline. The plant was designed to meet one-third of New Zealand's demand for transport fuels (Maiden 1988). The fuel was composed mainly of isoparaffins and aromatics with low benzene content and essentially zero sulfur (Spath and Dayton 2003).

The first coal-to-gasoline MTG plant, utilizing the second-generation MTG technology, was constructed by Jincheng Anthracite Mining Group (JAMG) in China (Dimitriou et al. 2018). The plant started up in 2009 and its current capacity is 2500 bpd (Dimitriou et al. 2018).

1.5.5.3 Topsoe Integrated Gasoline Synthesis (TIGAS)

The main principle of the TIGAS process is the incorporation of the methanol synthesis and the DME synthesis into a single process. It was developed by Haldor Topsoe to reduce investment costs and subsequently production costs of gasoline (Topp-Jorgensen 1988). The process has been demonstrated in Houston, Texas, using natural gas as feed to the process. The plant capacity was 1 t per day gasoline. The plant started working in early 1984 and terminated in January 1987 after

10,000 h of operation (Topp-Jorgensen 1988). The bioliq Process developed by Karlsruhe Institute of Technology (KIT), with a capacity of 1 t/day, is a similar process. It is in operation since 2014 and incorporates the following processing steps: decentralized fast pyrolysis to produce a pyrolysis bio-oil/char slurry, high-pressure entrained flow gasification of the pyrolysis slurry, hot gas cleaning, DME synthesis, and gasoline synthesis (Dimitriou et al. 2018).

The DME synthesis reactor operates at 250 °C (Larson et al. 2009). The gasoline synthesis reactor is quite similar to that of the MGT process. Then the gasoline product is separated from gas and water in a vapor-liquid separator.

1.6 Hydrotreated Vegetable Oils (HVO)

Saturating the double bonds present in a molecule through catalytic addition of hydrogen at certain temperature and pressure is known as "hydrogenation" (Hughes 1953). In the process known as "hydrotreatment" hydrogen, alongside a catalyst, is added after hydrogenation. After saturation is achieved, more hydrogen addition causes the breaking of the glycerol compound, forming propane and a chain of FFA. The carboxylic acid group of the FFA must be removed to form straight-chain alkanes. This can be performed through three ways:

- The hydrodeoxygenation (HDO) route, in which it reacts with hydrogen to produce a hydrocarbon with the same number of carbon atoms as the fatty acid chain and two moles of water
- The decarboxylation (DCOX) pathway, which yields a hydrocarbon with one carbon atom less than the fatty acid chain and a mole of CO_2
- The decarbonylation (DCO) route, which also produces a hydrocarbon with one carbon atom less, as well as a mole of CO and water

The hydrodeoxygenation and hydrodecarboxylation reactions shown in Figure 1.10 can be exemplified using a saturated molecule (palmitic triglyceride) in the next set of equations (Jeczmionek and Porzycka-Semczuk 2014):

$$\text{HDO}: C_{51}H_{98}O_6 + 12\ H_2 \rightarrow 3\ C_{16}H_{34} + C_3H_8 + 6H_2O \quad (1.3)$$

$$\text{DCOx}: C_{51}H_{98}O_6 + 3\ H_2 \rightarrow 3C_{15}H_{32} + C_3H_8 + 3CO_2 \quad (1.4)$$

$$\text{DCO}: C_{51}H_{98}O_6 + 6H_2 \rightarrow 3C_{15}H_{32} + C_3H_8 + 3CO + 3H_2O \quad (1.5)$$

The HDO reaction consumes 12 mol of H_2 per mole of required triglyceride, while DCOx reaction consumes 3 moles of H_2 and DCO reaction consumes 6 moles of H2. An additional mole of H_2 is required for each double bond that is present in the vegetable oil to grant saturation. The more saturated the feedstock is, the more it

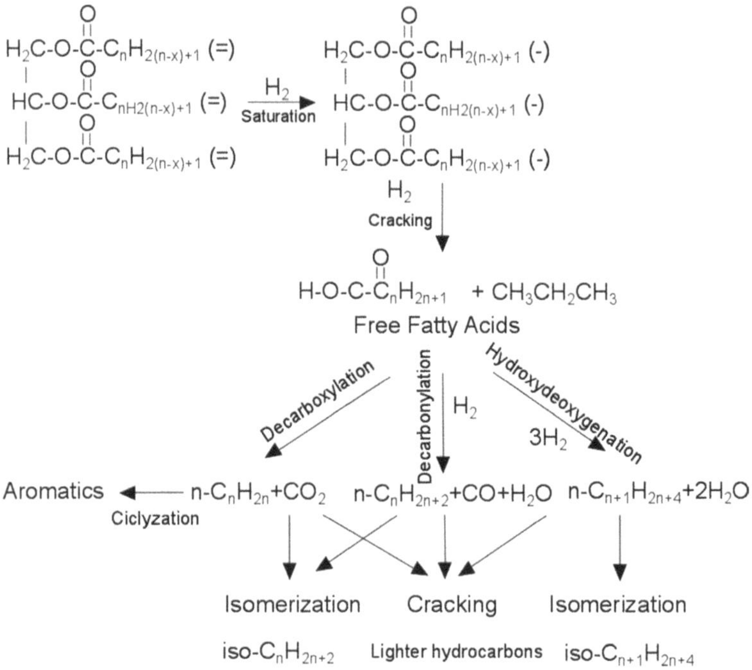

Fig. 1.10 Reactions happening during catalytic hydrotreatment (Vásquez et al. 2017); n, odd number; x,y,z, number of double bonds; =, double bonds; -, single bond; HDO, exothermic; DCOx, endothermic

is desirable, because less hydrogen will be needed during hydrogenation. The index determining unsaturation of fatty acids is known as iodine value (IV).

It has to be also considered that the CO and CO_2 formed during hydrodecarboxylation reactions may be converted into CH4 through a methanation reaction, and further addition of hydrogen would be necessary, as shown in equations (1.6) and (1.7) (Melero et al. 2012; Kaewmeesri et al. 2015):

$$CO + 3H_2 \rightarrow CH_4 + H_2O \tag{1.6}$$

$$CO_2 + 4\,H_2 \rightarrow CH_4 + 2H_2O \tag{1.7}$$

This implies that globally hydrodecarboxylation route will demand three more molecules of hydrogen than the hydrodeoxygenation pathway.

Depending of the composition of the final n-alkanes produced though the hydrotreatment process, they need to be subjected to either isomerization, cracking, or cyclization, to improve their combustion properties and obtain isoalkanes, lighter hydrocarbons, and aromatics, respectively (Veriansyah et al. 2012; Kiatkittipong et al. 2013).

During hydrotreatment, there are some variables that influence the process and the final composition of the product, including:

- Reaction conditions
- Type of catalyst used
- Selected feedstock

Dealing with reaction temperature, DCO and DCOX are more dominant over higher temperatures and moderate acidic catalyst than HDO reaction.

Two types of catalyst can be used for the hydrotreatment:

- Conventional bimetallic sulfide catalysts, such as $NiMoS_2$, $CoMoS_2$, and $NiWS_2$ supported on Al_2O_3
- Monometallic catalysts, in particular Ni, Pd, Pt, and Rh (Morgan et al. 2012; Rogers and Zheng 2016)

Nickel- and palladium-based catalysts are the most commonly used catalysts. Metal catalysts supported on activated carbon have been also tested for upgrading vegetable oils into hydrocarbon jet biofuels (Silva et al. 2016). Zeolite catalysts have been also studied for the hydrotreatment of vegetable oils (Zhao et al. 2015).

The process shown in Figure 1.11 consists mainly in a pretreatment of the raw material, a deoxygenation, a hydrocracking/isomerization, and a distillation (Hilbers et al. 2015). As it can be seen from Figure 1.11, the hydrotreating process is interesting for the production of both biodiesel and bio-jet fuel.

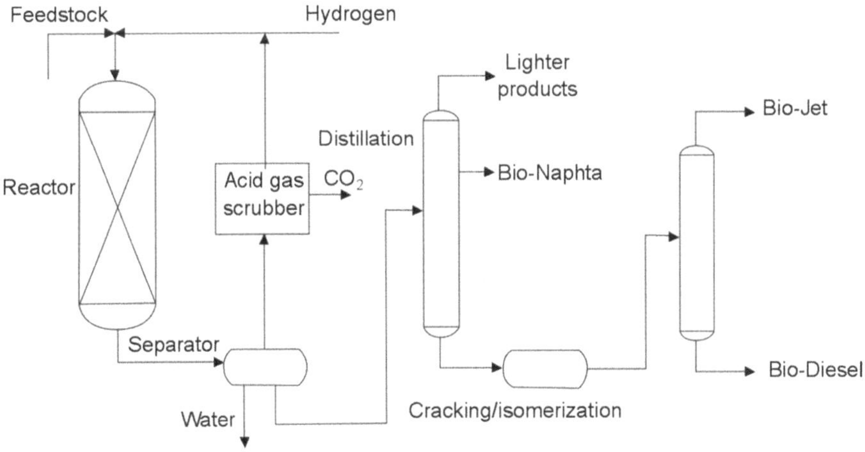

Fig. 1.11 Hydrotreatment of vegetable oils process diagram (Vásquez et al. 2017)

1.6.1 HVO Biofuel Plants

Neste Corporation is the leading company on HVO production, accounting for an annual production volume of more than 2 million tons of biofuel. Based in Finland, the company has two renewable refineries in Porvoo and two more abroad (i.e., in the Netherlands and in Singapore) (Vásquez et al. 2017). They developed the NExBTL technology, and currently, they process 10 types of raw materials including animal fats (food industry waste), fish fat (fish processing waste), vegetable oils, used cooking oil, and technical corn oil, though the focus is on waste and residue raw materials which account for an 80% of the feedstock.

Main competitor of NExBTL technology is the Ecofining Process, developed by the Honeywell UOP company, jointly with the Eni S.p.A (GREENEA 2015).

Some other technologies for the conversion of lipids through hydrotreatment are (Vásquez et al. 2017):

- The Vegan Technology marketed by the Axens Group, a French company
- Bio-Synfining process, patented by Syntroleum Corporation and bought by the Renewable Energy Group (United States) in 2014 (a plant with a capacity of 75 million gallons per year is operative)
- The UPM BioVerno technology, which converts crude tall oil into green diesel, developed by UPM Biofuels Company in Finland
- The Hydroflex technology created by the Haldor Topsøe group in Denmark, allowing its implementation as both stand-alone or co-processing unit (Vásquez et al. 2017)

1.6.2 HVO Jet Biofuel Plants

Most commercial applications of the hydrotreating process are optimized to produce green diesel; however, multiple agreements between airlines and refineries have boosted also jet biofuel production projects:

- SG Preston, in the United States, has signed a ten-year agreement with JetBlue Airways to deliver more than 33 million gallons of HEFA (hydroprocessed esters and fatty acids) jet per year.
- AltAir fuels have a dedicated capacity to produce jet biofuel to provide United Airlines 15 million gallons of sustainable biofuel over a three-year period contract. AltAir will also provide KLM Royal Dutch Airlines with sustainable jet fuel.
- Petrixo Oil & Gas is expected to be the most massive jet biofuel project, producing over 500,000 metric tons per year of jet biofuel at its new refinery to be built in Fujairah, United Arab Emirates (BiofuelsDigets 2017).

Table 1.6 Biofuel yields and costs for different feedstock and technologies (Festel et al. 2014)

Biofuel	Raw material	Yield (l/t)	Costs (€Cent/l) referred to 2015
Ethanol	Maize	400	105.85
Ethanol	Wheat	375	136.40
Ethanol	Lignocellulosic waste	250	157.34
Biodiesel	Rapeseed oil	1100	117.49
Biodiesel	Palm oil	1100	70.00
Biodiesel	Waste oil	1100	61.78
HVO	Palm oil	1100	216.18
BTL	Wood	158	827.95

1.7 Biofuel Yields and Costs

Table 1.6 shows the yields of ethanol, biodiesel, HVO, and BTL. Biodiesel and HVO have very higher yields. Bioethanol has less than half of the mass yield of biodiesel, while BTL has 14% of the yield of biodiesel.

Biofuels' costs have been calculated through the methodology presented in (Festel et al. 2014). This is based on the development of scenarios on future raw material prices and on the modeling of production costs. The cost of raw materials is obviously influenced by the price of oil (for the results shown in Table 1.6, a price of oil of 50 € per barrel is supposed).

If we consider that according to (Festel et al. 2014) a price of oil equal to 50 € per barrel corresponds to a cost of oil equal to 36.45 €/l, we see that no biofuel can be produce at competitive cost compared to fossil fuel. The biofuel with lower cost is biodiesel especially that produced from waste oil.

1.8 Biofuel Properties and Combustion Performance

Biofuels properties are shown in Table 1.7.

Table 1.7 shows that gasoline, fossil diesel, and HVO have very high heating values, compared to bioethanol and biodiesel. The main issue with use of ethanol comes from its lower energy density; in fact it contains only around two-thirds of the energy of a similar volume of gasoline (Gautam and Martin II 2000). This is not an issue during normal driving; however, it will result in reduced vehicle range and a lower peak power of the engine when the accelerator is fully pressed. An advantage is represented by the fact that it has higher octane rating of ethanol as compared to gasoline. This can allow higher compression ratio engines to be used (Bergthorson and Thomson 2015) increasing in this way the fuel efficiency. When blended with hydrocarbon fuels, ethanol acts as a sink of reactive species (OH radicals) that disrupts the chain branching of the hydrocarbon fuel under low-temperature chemistry conditions and slows ignition of the blend (Foong et al. 2014). It is, however, high-temperature flame chemistry which controls the combustion efficiency and

Table 1.7 Biofuels properties

Property	Fossil diesel (EN590)	Biodiesel (EN 14214)	HVO	Bioethanol	Gasoline
Density (kg/m^3 at 15°C)	820–845 (Vásquez et al. 2017)	860–900 (Vásquez et al. 2017)	775–785 (Vásquez et al. 2017)	785 (Ku and Tu 2005)	720–780 (Christensen et al. 2011)
Viscosity (mm^2/s at 40°C)	2–4.5 (Vásquez et al. 2017)	3.5–5.0 (Vásquez et al. 2017)	2.9–3.5 (Vásquez et al. 2017)	1.1 (Ku and Tu 2005)	0.37–0.44 (Christensen et al. 2011)
Heating value (MJ/kg)	43 (Vásquez et al. 2017)	38 (Vásquez et al. 2017)	44 (Vásquez et al. 2017)	26.87 (Agarwal 2007)	44 (Al-Hasan 2003)
Cetane number	51> (Vásquez et al. 2017)	51> (Vásquez et al. 2017)	84–99 (Vásquez et al. 2017)	6 (Yilmaz 2012)	n.a.
Sulfur content (mg/kg)	<10 (Vásquez et al. 2017)	<1 (Vásquez et al. 2017)	0 (Vásquez et al. 2017)	0 (Agarwal 2007)	7 (Rodríguez-Antón et al. 2015)
Oxygen content (wt%)	0 (Vásquez et al. 2017)	11 (Vásquez et al. 2017)	0 (Vásquez et al. 2017)]	5 (Rodríguez-Antón et al. 2015)	2.7 (Rodríguez-Antón et al. 2015)

*Calculated

pollutant emissions in SI engines. After ignition by the spark, a turbulent premixed flame propagates through the premixed fuel–air charge in the engine, rapidly converting the fuel into combustion products and producing the thermal energy and pressure that drive the engine. A recent study observing flames in optically accessible engines has shown that ethanol flames propagate faster than butanol, gasoline, and iso-octane (Aleiferis et al. 2013). Recent papers still do not produce consistent trends in relative NO_x potential of gasoline–alcohol blends (Karavalakis et al. 2014; Canakci et al. 2013; Gravalos et al. 2013; Balki et al. 2014), while agreement is growing on the risk of a potential increase in oxygenated emissions, such as formaldehyde, acetaldehyde, and ketones (Agarwal 2007; Lynd et al. 1991; Kohse-Höinghaus et al. 2010; Saxena and Williams 2007).

Dealing with biodiesel, when compared to diesel, it has a 9% lower volumetric energy content, due to its oxygen content (Agarwal 2007; Lapuerta et al. 2008). At high-temperatures, the reactivity of long-chain esters is nearly indistinguishable for saturated and unsaturated esters (Wang et al. 2013), while the unsaturated esters (with double bonds) have generally increased low-temperature ignition delay times and reduced cetane numbers (Westbrook 2013; Westbrook et al. 2013). Biodiesels have the positive aspect of reducing engine deposits and coking, compared to petroleum-derived fuels (Graboski and McCormick 1998; Xue et al. 2011). NO_x emissions have been observed to increase for biodiesel compared to petro-diesel for many engine tests (Lapuerta et al. 2008; Graboski and McCormick 1998; Xue et al. 2011; Coniglio et al. 2013; Lai et al. 2011; Sun et al. 2010; Giakoumis et al. 2012; Palash et al. 2013; Varatharajan and Cheralathan 2012; Szybist et al. 2007; Hoekman and Robbins 2012; Mueller et al. 2009; Szybist et al. 2005; Rajasekar et al. 2010), while others have shown no increase or even a decrease (Coniglio et al. 2013).

Renewable diesel fuel derived from either hydrotreating vegetable oils is more compatible with existing engine technology than first-generation biodiesels, thus leading to improved engine performance (Knothe 2010; Gill et al. 2011). These renewable fuels have effectively equivalent energy densities as petroleum-derived fuels, due to the lack of oxygen content and similar hydrogen-to-carbon ratios (Probstein and Hicks 2006). The cetane numbers are quite high, so that the straight-chain fuel must be blended with lower-quality fuels for use in diesel engines (Dry 2002a) or be branched via oligomerization reactions to a cetane number around 50 for use as a pure diesel fuel (Dry 2002b). Blending FT synthetic diesel with FAME biodiesel or petroleum diesel is used to improve also the lubricity of the fuel (Gill et al. 2011); otherwise to be used pure, the FT diesel needs to be mixed with specialized aromatic additives (Corporan et al. 2011). The high cetane number of FT or hydrotreated-renewable-diesel fuels and their lack of aromatic content are considered to be the primary factors responsible for the observed decrease in NOx, soot, unburned hydrocarbon, and CO emissions and increase in thermal efficiency, compared to conventional diesel (Szybist et al. 2005; Knothe 2010; Gill et al. 2011).

References

Abubackar HN, Veiga MC, Kennes C (2011) Biological conversion of carbon monoxide: rich syngas or waste gases to bioethanol. Biofuels, Bioprod Biorefin 5:93–114

Agarwal AK (2007) Biofuels (alcohols and biodiesel) applications as fuels for internal combustion engines. Progr Energy Comb Sci 33(3):233–271

Albers SC, Berklund AM, Graff GD (2016) The rise and fall of innovation in biofuels. Nat Biotechnol 34:814–821

Aleiferis PG, Serras-Pereira J, Richardson D (2013) Characterisation of flame development with ethanol, butanol, iso-octane, gasoline and methane in a direct-injection spark-ignition engine. Fuel 109:256–278

Al-Hasan M (2003) Effect of ethanol-unleaded gasoline blends on engine performance and exhaust emission. Energy Conver Manag 44(9):1547–1561

Arnold M, Tainter JA, Strumsky D (2019) Productivity of innovation in biofuel technologies. Energy Policy 124:54–62

Assman G, Blasey G, Gutsche B, Jeromin L, Rigal J, Armengand R, Cormary B (1996) Continuous progress for the production of lower alkyl esters. US Patent No. 5,514,820.

Balat M (2011) Production of bioethanol from lignocellulosic materials via the biochemical pathway: a review. Energy Convers Manag 52:858–875

Balki MK, Sayin C, Canakci M (2014) The effect of different alcohol fuels on the performance, emission and combustion characteristics of a gasoline engine. Fuel 115:901–906

Barbanera M, Lascaro E, Foschini D, Cotana F, Buratti C (2018) Optimization of bioethanol production from steam exploded hornbeam wood (Ostrya carpinifolia) by enzymatic hydrolysis. Renew Energy 124:136–143

Bergthorson JM, Thomson MJ (2015) A review of the combustion and emissions properties of advanced transportation biofuels and their impact on existing and future engines. Renew Sustain Energy Rev 42:1393–1417

BiofuelsDigets (2017) Ground delay: where are the sustainable aviation fuels? http://www.biofuelsdigest.com/bdigest/2017/02/20/ground-delaywhere-are-the-sustainable-aviation-fuels/

Boerrigter H (2006) Economy of Biomass-to-Liquids (BTL) plants – An engineering assessment. Energy research Centre of the Netherlands [Report ECN-C-06-019]
Bothast RJ, Schlicher MA (2005) Biotechnological processes for conversion of corn into ethanol. Appl Microbiol Biotechnol 67:19–25
Bridgwater AV, Maniatis K (2014) The production of biofuels by the thermochemical processing of biomass. In: Archer MD, Barber J (eds) Molecular to global photosynthesis. edIC Press, New York, pp 521–612
Buranov AU, Mazza G (2008) Lignin in straw of herbaceous crops. Ind Crop Prod 28:237–259
Buratti C, Barbanera M, Lascaro E (2015) Ethanol production from vineyard pruning residues with steam explosion pretreatment. Environ Progr Sustain Energy 34(3):802–809
Buratti C, Foschini D, Barbanera M, Fantozzi F (2018) Fermentable sugars production from peach tree prunings: response surface model optimization of NaOH alkaline pretreatment. Biomass Bioenergy 112:128–137
Caffall KH, Mohnen D (2009) The structure, function, and biosynthesis of plant cell wall pectic polysaccharides. Carbohydr Res 344:1879–1900
Canakci M, Ozsezen AN, Alptekin E, Eyidogan M (2013) Impact of alcohol-gasoline fuel blends on the exhaust emission of an SI engine. Renew Energy 52:111–117
Cardona C, Sanchez O (2007) Fuel ethanol production: process design trends and integration opportunities. Biores Technol 98:2415–2457
Cavalaglio G, Gelosia M, D'Antonio S, Nicolini A, Pisello AL, Barbanera M, Cotana F (2016) Lignocellulosic ethanol production from the recovery of stranded driftwood residues. Energies 9 (8):634
Cheng JJ, Timilsina GR (2011) Status and barriers of advanced biofuel technologies: a review. Renew Energy 36:3541–3549
Chiaramonti D (2007) Bioethanol: role and production technologies. In: Ranalli P (ed) Improvement of crop plants for industrial end uses. Springer, Dordrecht, pp 209–251
Christensen E, Yanowitz J, Ratcliff M, McCormick RL (2011) Renewable oxygenate blending effects on gasoline properties. Energy Fuels 25(10):4723–4733. https://doi.org/10.1021/ef2010089
Chundawat SPS, Beckham GT, Himmel ME, Dale BE (2011) Deconstruction of lignocellulosic biomass to fuels and chemicals. Annu Rev Chem Biomol Eng 2:121–145
Coniglio L, Bennadji H, Glaude PA, Herbinet O, Billaud F (2013) Combustion chemical kinetics of biodiesel and related compounds (methyl and ethyl esters): Experiments and modeling – Advances and future refinements. Progr Energy Combust Sci 39(4):340–382
Corporan E, Edwards T, Shafer L, MJ DW, Klingshirn C, Zabarnick S, West Z, Striebich R, Graham J, Klein J (2011) Chemical, thermal stability, seal swell, and emissions studies of alternative jet fuels. Energy Fuels 25(3):955–966. https://doi.org/10.1021/ef101520v
Costantini V, Crespi F, Curci Y (2013) BioPat: an investigative tool for analysis of industry evolution, technological paths and policy input in the biofuels sector. In: Costantini V, Mazzanti M (eds) The dynamics of environmental and economic systems: innovation, Environmental Policy and Competitiveness. Springer, Dordrecht, pp 203–226
Costantini V, Crespi F, Martini C, Pennacchio L (2015a) Demand-pull and technology-push public support for eco-innovation: the case of the biofuels sector. Res Policy 44:577–595
Costantini V, Crespi F, Curci Y (2015b) A keyword selection method for mapping technological knowledge in specific sectors through patent data: the case of biofuels sector. Econ Innov New Technol 24:282–308
Cotana F, Barbanera M, Foschini D, Lascaro E, Buratti C (2015) Preliminary optimization of alkaline pretreatment for ethanol production from vineyard pruning. Energy Proc 82:389–394
Datta R, Maher MA, Jones C, Brinker RW (2011) Ethanol – the primary renewable liquid fuel. J Chem Technol Biotechnol 86:473–480
Demirbas A (2009a) Biofuels securing the planet's future energy needs. Energy Conver Manag 50 (9):2239–2249

Demirbas MF (2009b) Biorefineries for biofuel upgrading: A critical review. Appl Energy 86(1): S151–S161

Dimitriou I, Goldingay H, Bridgwater AV (2018) Techno-economic and uncertainty analysis of Biomass to Liquid (BTL) systems for transport fuel production. Renew Sustain Energy Rev 88:60–175

Dry ME (2002a) The Fischer-Tropsch process: 1950–2000. Catal Today 71(3–4):227–241

Dry ME (2002b) High quality diesel via the Fischer-Tropsch process – a review. J Chem Technol Biotechnol 77(1):43–50

Dwivedi P, Alavalapati JRR, Lal P (2009) Cellulosic ethanol production in the United States: conversion technologies, current production status, economics, and emerging developments. Energy Sustain Dev 13:174–182

Eckey EW (1956) Esterification and interesterification. J Am Oil Chem Sot 33:575–579

Eilers J, Posthuma SA, Sie ST (1990) The Shell Middle Distillate Synthesis process (SMDS). Catal Lett 7:253–270

Fan Y, Wu G, Su F, Li K, Xu L, Han X, Yan Y (2016) Dendrimer-coated magnetic multiwalled carbon nanotubes: synthesis, characterization, and employed in immobilization of lipases toward catalyzing biodiesel production. Fuel 178:172–178

Fan Y, Su F, Li K, Ke C, Yan Y (2017) Carbon nanotube filled with magnetic iron oxide and modified with polyamidoamine dendrimers for immobilizing lipase toward application in biodiesel production. Sci Rep 7:45643. https://doi.org/10.1038/srep45643

Festel G, Würmseher M, Rammer C, Boles E, Bellof M (2014) Modelling production cost scenarios for biofuels and fossil fuels in Europe. J Clean Prod 66:242–253

Fleisch TH, Sills RA, Briscoe MD (2002) Emergence of the gas-to-liquids industry: a review of global GTL developments. J Nat Gas Chem 11:1–14

Foong TM, Morganti KJ, Brear MJ, da Silva G, Yang Y, Dryer FL (2014) The octane numbers of ethanol blended with gasoline and its surrogates. Fuel 115:727–739

Freedman B, Butterileld RO, Pryde EH (1986) Transesterification kinetics of soybean oil. J Am Oil Chem Sot 63:1375–1380

Fürnsinn S (2007) Outwitting the dilemma of scale: cost and energy efficient scale-down of the Fischer-Tropsch fuel production from biomass [Ph.D. Thesis]. Vienna University of Technology

Gautam M, Martin DW II (2000) Combustion characteristics of higher-alcohol/gasoline blends. Proc Inst Mech Eng, Part A: J Power Energy 214(5):497–511. https://doi.org/10.1243/095765000l538047

Giakoumis EG, Rakopoulos CD, Dimaratos AM, Rakopoulos DC (2012) Exhaust emissions of diesel engines operating under transient conditions with biodiesel fuel blends. Progr Energy Combust Sci 38(5):691–715

Gill SS, Tsolakis A, Dearn KD, Rodriguez-Fernandez J (2011) Combustion characteristics and emissions of Fischer-Tropsch diesel fuels in IC engines. Progr Energy Combust Sci 37 (4):503–523

Gírio FM, Fonseca C, Carvalheiro F, Duarte LC, Marques S, Bogel-Łukasik R (2010) Hemicelluloses for fuel ethanol: a review. Bioresour Technol 101:4775–4800

Graboski MS, McCormick RL (1998) Combustion of fat and vegetable oil derived fuels in diesel engines. Progr Energy Combust Sci 24(2):125–164

Gravalos I, Moshou D, Gialamas T, Xyradakis P, Kateris D, Tsiropoulos Z (2013) Emissions characteristics of spark ignition engine operating on lower-higher molecular mass alcohol blended gasoline fuels. Renew Energy 50:27–32

GREENEA (2015) Is HVO the Holy Grail of the World Biodiesel Market?

Gui MM, Lee KT, Bhatia S (2008) Feasibility of edible oil vs. non-edible oil vs. waste edible oil as biodiesel feedstock. Energy 33:1646–1653

Hahn-hägerdal B, Karhumaa K, Fonseca C, Spencer-Martins I, Gorwagrauslund MF (2007) Towards industrial pentose-fermenting yeast strains. Appl Microbiol Biotechnol 74:937–953

Henstra AM, Sipma J, Rinzema A, Stams AJM (2007) Microbiology of synthesis gas fermentation for biofuel production. Curr Opin Biotechnol 18:200–206

Hilbers TJ, Sprakel LMJ, van den Enk LBJ, Zaalberg B, van den Berg H, van der Ham LGJ (2015) Green diesel from hydrotreated vegetable oil process design study. Chem Eng Technol 38:651–657. https://doi.org/10.1002/ceat.201400648

Himmel ME, Ding SY, Johnson DK, Adney WS, Nimlos MR, Brady JW, Foust TD (2007) Biomass recalcitrance: engineering plants and enzymes for biofuels production. Science 315:804–807

Hoekman SK, Robbins C (2012) Review of the effects of biodiesel on NOx emissions. Fuel Process Technol 96:237–249

Hofbauer H, Rauch R, Ripfel-Nitsche K (2009) Gas cleaning for synthesis applications. In: Bridgwater AV, Hofbauer H, van Loo S (eds) Thermal biomass conversion. CPL Press, Newbury, pp 211–266

Huber G, Iborra S (2006) Synthesis of transportation fuels from biomass: chemistry, catalysts, and engineering. Chem Rev 106:4044–4098

Hughes JP (1953) Hydrogenation of fatty oils. J Am Oil Chem Soc 30:506–515. https://doi.org/10.1007/BF02641690

IEA (2018) World energy outlook 2018, available from: https://webstore.iea.org/world-energy-outlook-2018. Accessed 3 Feb 2019

Jacquet N, Quiévy N, Vanderghem C, Janas S, Blecker C, Wathelet B, Devaux J, Paquot M (2011) Influence of steam explosion on the thermal stability of cellulose fibres. Polym Degrad Stab 96 (9):1582–1588

Jeczmionek Ł, Porzycka-Semczuk K (2014) Hydrodeoxygenation, decarboxylation and decarbonylation reactions while co-processing vegetable oils over a NiMo hydrotreatment catalyst. Part I: thermal effects – theoretical considerations. Fuel 131:1–5. https://doi.org/10.1016/j.fuel.2014.04.055

Kaewmeesri R, Srifa A, Itthibenchapong V, Faungnawakij K (2015) Deoxygenation of waste chicken fats to green diesel over Ni/Al2O3: effect of water and free fatty acid content. Energy Fuels 29:833–840. https://doi.org/10.1021/ef5023362

Karavalakis G, Short D, Vu D, Villela M, Asa-Awuku A, Durbin TD (2014) Evaluating the regulated emissions, air toxics, ultrafine particles, and black carbon from SI-PFI and SI-DI vehicles operating on different ethanol and iso-butanol blends. Fuel 128:410–421

Keil FJ (1999) Methanol-to-hydrocarbons: process technology. Microporous Mesoporous Mater 29:49–66

Kessler J, Sperling D (2016) Tracking U.S. biofuel innovation through patents. Energy Policy 98:97–107

Kiatkittipong W, Phimsen S, Kiatkittipong K, Wongsakulphasatch S, Laosiripojana N, Assabumrungrat S (2013) Diesel-like hydrocarbon production from hydroprocessing of relevant refining palm oil. Fuel Process Technol 116:16–26

Kim S, Dale BE (2004) Global potential bioethanol production from wasted crops and crop residues. Biomass Bioenergy 26(4):361–375

Knothe G (2010) Biodiesel and renewable diesel: a comparison. Progr Energy Combust Sci 36 (3):364–373

Kohse-Höinghaus K, Oßwald P, Cool TA, Kasper T, Hansen N, Qi F, Westbrook CK, Westmoreland PR (2010) Biofuel combustion chemistry: from ethanol to biodiesel. Angew Chem – Int Ed 49(21):3572–3597

Kojima M, Johnson T (2005) Potential for biofuels for transport in developing countries, The International Bank for Reconstruction and Development/The World Bank, Energy Sector Management Assistance Programme Report

Köpke M, Mihalcea C, Bromley JC, Simpson SD (2011) Fermentative production of ethanol from carbon monoxide. Curr Opin Biotechnol 22:320–325

Ku HC, Tu CH (2005) Densities and Viscosities Of Binary And Ternary Mixtures Of Ethanol, 2-Butanone, And 2,2,4-Trimethylpentane At T = (298.15, 308.15, and 318.15) K. J Chem Eng Data 50(2):608–615. https://doi.org/10.1021/je049655w

Kumar S, Singh N, Prasad R (2010) Anhydrous ethanol: a renewable source of energy. Renew Sustain Energy Rev 14:1830–1844

Kurian JK, Nair GR, Hussain A, Vijaya Raghavan GS (2013) Feedstocks, logistics and pre-treatment processes for sustainable lignocellulosic biorefineries: a comprehensive review. Renew Sustain Energy Rev 25:205–219

Lai JYW, Lin KC, Violi A (2011) Biodiesel combustion: advances in chemical kinetic modeling. Progr Energy Combust Sci 37(1):1–14

Lam MK, Lee KT, Mohamed AR (2010) Homogeneous, heterogeneous and enzymatic catalysis for transesterification of high free fatty acid oil (waste cooking oil) to biodiesel: a review. Biotechnol Adv 28(4):500–518

Lapuerta M, Armas O, Rodriguez-Fernandez J (2008) Effect of biodiesel fuels on diesel engine emissions. Prog Energy Combust Sci 34(2):198–223

Larson ED, Jin H, Celik FE (2009) Large-scale gasification-based coproduction of fuels and electricity from switchgrass. Biofuels Bioprod Bioref 3:174–194

LeBlanc JR, Schneider RV, Strait RB (1994) Production of methanol. In: Cheng WH, Kung HH (eds) Methanol production and use. Marcel Dekker, New York

Lee S (1990) Methanol synthesis technology. CRC Press, Cleveland

Leung DYC, Wu X, Leung MKH (2010) A review on biodiesel production using catalyzed transesterification. Appl Energy 87(4):1083–1095

Lewis R, Datar R, Huhnke RL (2006) Biomass to ethanol. Encycl Chem Process 1:143–151

Li J, Henriksson G, Gellerstedt G (2005) Carbon reactions during high-temperature steam treatment of aspen wood. Appl Biochem Biotechnol 125:175. https://doi.org/10.1385/ABAB:125:3:175

Li K, Fan Y, He Y, Zeng L, Han X, Yan Y (2017) Burkholderia cepacia lipase immobilized on heterofunctional magnetic nanoparticles and its application in biodiesel synthesis. Sci Rep 7:16473. https://doi.org/10.1038/s41598-017-16626-5

Liu Y, Yan YJ, Zhang X, Tan H, Xin L, Yao A (2010) Combined lipases catalyzed transesterification for biodiesel production: optimization and kinetics. AIChE J 56 (6):1659–1665

Liu Y, Liu T, Li C, Yan Y (2011a) Biodiesel synthesis catalyzed by Burkholderia cenocepacia lipase supported on macroporous resin NKA in solvent-free and isooctane systems. Energy Fuels 25(3):1206–1212

Liu Y, Chen D, Yan Y, Peng C, Xu L (2011b) Biodiesel synthesis and conformation of lipase from Burkholderia cepacia in room temperature ionic liquids and organic solvents. Bioresour Technol 102(22):10414–10418

Lynd LR, Cushman JH, Nichols RJ, Wyman CE (1991) Fuel ethanol from cellulosic biomass. Science 251:1318–1323

Lynd LR, Weimer PJ, van Zyl WH, Pretorious IS (2002) Microbial cellulose utilization: fundamentals and biotechnology. Microbiol Mol Biol Rev 66:506–577

Maiden CJ (1988) The New Zealand gas-to-gasoline project. In: Bibby DM, Chang CD, Howe RF, Yurchak S (eds) Methane conversion, Studies in surface science and catalysis, vol 36. Elsevier, Amsterdam, pp 1–16

Mangena S (2012) Coal gasification and liquefaction – SA experiences and opportunities. In: 4th EU – South Africa Clean Coal Working Group Meeting, Sasol Technology (Pty) Ltd

May JB (1994) Wet milling: process and products. In: White PJ, Johnson LA (eds) Corn chemistry and technology. St. Paul, American Association of Cereal Chemist, pp 377–395

Melero JA, Iglesias J, Garcia A (2012) Biomass as renewable feedstock in standard refinery units. Feasibility, opportunities and challenges. Energy Environ Sci 5:7393. https://doi.org/10.1039/c2ee21231e

Mood SH, Golfeshan AH, Tabatabaei M, Jouzani GS, Najafi GH, Gholami M, Ardjmand M (2013) Lignocellulosic biomass to bioethanol, a comprehensive review with a focus on pretreatment. Renew Sustain Energy Rev 27:77–93

Morgan T, Santillan-Jimenez E, Harman-Ware AE, Ji Y, Grubb D, Crocker M (2012) Catalytic deoxygenation of triglycerides to hydrocarbons over supported nickel catalysts. Chem Eng J 189-190:346–355. https://doi.org/10.1016/j.cej.2012.02.027

Mueller CJ, Boehman AL, Martin GC (2009) An experimental investigation of the origin of increased NOx emissions when fueling a heavy-duty compression- ignition engine with soy biodiesel. SAE Paper 2009-01-1792

Neste Oil Corporation (n.d.). Annual report 2010. Retrieved from http://www.nesteoil.com. Accessed 10 Feb 2019.

Noureddini H, Zhu D (1997) Kinetics of transesterification of soybean oil. J Am Oil Chem Sot 74:1457–1463

Palash SM, Kalam MA, Masjuki HH, Masum BM, Rizwanul Fattah IM, Mofijur M (2013) Impacts of biodiesel combustion on NOx emissions and their reduction approaches. Renew Sustain Energy Rev 23:473–490

Pauly M, Keegstra K (2010) Plant cell wall polymers as precursors for biofuels. Curr Opin Plant Biol 13:305–312

Peng F, Ren JL, Xu F, Sun RC (2011) Chemicals from hemicelluloses: a review. In: Zhu JY, Zhang X, Pan X (eds) Sustainable production of fuels, chemicals, and fibers from forest biomass. ACS Symposium Series, Washington, DC, pp 219–259

Probstein RF, Hicks RE (2006) Synthetic fuels. Dover Publications Inc., Mineola

Rajasekar E, Murugesan A, Subramanian R, Nedunchezhian N (2010) Review of NOx reduction technologies in CI engines fuelled with oxygenated biomass fuels. Renew Sustain Energy Rev 14(7):2113–2121

Renewable Fuels Association (2015) Ethanol facts: environment. http://www.ethanolrfa.org/pages/ethanol-facts-environment. Accessed 3 Feb 2019

Rodríguez-Antón LM, Gutiérrez-Martín F, Martinez-Arevalo C (2015) Experimental determination of some physical properties of gasoline, ethanol and ETBE ternary blends. Fuel 156:81–86

Rogers KA, Zheng Y (2016) Selective deoxygenation of biomass-derived bio-oils within hydrogen-modest environments: a review and new insights. Chem Sus Chem 9:1750–1772. https://doi.org/10.1002/cssc.201600144

Ruane J, Sonnino A, Agostini A (2010) Bioenergy and the potential contribution of agricultural biotechnologies in developing countries. Biomass Bioenergy 34:1427–1439

Saxena P, Williams FA (2007) Numerical and experimental studies of ethanol flames. Proc Combust Inst 31(1):1149–1156

Shah V, Jurjevic M, Badia D (2007) Utilization of restaurant waste oil as a precursor for sophorolipid production. Biotechnol Prog 23:512–515

Silva LN, Fortes ICP, De Sousa FP, Pasa VMD (2016) Biokerosene and green diesel from macauba oils via catalytic deoxygenation over Pd/C. Fuel 164:329–338. https://doi.org/10.1016/j.fuel.2015.09.081

Sims R, Taylor M, Saddler J, Mabee W (2008) From 1st- to 2nd-generation biofuel technologies – full report- an overview of current industry and RD&D activities. International Energy Agency

Singh V, Rausch KD, Yang P, Shapouri H, Belyea RL, Tumbleson ME (2001) Modified dry grind ethanol process, University of Illinois at Urbana–Champaign. Report No UILU No. 2001-7021

Skutsch M, de los Rios E, Solis S, Riegelhaupt E, Hinojosa D, Gerfert S, Gao Y, Masera O (2011) Jatropha in Mexico: environmental and social impacts of an incipient biofuel program. Ecol Soc 16:4–11

Slade R, Bauen A (2013) Micro-algae cultivation and biofuels: cost, energy balance, environmental impacts and future prospects. Biomass Bioenergy 53:29–38

Solomon BD, Barnes JR, Halvorsen KE (2007) Grain and cellulosic ethanol: history, economics, and energy policy. Biomass Bioenergy 31:416–425

Somerville C (2011) Biofuels. Curr Biol 17:115–119

Spath PL, Dayton DC (2003) Preliminary screening – technical and economic assessment of synthesis gas to fuels and chemicals with emphasis on the potential for biomass- derived syngas. National Renewable Energy Laboratory [Report NREL/TP-510-34929]

Sticklen MB (2008) Plant genetic engineering for biofuel production: Towards affordable cellulosic ethanol. Nat Rev Genet 9:433–443

Stidham WD, Seaman DW, Danzer MF (2000) Method for preparing a lower alkyl ester product from vegetable oil. US Patent No. 6,127,560
Strumsky D, Lobo J, Tainter JA (2010) Complexity and the productivity of innovation. Syst Res Behav Sci 27:496–509
Su F, Li G, Fan Y, Yan Y (2016) Enhanced performance of lipase via microcapsulation and its application in biodiesel preparation. Sci Rep 6:29670. https://doi.org/10.1038/srep29670
Sun J, Caton JA, Jacobs TJ (2010) Oxides of nitrogen emissions from biodiesel fueled diesel engines. Prog Energy Combust Sci 26:667–695
Swanson RM et al (2010) Techno-economic analysis of biofuels production based on gasification. National Renewable Energy Laboratory [Report NREL/TP-6A20-46587]
Szybist JP, Kirby SR, Boehman AL (2005) NOx emissions of alternative diesel fuels: a comparative analysis of biodiesel and FT diesel. Energy Fuels 19(4):1484–1492. https://doi.org/10.1021/ef049702q
Szybist JP, Song J, Alam M, Boehman AL (2007) Biodiesel combustion, emissions and emission control. Fuel Process Technol 88(7):679–691
Tainter JA, Strumsky D, Taylor TG, Arnold M, Lobo J (2018) Depletion vs. innovation; the fundamental question of sustainability. In: Burlando R, Tartaglia A (eds) Physical limits to economic growth: perspectives of economic, social, and complexity science. Routledge, London, pp 65–93
Taschler D (2009) Optimization of a Biomass-based Fischer-Tropsch Synthesis – Location Güssing [Ph.D. Thesis]. Vienna University of Technology
The German Energy Agency (2006) Biomass to liquid – BtL implementation report, summary. Deutsche Energie-Agentur GmbH (DENA), Berlin
The Royal Society (2008) Sustainable biofuels: prospects and challenges. Policy document, 01/08. The Royal Society, London
Topp-Jorgensen J (1988) Topsoe integrated gasoline synthesis – the TIGAS process. In: Methane conversion, Studies in surface science and catalysis, vol 36. Elsevier, Amsterdam, pp 293–305
Ungerman AJ, Heindel TJ (2007) Carbon monoxide mass transfer for syngas fermentation in a stirred tank reactor with dual impeller configurations. Biotechnol Progr 23:613–620
Van der Drift A et al (2004) Entrained flow gasification of biomass. Ash behaviour, feeding issues, and system analyses. Energy research Centre of the Netherlands [Report ECN-C-–04-039]
Van Gerpen J (2005) Biodiesel processing and production. Fuel Process Technol 86 (10):1097–1107
Varatharajan K, Cheralathan M (2012) Influence of fuel properties and composition on NOx emissions from biodiesel powered diesel engines: A review. Renew Sustain Energy Rev 16 (6):3702–3710
Vásquez MC, Silva EE, Castillo EF (2017) Hydrotreatment of vegetable oils: a review of the technologies and its developments for jet biofuel production. Biomass Bioenergy 105:197–206
Veriansyah B, Han JY, Kim SK, Hong SA, Kim YJ, Lim JS, Shu YW, Oh SG, Kim J (2012) Production of renewable diesel by hydroprocessing of soybean oil: effect of catalysts. Fuel 94:578–585. https://doi.org/10.1016/j.fuel.2011.10.057
Vohra M, Manwar J, Manmode R, Padgilwar S, Patil S (2014) Bioethanol production: feedstock and current technologies. J Environ Chem Eng 2(1):573–584
WA A (1998) Report on biomass drying technology. National Renewable Energy Laboratory. [Report NREL/TP-570-25885]
Wang M (2000) Greet 1.5—transportation fuel-cycle model. Illinois: Argonne National Laboratory, Available at http://greet.anl.gov/publications.html
Wang W, Gowdagiri S, Oehlschlaeger MA (2013) Comparative study of the autoignition of methyl decenoates, unsaturated biodiesel fuel surrogates. Energy Fuels 27(9):5527–5532. https://doi.org/10.1021/ef4012593
Westbrook CK (2013) Biofuels combustion. Annu Rev Phys Chem 64:201–219

Westbrook CK, Pitz WJ, Sarathy SM, Mehl M (2013) Detailed chemical kinetic modeling of the effects of CC double bonds on the ignition of biodiesel fuels. Proc Combust Inst 34 (2):3049–3056

Wyman CE (1996) Ethanol production from lignocellulosic biomass: overview. In: Wyman CE (ed) Handbook on bioethanol: production and utilization. Taylor & Francis, Washington, DC, pp 1–18

Wyman CE (2004) Ethanol fuel. In: Cleveland CJ, Ayres RU, Costanza R, Goldemberg J et al (eds) Encyclopedia of energy, Elsevier science, vol 2. Elsevier, New York, pp 541–555

Xue J, Grift TE, Hansen AC (2011) Effect of biodiesel on engine performances and emissions. Renew Sustain Energy Rev 15(2):1098–1116

Yan Y, Li X, Wang G, Gui X, Li G, Su F, Wang X, Liu T (2014) Biotechnological preparation of biodiesel and its high-valued derivatives: a review. Appl Energy 113:1614–1631

Yilmaz N (2012) Comparative analysis of biodiesel-ethanol-diesel and biodiesel-methanol-diesel blends in a diesel engine. Energy 40(1):210–213

Yurchak S (1988) Development of mobil's fixed-bed methanol-to-gasoline (MTG) process. In: Methane conversion, Studies in surface science and catalysis, vol 36. Elsevier, Amsterdam, pp 251–272

Zabed H, Sahu JN, Suely A, Boyce AN, Faruq G (2017) Bioethanol production from renewable sources: Current perspectives and technological progress. Renew Sustain Energy Rev 71:475–501

Zhao X, Zhang L, Liu D (2012) Biomass recalcitrance. Part I: the chemical compositions and physical structures affecting the enzymatic hydrolysis of lignocellulose. Biofuels Bioprod Bioref 6:465–482. https://doi.org/10.1002/bbb.1331

Zhao X, Wei L, Julson J, Qiao Q, Dubey A, Anderson G (2015) Catalytic cracking of non-edible sunflower oil over ZSM-5 for hydrocarbon bio-jet fuel. Nat Biotechnol 32:300–312. https://doi.org/10.1016/j.nbt.2015.01.004

Biofuels Generation Based on Technical Process and Biomass Quality

2

Felipe Lange Shimizu, Hernan Dario Zamora Zamora, Alison Andrei Schmatz, Ranieri Bueno Melati, Danilo Bueno, and Michel Brienzo

Abstract

There is a wide variety of biomass types, implicating in biofuels and conversion process differences. Lignocellulosic biomass, for instance, can be converted into biofuels by biotechnology route. There are first-, second-, third-, and fourth-generation biofuels coming from different kinds of biomass and process. Besides biofuels, the carbohydrate and lignin of these biomasses can be used to generate other products of aggregated value. The biomasses have properties that resist the conversion processes, such as crystallinity and lignin contents. These difficulties are fought with genetic engineering and pretreatments to alter the material structure, decreasing the heterogeneity and recalcitrance, improving enzymatic hydrolysis and consequently the conversion into biofuels.

Keywords

Biofuels · Biomass · Renewable energy · Fossil fuels · Cellulose · Lignin

2.1 Introduction

Fossil fuels are a finite source of energy and its future increasing demand will raise CO_2 emissions and intensifies the greenhouse effect (Asumadu-Sarkodie and Owusu 2016). Alternatives to their use are necessary with urgency to stop environmental damage. Recently, biomass has been studied and used in biorefinery to provide these renewable alternatives (Zhao et al. 2012). Besides biodiesel, bioethanol is one strong candidate to replace or be combined with fossil fuel uses (Fig. 2.1). Lignocellulosic

F. L. Shimizu · H. D. Z. Zamora · A. A. Schmatz · R. B. Melati · D. Bueno · M. Brienzo (✉)
Institute for Research in Bioenergy (IPBEN), São Paulo State University (Unesp), Rio Claro, São Paulo, Brazil
e mail: michel.brienzo@unesp.br

N. Srivastava et al. (eds.), *Biofuel Production Technologies: Critical Analysis for Sustainability*, Clean Energy Production Technologies,
https://doi.org/10.1007/978-981-13-8637-4_2

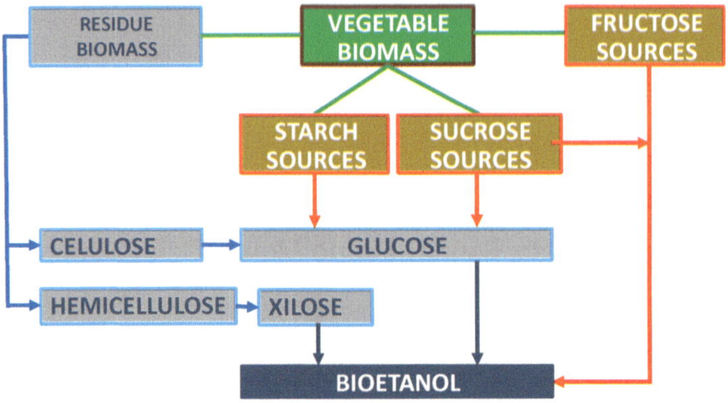

Fig. 2.1 Bioethanol production from carbohydrate sources of biomass

materials are available in large quantities and their products are all over our daily lives (Kamm and Kamm 2007). For instance, the 2018/2019 Brazilian sugarcane harvest was estimated at 615.84 million tons (CONAB 2018). The annual production of lignocellulosic material around the world hovers from 10 to 50 billion tons, representing 50% of the biomass materials on Earth (Zhao et al. 2012).

Lignocellulosic biomasses are organized into different categories based on their sources, which include crops destined to energy generation, aquatic plants, forest biomass and wastes, agricultural, and organic materials obtained from solid waste (Limayem and Ricke 2012). The lignocellulosic materials possess large quantities of carbohydrates, reaching more than 70% of its weight. These carbohydrates are an important source of sugar for biofuel generation and biodegradable products (Jørgensen et al. 2007). Unfortunately, the highly organized and resistant framework present in these materials results in protection from biological and chemical processes, a phenomenon called recalcitrance (Zhao et al. 2012). Recalcitrance is associated with the physicochemical characteristics of the plant cell walls. The presence of hemicellulose and lignin in the cell structure builds natural barriers that protect cellulose from hydrolysis or chemical aggression, maintaining its structure. Besides lignin and hemicelluloses, other factors affect biomass recalcitrance, such as physicochemical properties of the cellulose.

Several physical, chemical, biological, and combination of these pretreatments have been studied with the goal of enhancing the access to lignocellulosic biomass compounds, eliminating/modifying hemicelluloses and lignin, diminishing the cellulose crystallinity, improving cellulose accessibility and porosity, minimizing the sugars loss, and limiting the inhibitor formation (Kumar and Sharma 2017). Genetic modifications of the plant cell wall are being studied to alter the wall component interconnections and cellulose crystallinity, reducing lignin levels and increasing carbohydrate content.

2.2 Biomass Types for Biofuel Production

Biomass is the general name used for all organic materials derived from plants such as grasses, aquatic plants, woody plants, and manure (McKendry 2002). When subjected to pyrolysis, biomass converts into a dark-colored liquid called bio-oil, a biofuel precursor. Biofuels such as biodiesel can be obtained from rapeseed (*Brassica napus)*, canola (*Brassica* sp.), soybean (*Glycine max)*, sunflower (*Helianthus annuus)*, palm (*Elaeis* sp.), used cooking oils, and other oil sources. Looking at other sources, beef and sheep tallow also act as viable fat sources (Lü et al. 2011). Among these materials, soybean, rapeseed, and palm oils are commonly employed in biodiesel generation. However, soybean oil is valuable to the food industry, making it a costly alternative. Beef tallow, yellow grease, and pork lard are an abundant restaurant waste, providing viable biomass to biodiesel production (Demirbas 2011).

On the other hand, bioethanol can be obtained from carbohydrates found in lignocellulosic biomass such as sugarcane (*Saccharum officinarum*), algae, cotton (*Gossypium* sp.), wheat straw (*Triticum sp.*), corn (*Zea mays*), and others (Demirbas 2009). First-, second-, third-, and fourth-generation fuels can be obtained from various types of biomass (Table 2.1). Biofuels pertaining to the first generation come from edible agricultural products such as sugarcane, corn, and sugar beet, using these materials specifically for fuel production. Second-generation fuels come from many sources of cellulose, such as sugarcane bagasse and corn straw, using what remains from already used biomasses. Third-generation fuels come from processing algae into biofuels, while the fourth generation comes from bioengineered algae designed to produce biofuels and other vegetable oils (Lü et al. 2011).

Table 2.1 Biomasses used to produce different generations of biofuels

Biofuel types	
1st generation	
Biomass	Biofuel
Vegetable oil, sucrose, and starch	Bioethanol, biogas, biokerosene, and biodiesel
2nd generation	
Biomass	Biofuel
Solid waste, wood, sugarcane bagasse and straw, corn and wheat straw	Bioethanol, bio-oil, biohydrogen, biobutanol, biokerosene, and biodiesel
3rd generation	
Biomass	Biofuel
Algae	Biodiesel, vegetable oil
4th generation	
Biomass	Biofuel
Metabolic engineered biomass and vegetable oil	Biogasoline and biobutanol

2.3 Biomass Availability

Biomass is an organic matter that consists of cellulose, hemicelluloses, and lignin. In the world, biomasses such as softwood, hardwood, grasses, food waste, among others are used for the production of bioenergy and chemical in a biorefinery platform. The world widely needs to produce energy from renewable sources such as vegetable biomasses, an important source for bioenergy production. Therefore, biogas, biodiesel, aviation biokerosene, and bioethanol are gaining prominence in the global bioeconomic scenario.

However, the lignocellulosic materials represent a raw material source meager exploited in biotechnological processes. Its use is mainly based on energy generation through burning process and animal feed. However, new applications are emerging and can provide a better use of these materials. There are numerous researches with the objective of using the biomass macromolecules, cellulose, hemicelluloses, and lignin.

Among the lignocellulosic materials, sugarcane stands out, with Brazil being the largest producer and consumer in the world. The 2018/2019 sugarcane harvest was estimated at 615.84 million tons, with a projected production of 32.3 million liters of ethanol (CONAB 2018). Each ton generated from processed sugarcane obtained 250–280 kg of bagasse (Aguiar et al. 2010).

Another waste biomass resulting from sugarcane crops is the straw. As with the bagasse, the straw is rich in cellulose (40–44%), hemicelluloses (30–32%), and lignin (22–25%) (Santos et al. 2012). Each ton of cane used in Brazil produced 140 kg of straw (Santos et al. 2013). Sugarcane straw has been used to make second-generation ethanol like bagasse. One study has shown by estimates that 444 kg of glucose can be obtained for each ton of straw, and through a fermentation process, it would be possible to obtain 226.4 kg of ethanol, and to each ton of straw, 287 liters of ethanol are produced (Santos et al. 2012).

In addition to sugarcane, Brazil produces other biomass types, and one of them is forest biomass. The supply of forest biomass is by residues or energy forests (BNDES 2011) and may be composed of softwood or hardwood. In Brazil, the cultivation and production of softwood is highlighted by the pulp and paper industry. Among the softwoods produced in Brazil are the *Eucalyptus* ssp. and *Pinus* ssp. Brazil had 9.85 million hectares of forest plants, among which the *Eucalyptus* plantation was equivalent to about 75.2% and Pinus was equivalent to 20.6% of that amount (IBGE 2017). Brazil produced 19.85 thousand tons of cellulose in 2018 (IBÁ 2018). The production of the pulp and paper industry represents an important economic sector in Brazil. Softwood biomass from *Pinus* ssp. is prominent in Brazil due to its production amount. The *Pinus taeda* species is the most cultivated, with an extension of one million hectares planted in the southern region of the country (EMBRAPA 2014), with average yields varying from 18 to 28 m^3 per hectare per year (IPEF 2018).

Furthermore the biomass types already mentioned, such as sugarcane bagasse, hardwood, and softwood, soybeans are also considered in Brazil, one of the main raw materials for Brazilian bioenergy, since it is utilized in the production of Brazilian biodiesel.

Brazil has a prominence in the world soybean production, with around 116.996 million tons in the 2017/2018 harvest, with a 35.100 million hectares planted and a productivity of around 3333 kg per hectare, being the state of Mato Grosso the main producer of Brazilian soybean (EMBRAPA 2018).

Algae are another potential biomass for the production of biodiesel. In one hectare of algae cultivation, it would be possible to produce 100 to 237 thousand liters of oil (Defanti et al. 2010). Algae capture atmospheric carbon dioxide and produce bio-oils and products with high-added value (Espinosa et al. 2014). Compared to biomasses such as sugarcane, corn, soybean, and sunflower, algae are the ones with the highest productivity and photosynthetic efficiency, with 390–700 bep. $ha^{-1}.year^{-1}$ and 4–7%, respectively (Franco 2013).

Biogas is another promising biofuel, due to its potential for reducing organic methane and carbon dioxide released into the atmosphere (BNDES 2018). This biofuel is produced by the action of anaerobic microorganisms that decompose the organic matter. Among the main sources of organic matter abundant in Brazil is the waste from landfills (BNDES 2018). As an example of its potential, in 2015 Brazil produced about 705 thousand $Nm^3 \cdot day^{-1}$ of biogas from the landfills biomass and 349.705 thousand $Nm^3 \cdot day^{-1}$ from the waste biomass from the food and beverage industry (Brazilian Ministry of Mines and Energy 2017).

In addition to all biomasses described so far that are of great importance for the Brazilian bioenergy matrix, other local biomasses are being exploited for the production of aviation biokerosene, such as Macaúba (*Acrocomia aculeata*), a palm tree of great occurrence in Brazil, besides Babaçu (*Attalea speciosa*), Buriti (*Mauritia flexuosa*), and sunflower, from which oils can be extracted for biokerosene production. In Brazil, Macaúba occupies a planted area of 12 million hectares, Babaçu occupies 18 million hectares in the Brazilian territory, and the buriti 4 million hectares, and the sunflower 111.8 thousand hectares (UBRABIO 2017).

According to data released by the Brazilian government, biomass became the second most abundant energy source in the country in 2016 in the Internal Energy Offering (OIEE), with an energy generation equivalent to 54 tons of Wh, of which 67% correspond to biomass from sugarcane bagasse and straw. The survey demonstrates how important biomass is in the country's energy generation.

In the world, cultivars dedicated to energy production are called the energy crops. Besides Brazil, India is the second major producer of ethanol from sugarcane energy crops. India produced in 2017/2018 around 1.4 billion liters of ethanol (Reuters 2017).

In addition to sugarcane cultivars, elephant grass (*Pennisetum purpureum*) has been prominent in recent years, since ethanol produced from sugarcane has not shown significant increase in recent years (Campos 2015). Elephant grass has been evaluated as a high productivity biomass for second-generation ethanol, and its production per hectare can reach 150–200 tons of fresh mass (EMBRAPA 2015), and in the next few years this potentiality can be exploited, which can make elephant grass an important global energy crop.

Sorghum ssp. is also considered an energy crop, since it has potential to produce second-generation (2G) ethanol. The United States produced 9.24 million tons in the 2017/2018 harvest, followed by Nigeria, with 6.55 million tons produced (CONAB 2018). When cultivated, sorghum can produce about 35 to 70 tons of stems per hectare, and the production of ethanol from this biomass can reach 1400 to 2800 liters per hectare.

Some countries have the advantage of producing biofuels from waste, while others cultivate biomass to have feedstock, that is, production of energy crops. Brazil has excelled in both sources of biomass and presents a potential to produce biofuels from the vegetal biomass.

2.4 Carbohydrate Content in Biomass

In its natural form, the lignocellulosic biomass is composed of organic compounds which are considered to be a renewable kind of feedstock for producing inputs of industrial interest and biofuels (Kamm 2012). Cellulose is a macromolecule insoluble in water, made up of glucose units connected by β-1,4 glycoside connections, while hemicelluloses are a heteropolymer aggregate that includes several pentose and hexose sugars connected by several kinds of glycoside links. Lignin is a tridimensional macromolecule based on phenylpropane units (Melati et al. 2019), protecting and difficulty polysaccharides solubilization. The mass composition of these three components in the lignocellulosic biomass depends on the kind of plants and regions, as it is possible to see in Table 2.2. However, the entire mass proportion found in cellulose and hemicelluloses usually represents more than 60% (dry basis). Thus, the sugars derived from lignocellulosic biomass would be the first chemical products from this vegetal platform to be used by biorefineries of the next generation (Zakzeski et al. 2010).

It is possible to classify the sources of lignocellulosic biomass into several groups. They account for energy crops (the grasses that are perennial and other energy crops grown exclusively for this purpose), water plants (e.g., water hyacinth), wastes and biomasses from forests (soft- and hardwood, sawdust, pruning, and residues from bark thinning), residues from agriculture (straws from cereals, stovers, and bagasse), as well as organic fractions found in municipal solid wastes (MSW) (Kumar et al. 2009; Limayem and Ricke 2012). These resources from biomass appear to be the largest, most promising, and most abundant ones, considering that it is possible to find them all over the world (Table 2.2). One of the uses possible for the lignocellulosic biomass is as ethanol feedstock, and it practically does not need any extra requirements or any interference on the production of food and fiber crops (Sims et al. 2010). Roughly, 200 tons·year^{-1} of plant biomass are produced in the world. Nearly 109 tons of the primary biomass continues to be potentially accessible for the production of biofuels (Saini et al. 2015). There are predictions that claim that around 442 billion liters of bioethanol could be produced a year through the use of lignocellulosic biomass, if total crop waste and wasted crops were to be considered (Kim and Dale 2004; Sarkar et al. 2012).

The United States alone generate 1368 million tons of biomass that can be used for producing bioethanol. Of these, 428 million tons are derived from agriculture waste. Forestry wastes, energy crops, grains and corn, municipal and industrial wastes, and other kinds of waste contribute with 370, 377, 87, 58, and 48 million tons, respectively (Saini et al. 2015).

Table 2.2 Polysaccharide composition (cellulose and hemicelluloses) in some kinds of vegetable biomasses

Biomass	Cellulose (%)	Hemicelluloses (%)	Reference
Switch grass	5–20	30–50	McKendry (2002)
Miscanthus (spp. ?)	38.2–40	18–24.3	Brosse et al. (2010)
Grass esparto	33–38	27–32	Sánchez (2009)
Elephant grass	22	24	Sánchez (2009)
Bermuda grass	25	35.7	Prasad et al. (2007)
Grasses (general)	25–40	25–50	Saini et al. (2015)
Alfalfa stalks	48.5	6.5	Chandel et al. (2007)
Sugarcane whole	25	17	Saxena et al. (2009)
Napier grass	32	20	Saxena et al. (2009)
S32 rye grass (early leaf)	21.3–26.7	15.8–25.7	Sánchez (2009)
Orchard grass	32	40	Sánchez (2009)
Water hyacinth	18.2–18.4	48.7–49.2	Kumar et al. (2016), Nigam (2002)
General MSW	33–49	9–16	Li et al. (2012), Saxena et al. (2009)
Kraft paper	57.3	9.9	Schmitt et al. (2012)
High-grade paper	87.4	8.4	Schmitt et al. (2012)
Mixed- or low-grade paper	42.3	9.4	Schmitt et al. (2012)
Food waste	55.4	7.2	Schmitt et al. (2012)
Office paper	68.6	12.4	Mosier et al. (2005)
Newspaper	40–55	25–40	Howard et al. (2003)
Waste papers from chemical pulps	60–70	10–20	Prasad et al. (2007)
Sorted refuse	60	20	Prasad et al. (2007)
Primary waste water solids	8–15	0	Sánchez (2009)
Leaves and grass	15.3	10.5	Schmitt et al. (2012)
Solid cattle manure	16–4.7	1.4–1.33	Sánchez (2009)
Coffee husk	43	7	Gouveia et al. (2009)
Nut shells	25–30	25–30	Howard et al. (2003)
Corn cob	42–45	35–39	Prasad et al. (2007), Kuhad et al. (1993)
Cotton seed hairs	80–90	5–20	Prasad et al. (2007)
Corn stover	38–40	24–26	Saini et al. (2015), Zhu et al. (2005)
Corn fiber	14.28	16.8	Mosier et al. (2005)
Coir	36–43	0.15–0.25	Saini et al. (2015)
Sugarcane bagasse	42–48	19–25	Saini et al. (2015), Kim and Day (2011)
Rice straw	28–36	23–28	Saini et al. (2015)
Wheat straw	33–38	26–32	Saini et al. (2015)
Barley straw	31–45	27–38	Saini et al. (2015)
Sweet sorghum bagasse	34–45	18–27	Saini et al. (2015)

(continued)

Table 2.2 (continued)

Biomass	Cellulose (%)	Hemicelluloses (%)	Reference
Banana waste	13.2	14.8	Medina and Colorado (2006)
Sponge gourd fibers	66.6	17.4	Guimarães (2009)
Pineapple leaf fiber	70–82	18	Reddy and Yang (2005)
Oat straw	31–37	27–38	Sánchez (2009)
Rye straw	33–35	27–30	Sánchez (2009)
Bamboo	26–43	15–26	Sánchez (2009)
Bast fiber seed flax	47	25	Sánchez (2009)
Bast fiber kenaf	31–39	22–23	Sánchez (2009)
Bast fiber jute	45–53	18–21	Sánchez (2009)
Leaf fiber abaca (Manila)	60.8	17.3	Sánchez (2009)
Leaf fiber sisal (agave)	43–56	21–24	Sánchez (2009)
Leaf fiber henequen	77.6	4–8	Sánchez (2009)
Coffee pulp	35	46.3	Sánchez (2009)
Banana waste	13.2	14.8	Sánchez (2009)
Rice husk	25–35	18–21	Ludueña et al. (2011)
Softwood	27–30	35–40	McKendry (2002)
Softwood bark	18–38	15–33	Saini et al. (2015)
Softwood stem	45–50	25–35	Sánchez (2009)
Pine	44–46.4	8.8–26	Mosier et al. (2005), Olsson and Hahn-Hägerdal (1996)
Hardwood	20–25	45–50	McKendry (2002)
Hardwood bark	22–40	20–38	Saini et al. (2015)
Hardwood stem	40–55	24–40	Sánchez (2009)
Poplar	47.6–49.9	27.4–28.7	Mosier et al. (2005), Olsson and Hahn-Hägerdal (1996)
Swine waste	6	28	Sun et al. (2002)
Silver grass	37	25-35	Sun and Cheng (2002)
Beet pulp	23	36	Zheng et al. (2013)
Coconut husk	44	12	Goh et al. (2010)

2.5 Cellulose

Cellulose is the renewable polysaccharide that is most abundant in the world. It is composed of chains of homopolysaccharides, which are formed by units of anhydrous glucose and connected by β-1,4-D-glucose bonds, making up a crystalline structure. This happens because of its long intra- and intermolecular hydrogen links, facilitating its aggregation in fibrils (Pu et al. 2008). The microfibrils from cellulose have highly organized regions (crystalline regions) and less organized ones (amorphous). Native cellulose is a polymorphic structure, defined as cellulose I, that can suffer some chemical/thermal treatments converting it into polymorphous II, III, and

IV. Native cellulose can be found in two crystalline phases, known as Iα e Iβ (Atalla and Vander Hart 1984).

The average molecular mass of celluloses from sugarcane bagasse is around 157.800–168.400 g mol^{-1} and the cellulose fibers vary from 1.0 to 1.5 mm in size. One unit of cellulose, which is called elementary fibril, goes through self-assembly, so as to compose microfibrils that can be englobed by hemicellulose matrices, in order to produce macrofibrils, becoming resistant to chemical and enzymatic degradation (Bezerra and Ragauskas 2016). The average polymerization degree of cellulose from sugarcane bagasse varies from 974 to 1039 glucose monomers (Bian et al. 2014).

The polymerization degree refers to the amount of glucose monomers present in the chain, besides being able to influence in how efficient the enzymatic hydrolysis is. The natural structure of cellulose is firstly resistant to enzymatic hydrolysis, a condition that is a result of its high crystallinity and polymerization degree; moreover cellulose has the characteristic to be insoluble in water (Sant'anna et al. 2014).

2.6 Hemicelluloses

Hemicellulose compound is the second most abundant renewable polysaccharide in the planet, only behind cellulose (Sun et al. 2001; Peng 2009). They are classified as heterogeneous polysaccharides, found in big quantities in wood and perennial plants, forming the plant cell wall, along with cellulose and lignin (Peng 2009). However, hemicelluloses are closely associated with celluloses via physical and hydrogen connections; meanwhile, when it comes to lignin, they are gathered by covalent connections, which are characterized by sharing of one or more pairs of electrons between atoms, being them mainly R(radical)-benzyl and ether connections (Freudenberg 1965; Peng 2009).

In general, hemicelluloses are polysaccharides composed of 80 to 200 units of low molecular mass sugar residues. The general chemical formula for hemicelluloses is (C5H8O4) and they are organized into oses and hexoses (Paszner 1988). Among pentoses, D-xylose, L-arabinose, and L-rhamnose are highlighted, and among hexoses more common are the following sugars: D-glucose, D-mannose, and D-galactose. Hemicelluloses may also contain uronic acids, being the most important ones the 4-omethyl-D-glucoronic and the D-galacturonic acids (Timell 1964).

In annual plants, the most abundant hemicellulose kind is arabinoxylan, which contains D-xylopyranosyl molecules connected by β-(1-4) glycoside bonds. The hemicellulose from the sugarcane bagasse, for example, is defined as L-arabino-(4-O-methyl-D-glucuron)-D xylan (Brienzo et al. 2009). α-L-arabinofuranose and α-D-glucuronic acid (or its derivative 4-O-methyl) molecules are connected to them, in positions C-2, C-3, or both, as simple unit chains (Macgregor and Fincher 1993; Peng 2009). A natural compound, xylan, contains O-acetyl groups in some of the hydroxyl groups, at carbons 2 or 3, in its main chain (Peng 2009).

In opposition to cellulose, hemicelluloses are not chemically homogeneous and its compositions present great variation in proportion and content, depending on species, tissue kind, growth phase, and environmental and physiological conditions (Fengel and Wegner 1984; Brienzo et al. 2010; Brienzo et al. 2014). There is a great variety in content and composition of hemicelluloses in stalks, leaves, roots, and rinds. For example, the wheat hemicelluloses present different degrees of substitution between their tissues. The relation arabinose/xylose equals 1, when the external part of the grain (pericarp) is analyzed, while the internal part presents low substitution degree (0.3) (Brillouet and Joseleau 1987). Xylan is the main polysaccharide found in hemicelluloses from hardwoods, while mannan prevails in softwoods (Gao et al. 2013).

Hemicelluloses present an amorphous and hydrophilic structure and, thus, can be removed from cell walls with more ease than cellulose. Xylose and xylooligomers are frequently the main products obtained by pretreating and by enzymatic hydrolysis of hemicelluloses, being able to be used as fermentable sugars for bioethanol and organic acids generation, for example (Gao et al. 2013).

Other products of aggregated value of intermediary compounds, such as xylitol, furfural, and levulinic acid, used as producing chemicals and polymers, can also be generated from hemicelluloses through appropriate catalytic approaches (Werpy et al. 2004; Alonso et al. 2010). Other authors have suggested that the hemicellulose content and composition may also affect the recalcitrance of the cell wall (York and O'neill 2008).

It is considered that interactions between microfibers from hemicelluloses and cellulose and the lignin-carbohydrate connections can stop enzyme attack (Chundawat et al. 2011; HSU 1996). The transgenic *Arabidopsis*, with less content of methyl groups in the lateral glucuronoxylan chains, has released more xylose than the type of wild control in less severe conditions after the enzymatic hydrolysis (Urbanowicz et al. 2012).

2.7 Biomass Quality

Biomass quality can be classified by its physical properties and its component characteristics, such as density, particle size, calorific value, specific gravity, moisture content, extractive content, ash (as mineral content), and lignocellulosic composition (Kenney et al. 1990). To exemplify some of these properties, let us start with cellulose crystallinity. More organized regions of cellulose chain are called crystalline regions, differing them from the less systematic amorphous regions. Formed by microfibrils, crystalline regions present a more coordinated cellulose chain than the amorphous regions (Nakamura et al. 2014). These crystalline regions provide a higher recalcitrance to the biomass, making it harder for enzymes to reach cellulose and diminishing digestibility and accessibility (Zhao et al. 2012). Moving on, the polymerization degree impacts accessibility and can raise recalcitrance. Carbohydrates, more specifically glucose units, form cellulose. The quantity of these units in a cellulose chain determines the degree of polymerization. If cellulose has a high polymerization degree, then a bigger and stronger structure gets interrupting any enzymatic action (Zhao et al. 2012).

Accessibility is a special variable among factors that influence the biomass conversion into biofuels. It is very important due to its correlation to glucose yield (Crowe et al. 2017). To put it in a simple way, accessibility shows how much cellulose content is accessible to enzymatic hydrolysis. Cellulose attributes can affect biomass quality, but hemicelluloses and lignin content add heterogeneity to material structure and therefore they are taken into consideration too (Chandra et al. 2008). A strong and organized microfibril structure (surrounded by hemicelluloses and lignin) produces a protecting barrier in cellulose (Arantes and Saddler 2010).

The exposed surface areas of cellulose are commonly divided into internal and external surfaces. The external surface has bigger pores and increases in smaller particle sizes. The internal surface, on the other hand, has smaller pores and seems to be related to how effective enzymatic hydrolysis can be (Cosgrove 2005). It happens due to pore size distribution, key factor that suggest more importance than external surface area (Wang et al. 2018). Some enzymes can arrive to internal surface by entering the bigger pores on external surface area, but they can't reach the internal surface, which has smaller pores (Harmoko et al. 2016). Pores have to be bigger than 5.1 nm to allow the enzymes to reach the lignocellulosic material, demanding alterations in the biomass structure to increase pore size (Chandra et al. 2007). Adding to this problem, most studies dry and pretreat their materials, making smaller pore sizes (Luo and Zhu 2011).

Cellulose with low polymerization degree offers a larger number of settlement sites for enzyme attack, collaborating substantially with hydrolysis of cellulose chains. When materials are milling, it speeds up the process involving changes of the degree of polymerization, porosity, pore size distribution, and crystallinity (Zhao et al. 2012). Also worth noting, depolymerization is a mechanism that breaks polysaccharides into monomers (Goufo and Mugisha 2018). This is important because the depolymerized cellulose chains are formed in one stage of the hydrolysis. Cellulases are enzymes responsible for depolymerization, as they convert polysaccharides (cellulose) into monomers (glucose) (Yücel and Göycıncık 2015). This system is composed of three enzymes acting in a synergic way, meaning their effects combined are stronger than the effects of them working on their own (Hideno et al. 2009). Exposing cellulose by removing hemicelluloses and lignin, plus altering the material, results in more efficient enzyme action, enhancing glucose yield (Brienzo et al. 2017).

Particle size can also be considered to assess biomass quality (Chandra et al. 2008), as it influences in biomass conversion. The biomass particles have different sizes and shapes, often forming agglomerates that make measuring their exact sizes a difficult endeavor (Karimi and Taherzadeh 2016). It is assumed that smaller particles provide a larger exposed surface area, increasing cellulose accessibility, which consequently improves glucose yield (Chandra et al. 2008).

The hemicellulose presence interferes with cellulose accessibility, messing with pore distribution and pretreatments and enzymatic hydrolysis efficiencies (Zhao et al. 2017). Hemicelluloses occupy the spaces among and around cellulose fibers in the plant cell walls, acting as barrier against enzymatic hydrolysis and other external threats to its structure (Zhu et al. 2010). It has been confirmed that hemicellulose removal improves glucose yield after enzymatic hydrolysis of material (Chen et al. 2015). Lignin causes another problem besides providing a physical barrier like hemicelluloses. Lignin causes a non-productive adsorption of enzymes employed in cellulose hydrolysis, deactivating them along the process (Ximenes et al. 2011). Lignin removal with pretreatments can reduce loss of active enzymes (Yang and Pan 2016). Changes in biomass structure are observed by looking at superficial cavities, gaps, and hollow spaces created by hemicellulose elimination and/or lignin amendment by pretreatments like dilute acid, peroxide, steam explosion, and alkaline methods (Karimi and Taherzadeh 2016). A quantification of these components (cellulose, hemicelluloses, and lignin) can determine biomass quality and drive its application. The pretreatment processes improve lignocellulosic biomass quality allowing its conversion into biofuel, organic acids, and other chemicals with high industrial interest.

2.8 Biomass Genetic Improvement

Modification of cell wall structure of lignocellulosic biomass is the key step in improving the quality of bioenergetic crops. However, genetic improvement of energy crops is still a challenge to increase the production of biomass and biofuels on large scale. A genetic modification in the plant cell wall structure is desired to specifically alter the cell wall interconnections (cellulose crystallinity) and reducing lignin levels and increasing carbohydrate content. The genetic manipulation must select appropriate genes, which is a fundamental step to meet the expected objectives. However, there are more than 1,000 genes related to biosynthesis, degradation, and regulation of the plant cell wall (Xie and Peng 2011). As complementary techniques for genetic improvement of biomass quality, several tools have been developed: high-throughput genotyping, generation sequencing, and molecular breeding techniques such as marker assisted selection and genomic selection. These modern tools have been applied in a wide variety of bioenergetic species such as sugarcane, elephant grass, sorghum, and miscanthus (Allwright and Taylor 2016).

In order to optimize the biomass conversion from sugarcane into bioethanol, it is imperative to use genetically improved hybrids with better biomass degradability. However, genetic engineering is still a challenge due to limited understanding of the large and complex genome of this plant (Hoang et al. 2015).

Modern sugarcane represents the extreme example of polyploidy, with a double-genome structure, where genotypes resulting from the same crossover may vary in amount of homologous and homologous chromosomes. Mapping of genetic linkages is particularly difficult to construct in polyploid species because (i) the statistics are much more complicated for polyploids than for diploids, (ii) a wide variety of genotypes are expected in a segregating population, (iii) there are several ways of forming gametes and random pairs of multiple homologous chromosomes, (iv) alleles segregation with different dosage levels makes it impossible for the different frequencies of gametes to be identified with absolute certainty, and (v) the genome constitution of some polyploids are unclear, making their inheritance pattern difficult to determine (Kole 2010).

Modern hybrids of sugarcane are results of crosses between varieties and/or clones, making the combination of chromosomes unique among each offspring, due to random classification of chromosomes. Genetic and biotechnological breeding approaches in sugarcane germplasm can play a key role in improving this biomass in the biofuel production. Variations in the phenotypes can be found in the biomass yield, fiber content, and sugar composition (Hoang et al. 2015).

To produce bioethanol from sugarcane, transcription factors (TFs) that regulate monolignol biosynthesis in lignin pathway have been studied, since the understanding of this metabolic pathway reduces and modifies the lignin content, which is essential to address the biomass reclassification problem. The lignin biosynthesis pathways are complex and involve at least 10 enzymes and 28 associated unigenes identified in sugarcane. There are some key genes in lignin pathway that encode terminal enzymes such as caffeic acid O-methyltransferase (COMT) and cinnamyl alcohol dehydrogenase (CAD) (Hoang et al. 2015).

Lignin synthesis is like metabolism of phenylpropanoids in plants, containing a series of enzymes common to other processes (phenylalanine ammonia-lyase (PAL), cinnamate-4-hydroxylase (C4H), and COMT) and specific enzymes such as cinnamoyl-CoA reductase (CCR) and cinnamyl alcohol dehydrogenase (CAD), among others like ferulate-5-hydroxylase (F5H), caffeoyl CoA O-methyltransferase (CCoAOMT), and hydroxycinnamate CoA ligase (4CL). Studies have shown that maize and sorghum mutants that were deficient in CAD and/or COMT showed an easy digestibility due to the reduced lignin content (Ramos et al. 2001). Therefore, understanding enzymes and mechanisms involved in lignin biosynthesis is fundamental in genetic improvement of biomass to produce biofuels.

Downregulation of enzymes (COMT or CAD) in ending steps of the monolignol biosynthetic pathway has little or no negative effect on plant growth. A moderate reduction in lignin content (3.9% to 8.4%) can significantly reduce recalcitrance of sugarcane biomass without affecting plant growth under indoor controlled environmental conditions. Genetic studies demonstrate that downregulation of sugarcane COMT gene by 67% to 97% reduced lignin content by 3.9% to 13.7%, respectively. On reflection, the syringyl/guaiacyl (S/G) ratio in lignin was reduced from 1.47 in wild type to values between 1.27 and 0.79, and fermentable glucose increased 29% without pretreatment and up to 34% with pretreatment, both after enzymatic hydrolysis (Jung et al. 2012).

The suppression of COMT on lignin biosynthesis has shown that transgenic lines showed a significant reduction of S monomers and similar amount of G monomers in relation to wild type. The S/G ratio was also lower in the transgenic lines, which in turn had improvements in enzymatic digestibility. However, a microscopic analysis demonstrated that vascular bundle tissues and sclerenchyma fiber cells were intact in varieties with reduced lignin concentration. A histochemical analysis revealed the reduction of S lignin units in sclerenchyma fiber cells (Fig. 2.2). However, the transgenic lines showed thinner stems and lower biomass production in relation to wild plants (Jung et al. 2012).

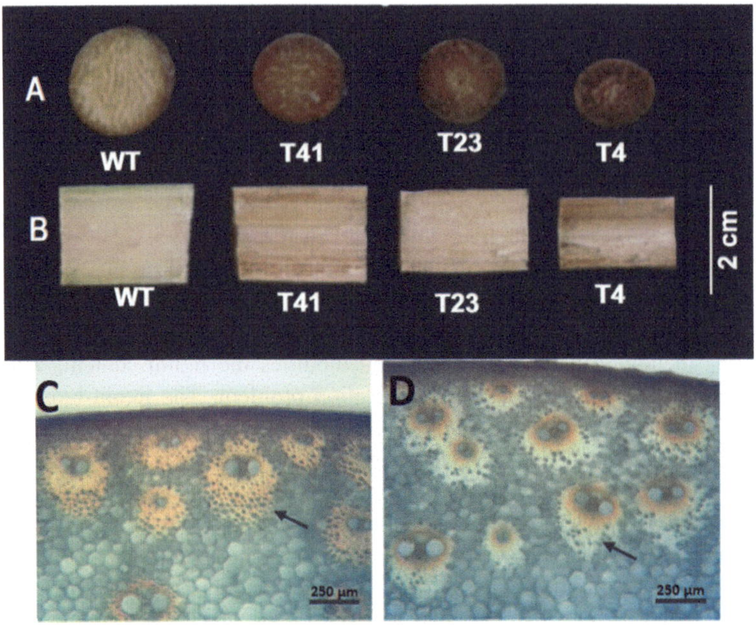

Fig. 2.2 Phenotypic differences in sugarcane stalks through suppression of the COMT gene and reduction in lignin content. WT: wild type. T41, T23, and T4: transgenic plants with total lignin content reduced by 3.9, 8.4, and 13.7%, respectively, when compared to wild type. (**a**) Transversal cuts. (**b**) Longitudinal cuts. (**c**) and (**d**) represent wild-type and T4 lineage, respectively, stained with Mäule reagent indicating a reduction in lignin S units in sclerenchyma fiber cells (Jung et al. 2012)

Breeding strategies also play a key role in improving species of raw materials in production of cellulosic ethanol, as have been done with food crops from their wild ancestors for decades, resulting in seeds more resistant to biotic and abiotic factors (Sticklen 2008). Important features should be improved in biomass for biofuel production, such as yield, adaptability, resistance to drought and diseases, and high biomass yield per hectare (classical breeding) (Hoang et al. 2015). By changing DNA code of ancestral sugarcane species together with traditional breeding methods, it was possible to develop sugarcane energy (energy cane). The main characteristic of this variety is the significant increase of biomass, containing more fibers than sucrose in its composition (Matsuoka et al. 2014).

In studies carried out with 115 varieties of sugarcane (100 of classic breeding and 15 of precision breeding), it was observed the potential of genotypes for fiber processing capacity and yield of fermentable sugars, obtained from acid pre-treatment and enzymatic hydrolysis (Benjamin et al. 2013). Varieties of genetic improvement showed higher levels of arabinoxylan and lower lignin and ash content, whereas classical breeding varieties that presented low lignin content did not present higher amounts of structural carbohydrates. The highest glucose yields after enzymatic hydrolysis were observed in samples that had lower lignin contents.

Most precision breeding variety exhibited higher digestibility than many classical breeding varieties (Benjamin et al. 2013).

Sugarcane clones (transgenic and wild type) showed differences between agronomic properties and sugar conversion, since transgenic specimens showed higher stature, shoot population, and soluble sugars in juice (144 cm, 129.603 e 147 kg·ton^{-1} of sugarcane, respectively) when compared to wild type (123 cm in height, population of 104,039 stems and 140 kg of soluble sugars in juice per ton of sugarcane). As for ethanol production, transgenic variety yielded 29.6 g·L^{-1}, corresponding to 75.8% potential glucose in pretreated material, while wild variety yielded 26.8 g·L^{-1}, corresponding to 69.2 % of potential glucose. These variations can be attributed to different lignin contents found in transgenic and wild-type genotypes (16.4% and 21.3%, respectively) (Benjamin and Görgens 2015).

Among various biomass used in biofuel production, elephant grass has potential for coal, alcohol, and methane production or even for direct burning in the boiler feed, due to extremely high positive energy balance and efficiency in fixation of atmospheric CO_2. Elephant grass genotypes (Paraíso and Roxo) can accumulate dry biomass varying from 30 Mg·ha^{-1} to 42 Mg·ha^{-1}, respectively (Flores et al. 2012). Elephant grass presents potential for bioenergy by direct combustion. Among five evaluated varieties, it was observed that yield of dry mass ranged from 45 to 66.6 Mg·ha^{-1}. In addition, the recalcitrance of this biomass can vary according to genotypes (Morais et al. 2009).

The genetic diversity of 100 varieties of elephant grass was quantified in relation to their bioenergetic use. The study classified varieties into 6 groups according to agronomic characteristics, which could be employed, to produce cellulosic ethanol, forage, or direct combustion. Elephant grass can contribute to diversification of sustainable energy matrix. To reach energy goal, some actions must be taken to enhance their bioenergetic potential, as crosses between individuals diverging from different clusters. The genetic variability of the Active Elephant Grass Germplasm Bank (BAGCE) can be exploited to produce higher quality combinations that maximize conversion to second-generation ethanol and direct biomass combustion (Rocha et al. 2017).

Sorghum sp. was evaluated as potential and yield (of six genotypes of sorghum, one cultivar and five hybrids) during five years in several areas of the United States (six different states). There was a significant variation between genotypes for fresh and dry weight, moisture, and brix, where some hybrids are more tractable to produce bioenergy and biomass. The results showed that one of the hybrids produced the highest fresh and dry average weights (58.6 and 17.9 tons per hectare, respectively), while the lowest averages were 35.1 (fresh weight) and 10.1 (dry weight) metric tons per hectare, but the consistency of production may vary year by year according to each locality. This study demonstrated that sorghum can yield sufficient biomass amounts to meet needs of developing lignocellulosic industry, allowing for breakthroughs in cellulosic ethanol industries (Gill et al. 2014).

Miscanthus sinensis (102 varieties) is used to demonstrate the genetic variety distribution, relating genotype variation of cell wall composition with lignocellulose degradation rate and its relationships with biohydrogen production. Lignin contents were found between 6.96 and 20.75%, while cellulose and hemicellulose degradation ranged from 2.08% to 37.87% and from 14.71% to 52.50%, respectively. After fermentation, biohydrogen production ranged from 14.59 to 40.66 mL per gram of *Miscanthus* biomass. Improvement of varieties in their composition and cell wall structure with high rates of degradability can significantly increase the sustainable production of bioenergy (Zhao et al. 2014).

Switchgrass (*Panicum virgatum*) was genetically engineered and demonstrated to be possible to increase cellulosic ethanol production by up to 2.6-fold (by overexpressing the PvMYB4 transcription factor). This strategy decreased carbon deposition into phenolic fermentation inhibitors and lignin and maintains the propitiation of potentially fermentable soluble sugars. Overexpression of the switchgrass R2-R3 MYB transcription factor PvMYB4 restricts genes from biosynthetic lignin pathway and directly affects saccharification without acid pretreatment, directly affecting recalcitrance. This variety provides new germplasm for development of cultivars as a raw material to produce biomass for biofuels (Shen et al. 2013).

Hybrid poplars (genus *Populus*) are among the fastest growing temperate climate trees in the world and are also considered one of the most promising raw materials for biofuels and other value-added products. The estimated nominal yield (including the moisture content at harvest) of hybrid poplar species grown in North America is between 14 $Mg{\cdot}ha^{-1}{\cdot}year^{-1}$ and can be compared to switchgrass (14 $Mg{\cdot}ha^{-1}{\cdot}year^{-1}$) and higher than wheat straw and corn stover. The species and hybrids of the genus *Populus* have cellulose contents between 42 and 49% (higher than that of switchgrass), hemicelluloses of 16 to 23% and lignin of 21 to 29%. Studies performed on poplar clones demonstrated that a small decrease in S/G ratio resulted in a statistical increase in releasing of xylose after hydrolysis with dilute sulfuric acid. Suppression of the COMT gene in poplar does not result in a reduction in lignin content, but S/G ratios were decreased. Sequencing of poplar genome paved ways for development of new cultivars and clones optimized to produce biofuels (Sannigrahi and Ragauskas 2010).

Basically, the cell wall of plants consists of cellulose, non-cellulosic polysaccharides, and lignin. Manipulations of molecular and structural levels of wall components can improve performances of plants in biofuel production. This can be done through genetic improvement or through methods such as screening of natural variation and random mutagenesis. Altering genes encoding enzymes responsible for lignin biosynthesis or manipulating genes involved in cell wall polysaccharide biosynthesis may facilitate their deconstruction and improve mass/energy balances during biofuel production (Burton and Fincher 2014). Plant crops for biofuels that hold high yields and efficiency in conversion allow crops in smaller areas, minimizing competition for food crops and conserving biodiversity. In this way, development of genetically modified plants is key to biofuel production, improving the efficiency and viability of sustainable production processes (Furtado et al. 2014).

2.9 Biomass Conversion Processes

As mentioned earlier, nowadays the renewable biomass is considered a real source to obtain many types of products like fuels (bioethanol, biodiesel, biohydrogen, biobutanol, biomethane, etc.) and platform chemicals (organic acids, monomers, etc.) (Müller-Langer and Kaltschmitt 2015; Du Preez 2016), which have been included inside civilization needs, taking into account the humanity sustainable development (Solomon et al. 2015). Thus, in this part, what kind of fuels and chemicals are obtained from renewable resources and the principal and current processes used to goal the conversions will be discussed.

2.10 Fuels and Chemicals from Lignocellulosic Biomass

At present day, the society has basic needs of food, clothes, fuels, and other chemicals, which are dependent principally on oil, natural gas, and coal. Among them, the oil is the number one precursor for different industries, which has generated a lot of direct negative alterations to natural ecosystems and adverse effects to the human health, mainly by green gas emissions. Therefore, to replace fossil-based products by bio-based products is an attractive issue in the current world (De Bhowmick et al. 2018).

Taking specifically about agricultural waste biomass, lignocellulosic raw material has been studied with the aim to produce different chemicals, since this material is considered clean, renewable, abundant, and available. However, these lignocellulosic biomass must be conditioned through pretreatment and hydrolysis processes in order to generate an unstructured material, thus allowing cellulose, hemicelluloses, and lignin to be more exposed and to obtain simpler molecules (De Bhowmick et al. 2018). To simply explain the difference between pretreatment and hydrolysis, it could imagine a wool scarf, when it is submitted to stretch, the wood fibers are more visible (pretreatment), and when these stretched fibers are cut with scissors, the wool fibers will be broken into smaller pieces (hydrolysis). In general, pretreatments are processes applied to lignocellulosic biomass to promote access to cellulose, hemicelluloses, and removal or modification of lignin, and hydrolysis is a process to break cellulose and hemicelluloses into smaller molecules (Kumar et al. 2018). Both processes (pretreatment and hydrolysis) are used to obtain other bio-based products known like platform chemicals or intermediate products (Du Preez 2016; Yung 2016). Furfural, hydroxymethylfurfural, levulinic acid, furfuryl alcohol, sorbitol and others are examples of bio-based products different to biofuels (Iroegbu and Hlangothi 2018; Steinbach et al. 2017; Hartono et al. 2016).

2.11 Lignocellulosic Biomass Pretreatments

Without pretreatment, it is considered that just 20% of lignocellulosic biomass in natura could be hydrolyzed, and this low percentage is because the chemical structure is recalcitrant, which justifies the importance to carry out a pretreatment stage on lignocellulosic material before subsequent processes (Rocha et al. 2012). Several physical, chemical, biological, and any combination of these pretreatments (Table 2.3) have been researched to enhance the access to lignocellulosic biomass

Table 2.3 Benefits and disadvantages of chemical, physical, physicochemical, and biological pretreatments applied on lignocellulosic biomass

Pretreatment	Example	Benefits	Disadvantages
Physical	Milling, microwave irradiation and thermal	Increase superficial area	High energetic expense
		Reduce cellulose crystallinity	Not remove lignin
		Reduce cellulose polymerization index	Not remove hemicelluloses
Chemical	Acid, alkaline, and solvent	Increase superficial area	High price of acids, alkalis, and solvents
		Reduce cellulose crystallinity	Difficulty to recover the acids, alkalis, and solvents
		Reduce cellulose polymerization index	Formation of inhibitors for hydrolysis and fermentation processes (depending on acid or alkali concentration)
		Partial or total solubilization of lignin (depending on alkali concentration)	Corrosion (concentrated acid)
		Partial or total solubilization of hemicelluloses (depending on acid concentration)	
		Produce glucose (diluted acid)	
Physicochemical	Steam explosion and hydrothermal	Partial or total solubilization of hemicelluloses (depending on acid concentration)	Not act on lignin
		Produce sugars	Require devices equipped with heat and pressure
Biological	White-rot fungi and brown-rot fungi	Remove lignin (white rot)	High time of operation
		Remove hemicelluloses and cellulose (brown rot)	Fungi consume monosaccharides
		Not produce inhibitors	Special conditions to use fungi

compounds (cellulose and hemicelluloses), to remove or modify lignin or hemicelluloses, to diminish the cellulose crystallinity, to enlarge the lignocellulosic biomass accessibility and porosity, to minimize the sugars loss, and to limit the inhibitor formation (Kumar and Sharma 2017). The principal path to assess the pretreatment effect on lignocellulosic materials is based on determination of mass composition, x-ray crystallography, scanning electron microscope, high-performance liquid chromatography (HPLC), nuclear magnetic resonance (RMN), and Fourier-transform infrared spectroscopy (FT-IR).

Among pretreatments, the steam explosion is a technique utilized on lignocellulosic biomass, which promotes chemical modifications in its composition and fracks the cellular wall structure, guaranteeing more accessibility to cellulose and hemicelluloses (Pielhop et al. 2016). Steam explosion is a practice where lignocellulosic biomass is submitted to conditions of temperature (160–240 °C), pressure (0.7–4.8 MPa) and superheated or saturated vapor (it can use water or prepared solutions with acid or alkali) (Han et al. 2018; Silva et al. 2018). Hemicellulose solubilization and lignin structure modifications are caused by a sudden decompression, which lets a higher accessibility to cellulose and higher digestion of lignocellulosic biomass (Kumar et al. 2018).

The method takes place in two steps, the first named autohydrolysis and the second one known as decompression. At the first stage, the substance changes the phase (from liquid to vapor) and gets into lignocellulosic structure, and it allows to remove acetyl, carbonyl, and carboxyl groups present in hemicellulose chains and to obtain acetic and uronic acid, which act in hemicellulose hydrolysis. In the second stage, condensed steam goes from lignocellulosic biomass to outside, causing a mechanical breaking (explosion) into lignocellulosic structure (Martino et al. 2017; Asada et al. 2018; Wang et al. 2018). The steam explosion pretreatment can be ameliorated by using acid catalysts, which help to reduce the operation time of the whole process, because hemicellulose hydrolysis (the first stage) occurs faster than without catalyst process (Silva et al. 2018).

The gains of using steam explosion as pretreatment are limited use of chemical products (generally water or dilute solutions), low energy expense compared to mechanical pretreatment, hemicelluloses that are partially hydrolyzed, and more accessibility to cellulose and lignocellulosic biomass after process is more susceptible to enzymes action (Chen et al. 2015). The disadvantages of steam explosion as pretreatment are obtaining by-products as furfural and hydroxymethylfurfural, partial solubilization of hemicelluloses, and the necessity to wash the lignocellulosic biomass after process and not acting on lignin (Bhagwat et al. 2015).

On the other side, thermal pretreatment with dilute acid is another kind of process that has been considered in the last years as a method to get unstructured lignocellulosic biomass. The substances generally employed are the hydrochloric and sulfuric acid, which are prepared in low concentration solutions (1–3% w/v). The dilute acid pretreatment is selected over the concentrated acid process, because it reduces the inhibitor formation and minimizes the corrosion problems (Loow et al. 2016).

The temperature conditions of this pretreatment generally are between 90 and 140 °C, and it is selected when it is wanted to act on hemicelluloses and low degradation of polysaccharides. The dilute acid pretreatment is recommended over lignocellulosic biomass, whose lignin composition is lower than 15%. Generally, dilute acid processes do not remove lignin (but do modify its chemical structure, promoting more accessibility) and remain the cellulose, letting elevated enzymatic hydrolysis rates as result of amplified accessibility of the exposed cellulose (Karapatsia et al. 2017; Kumar et al. 2018).

2.12 Enzymatic Hydrolysis

The conversion process of cellulose and hemicelluloses into smaller molecules (commonly called total reducing sugars or fermentable sugars with 6 and 5 carbons, respectively) is named hydrolysis or saccharification. This process can be developed in the presence of acids, enzymes, or both simultaneously (Loow et al. 2016; Sharma et al. 2017). Acid hydrolysis involves organic or inorganic acids (such as sulfuric, hydrochloride, oxalic, acetic, and uronic). Enzymatic hydrolysis includes different type of enzymes acting on cellulose and hemicellulose chains (Loow et al. 2016). Cellulose hydrolysis occurs by action of cellulases, called cellulolytic enzymes too, and hemicelluloses hydrolysis occurs by action of hemicellulases, called hemicellulolytic enzymes too (Bhattacharya et al. 2015).

Cellulases are grouping in three enzymes categories. Endoglycanases, carboxylmethylcellulases, or CMCases (EC 3.2.1.4) are enzymes that hydrolyze randomly the non-crystalline sections of cellulose generating new non-reducing and reducing ends. Exoglycanases, cellobiohydrolases, CBH, or Avicelases (EC 3.2.1.91) are enzymes that act over ends of cellulose chains and the new non-reducing and reducing ends to produce cellobiose molecules. And β-glycosidases (EC 3.2.1.21) are responsible to break cellobiose and thus to release glucose (Dotsenko et al. 2018) (Fig. 2.3).

According to chemical composition of hemicelluloses, its biological breaking is completed by several groups of enzymes. Endoxylanases, β-xylosidases, α-L-arabinofuranosidases, and β-galactosidases are the hemicellulolytic enzymes most relevant in hydrolysis process of xylan, the principal polymer that makes up hemicelluloses. Endoxylanases (EC 3.2.1.8) are enzymes that attack the principal chain of xylan to generate xylobiose. β-Xylosidases (EC 3.2.1.37) are responsible to break xylobiose and thus to release xylose. α-Arabinofuranosidase (EC 3.2.1.55) and β-galactosidase (EC 3.2.1.23) enzymes attack the xylan branches to release arabinose and galactose molecules, respectively (Quiroz-Castañeda and Folch-Mallol 2011).

2.13 Conclusion

Fossil fuels are a finite resource and damage the environment, turning their use an unsustainable practice if new technologies continue to advance demanding more and more energy. Sugarcane, corn, wood, algae, and many more biomasses are

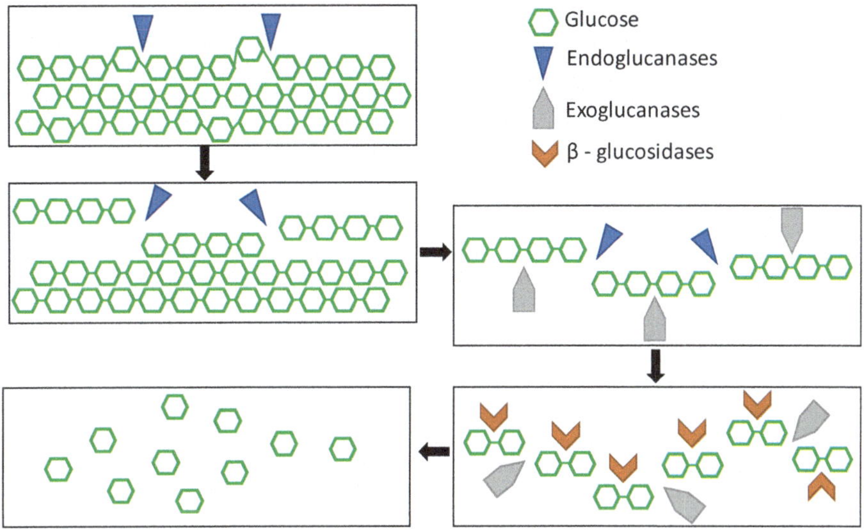

Fig. 2.3 Cellulases action on cellulose hydrolysis into glucose

employed to produce biofuels. Besides converting biomass into biofuels, their components are individually useful for the assembly of high-value-added products.

Recalcitrance caused by the material properties slows progress toward a better and refined conversion. Genetically engineering crops is not a new process, which means that investing and studying ways to genetically improve the biomasses regarding their resistance to pretreatments and conversion is already a reality. These pretreatments are also constantly being updated to offer different ways of altering the biomass. Biomass conversion is a modern necessity and will be increasingly vital.

Acknowledgment Authors are thankful for the support of the Brazilian Council for Research and Development (CNPq, process 401900/2016-9) and São Paulo Research Foundation (FAPESP, process 2017/11345-0).

References

Aguiar MM, Ferreira LFR, Monteiro RTR (2010) Use of vinasse and sugarcane bagasse for the production of enzymes by lignocellulolytic fungi. Braz Arch Biol Technol 53(5):1245–1254

Alonso DM, Bond JQ, Dumesic JA (2010) Catalytic conversion of biomass to biofuels. Green Chem 12(9):1493–1513

Allwright MR, Taylor G (2016) Molecular breeding for improved second generation bioenergy crops. Trends Plant Sci 21(1):43–54

Arantes V, Saddler JN (2010) Access to cellulose limits the efficiency of enzymatic hydrolysis: the role of amorphogenesis. Biotechnol Biofuels 3(1):4

Asada C, Sasaki C, Suzuki A, Nakamura Y (2018) Total biorefinery process of lignocellulosic waste using steam explosion followed by water and acetone extractions. Waste Biomass Valoriz 9:2423–2432. https://doi.org/10.1007/s12649-017-0157-x

Asumadu-Sarkodie S, Owusu PA (2016) Feasibility of biomass heating system in Middle East Technical University, Northern Cyprus Campus. Cogent Eng 3(1):1134304

Atalla RH, Vanderhart DL (1984) Native cellulose: a composite of two distinct crystalline forms. Science 223(4633):283–285

Benjamin Y, Görgens JF (2015) Improving sugar conversion and bioethanol yield through transgenic sugarcane clone in South Africa. Int J Environ Bioenergy 10(1, October):9–25

Benjamin Y, Cheng H, Görgens JF (2013) Evaluation of bagasse from different varieties of sugarcane by dilute acid pretreatment and enzymatic hydrolysis. Ind Crops Prod Elsevier B.V. 51:7–18. https://doi.org/10.1016/j.indcrop.2013.08.067

Bezerra TL, Ragauskas AJ (2016) A review of sugarcane bagasse for second-generation bioethanol and biopower production. Biofuels Bioprod Biorefin 10(5):634–647

Bhagwat S, Ratnaparkhe S, Kumar A (2015) Biomass pre-treatment methods and their economic viability for efficient production of biofuel. Br Biotechnol J 8:1–17. https://doi.org/10.9734/BBJ/2015/18284

Bhattacharya AS, Bhattacharya A, Pletschke BI (2015) Synergism of fungal and bacterial cellulases and hemicellulases: a novel perspective for enhanced bio-ethanol production. Biotechnol Lett 37:1117–1129. https://doi.org/10.1007/s10529-015-1779-3

Bian J, Peng F, Peng XP, Xiao X, Peng P, Xu F, Sun RC (2014) Effect of [Emim] ac pretreatment on the structure and enzymatic hydrolysis of sugarcane bagasse cellulose. Carbohydr Polym 100:211–217

BNDES, National Development Bank (2011) Perspectivas do setor de biomassa de madeira para a geração de energia. https://web.bndes.gov.br/bib/jspui/bitstream/1408/2523/1/A%20BS%2033%20Perspectivas%20do%20setor%20de%20biomassa%20de%20madeira%20para%20a%20gera%C3%A7%C3%A3o%20de%20energia_P.pdf. Accessed 8 Jan 2019.

BNDES, National Development Bank (2018) Biogás de resíduos agroindustriais: panorama e perspectivas. https://web.bndes.gov.br/bib/jspui/bitstream/1408/15384/1/BS47__Biogas__FECHADO.pdf. Accessed 29 Jan 2019.

Brazilian Ministry of Mines and Energy (2017) Impactos da participação do biogas e do biometano na matriz brasileira. http://www.epe.gov.br/sites-pt/publicacoes-dados-abertos/publicacoes/PublicacoesArquivos/publicacao-244/topico-257/EPE_IV%20FORUM%20BIOGAS_JOSE%20MAURO_2017_1710.pdf. Accessed 10 Jan 2019.

Brienzo M, Siqueira AF, Milagres AMF (2009) Search for optimum conditions of sugarcane bagasse hemicellulose extraction. Biochem Eng J 46(2):199–204

Brienzo M, Carvalho W, Milagres AM (2010) Xylooligosaccharides production from alkali-pretreated sugarcane bagasse using xylanases from Thermoascus aurantiacus. Appl Biochem Biotechnol 162(4):1195–1205

Brienzo M, Ferreira S, Vicentim MP, de Souza W, Sant'Anna C (2014) Comparison study on the biomass recalcitrance of different tissue fractions of sugarcane culm. Bioenergy Res 7(4):1454–1465

Brienzo M, Fikizolo S, Benjamin Y, Tyhoda L, Görgens J (2017) Influence of pretreatment severity on structural changes, lignin content and enzymatic hydrolysis of sugarcane bagasse samples. Renew Energy 104:271–280

Brillouet JM, Joseleau JP (1987) Investigation of the structure of a heteroxylan from the outer pericarp (beeswing bran) of wheat kernel. Carbohydr Res 159(1):109–126

Brosse N, El Hage R, Chaouch M, Pétrissans M, Dumarçay S, Gérardin P (2010) Investigation of the chemical modifications of beech wood lignin during heat treatment. Polym Degrad Stab 95(9):1721–1726

Burton RA, Fincher GB (2014) Plant cell wall engineering: applications in biofuel production and improved human health. Curr Opin Biotechnol 26:79–84

Campos BB (2015) Produção de etanol em biomassa de capim-elefante por Kluyveromyces marxianus CCT 7735. Dissertation, Federal University of Viçosa

Chandra RP, Bura R, Mabee WE, Berlin DA, Pan X, Saddler JN (2007) Substrate pretreatment: The key to effective enzymatic hydrolysis of lignocellulosics? In: Biofuels. Springer, Berlin Heidelberg, pp 67–93
Chandra RP, Esteghlalian AR, Saddler JN (2008) Assessing substrate accessibility to enzymatic hydrolysis by cellulases. In: Characterization of lignocellulosic materials, pp 60–80. https://doi.org/10.1002/9781444305425.ch4
Chandel AK, Kapoor RK, Singh A, Kuhad RC (2007) Detoxification of sugarcane bagasse hydrolysate improves ethanol production by Candida shehatae NCIM 3501. Bioresour Technol 98(10):1947–1950
Chen JH, Wang K, Xu F, Sun RC (2015) Effect of hemicellulose removal on the structural and mechanical properties of regenerated fibers from bamboo. Cellulose 22(1):63–72
Chundawat SP, Beckham GT, Himmel ME, Dale BE (2011) Deconstruction of lignocellulosic biomass to fuels and chemicals. Annual Review Chem Biomolecular Eng 2:121–145
CONAB, National Supply Company (2018) Análise mensal sorgo Março de 2018. https://www.conab.gov.br/info-agro/analises-do-mercadoagropecuarioextrativista/analises-do-mercado/historico-mensal-de-sorgo. Accessed 5 Jan 2019
CONAB, National Supply Company. Produção total de etanol no Brasil bate recorde com 32,3 bilhões de litros (2018). https://www.conab.gov.br/ultimas-noticias/2630-producao-total-de-etanol-no-brasil-bate-recorde-com-32-3-bilhoes-de-litros. Accessed 3 Jan 2019
Cosgrove DJ (2005) Growth of the plant cell wall. Na Rev Mol Cell Biol 6(11):850
Crowe JD, Zarger RA, Hodge DB (2017) Relating nanoscale accessibility within plant cell walls to improved enzyme hydrolysis yields in corn stover subjected to diverse pretreatments. J Agric Food Chem 65(39):8652–8662
De Bhowmick G, Sarmah AK, Sen R (2018) Lignocellulosic biorefinery as a model for sustainable development of biofuels and value added products. Bioresour Technol 247:1144–1154. https://doi.org/10.1016/j.biortech.2017.09.163
Defanti LS, Siqueira NS, Linhares PC (2010) Produção de biocombustíveis a partir de algas fotossintetizantes. Bolsista de Valor 1:11–21
Demirbas A (2009) Progress and recent trends in biodiesel fuels. Energy Convers Manag 50 (1):14–34
Demirbas A (2011) Competitive liquid biofuels from biomass. Appl Energy 88(1):17–28
Dotsenko AS, Gusakov AV, Rozhkova AM et al (2018) Enzymatic hydrolysis of cellulose using mixes of mutant forms of cellulases from Penicillium verruculosum. Moscow Univ Chem Bull 73:58–62. https://doi.org/10.3103/S0027131418020037
du Preez JC (2016) Editorial: chemicals and bioproducts from biomass. Biotechnol Biofuels 9:233. https://doi.org/10.1186/s13068-016-0637-4
EMBRAPA, Brazilian Agricultural Research Corporation (2014) Cultivo de Pínus. https://www.spo.cnptia.embrapa.br/conteudo?p_p_id=conteudoportlet_WAR_sistemasdeproducaolf6_1ga1ceportlet&p_p_lifecycle=0&p_p_state=normal&p_p_mode=view&p_p_col_id=column1&p_p_col_count=1&p_r_p_76293187_sistemaProducaoId=3715&p_r_p_-996514994_topicoId=3229. Accessed 17 Jan 2019
EMBRAPA, Brazilian Agricultural Research Corporation (2015) Pesquisa investe em capim como fonte de energia. https://www.embrapa.br/busca-de-noticias/-/noticia/2422024/pesquisa-investe-em-capim-como-fonte-de-energia. Accessed 5 Feb 2019
EMBRAPA, Brazilian Agricultural Research Corporation (2018) Embrapa soja. https://www.embrapa.br/soja/cultivos/soja1/dados-economicos. Accessed 19 Jan 2019
Espinosa L, Tapanes NLCOM, Aranda DAG, Cruz YR (2014) As microalgas como fonte de produção de biodiesel: discussão de sua viabilidade. Acta Sci Techn 2(1):1–10
Fengel D, Wegener G (1984) Wood: Chemistry, Ultrastructure, React 613:1960–1982
Flores RA et al (2012) Yield and quality of elephant grass biomass produced in the cerrados region for bioenergy. Eng. Agríc 32(5):831–839
Franco ALC (2013) Biodiesel de microalgas: avanços e desafios. Química Nova 36(3):437–448

Freudenberg K (1965) Lignin: its constitution and formation from p-hydroxycinnamyl alcohols. Science 148(3670):595–600

Furtado A et al (2014) Modifying plants for biofuel and biomaterial production. Plant Biotechnol J 12:1246–1258. https://doi.org/10.1111/pbi.12300

Gao J, Chen L, Yuan K, Huang H, Yan Z (2013) Ionic liquid pretreatment to enhance the anaerobic digestion of lignocellulosic biomass. Bioresour Technol 150:352–358

Gill JR et al (2014) Yield Results and Stability Analysis from the Sorghum Regional Biomass Feedstock Trial. Bioenerg. Res 7(3):1026–1034. https://doi.org/10.1007/s12155-014-9445-5

Goh CS, Tan KT, Lee KT, Bhatia S (2010) Bio-ethanol from lignocellulose: status, perspectives and challenges in Malaysia. Bioresour Technol 101(13):4834–4841

Goufo ED, Mugisha S (2018) Complex harmonic poles in the evolution of macromolecules depolymerization. J Comput Anal Appl 18:1490

Gouveia L, Oliveira AC (2009) Microalgae as a raw material for biofuels production. J Ind Microbiol Biotechnol 36(2):269–274

Guimarães CC (2009) Experimentação no ensino de química: caminhos e descaminhos rumo à aprendizagem significativa. Química nova na escola 31(3):198–202

Han Y, Xu J, Zhao Z, Zhao J (2018) Analysis of enzymolysis process kinetics and estimation of the resource conversion efficiency to corn cobs with alkali soaking, water and acid steam explosion pretreatments. Bioresour Technol 264:391–394. https://doi.org/10.1016/j.biortech.2018.06.045

Harmoko C, Sucipto KI, Retnoningtyas ES, Hartono SB (2016) Vinyl functionalized cubic mesoporous silica nanoparticles as supporting material to enhance cellulase enzyme stability. ARPN J Eng Appl Sci 11:2981–2992

Hartono CD, Marlie KJ, Putro JN et al (2016) Levulinic acid from corncob by subcritical water process. Int J Ind Chem 7:401–409. https://doi.org/10.1007/s40090-016-0086-8

Hideno A, Inoue H, Tsukahara K, Fujimoto S, Minowa T, Inoue S, Sawayama S (2009) Wet disk milling pretreatment without sulfuric acid for enzymatic hydrolysis of rice straw. Bioresour Technol 100(10):2706–2711

Hoang NV et al (2015) Potential for genetic improvement of sugarcane as a source of biomass for biofuels. Front Bioeng Biotechnol 3(November):1–15. https://doi.org/10.3389/fbioe.2015.00182

Howard RL, Abotsi ELJR, Van Rensburg EJ, Howard S (2003) Lignocellulose biotechnology: issues of bioconversion and enzyme production. Afr J Biotechnol 2(12):602–619

Hsu T-A (1996) Pretreatment of biomass. In: Wyman CE (ed) Handbook on bioethanol production and utilization. Applied Energy Technology Series. Taylor & Francis, Washington DC, Chapter 10

IBÁ, Brazilian Industry of Trees (2018) Sumário executivo 2018. https://iba.org/datafiles/publicacoes/relatorios/digital-sumarioexecutivo-2018.pdf. Accessed 17 Jan 2019

IBGE, Geographic Statistical Brazilian Institute (2017) Produção da Extração Vegetal e da Silvicultura 2017. https://agenciadenoticias.ibge.gov.br/media/com_mediaibge/arquivos/15f538e9095614fc3204f828b22fa714.pdf. Accessed 20 Dec 2019

IPEF, Institute for Research and Forest Studies (2018) Programa cooperativo sobre produtividade potencial do Pinus no Brasil. https://www.ipef.br/pppib/. Accessed 10 Jan 2019

Iroegbu AO, Hlangothi SP (2018) Furfuryl alcohol a versatile, eco-sustainable compound in perspective. Chem Africa. https://doi.org/10.1007/s42250-018-00036-9

Jørgensen H, Kristensen JB, Felby C (2007) Enzymatic conversion of lignocellulose into fermentable sugars: challenges and opportunities. Biofuels Bioprod Biorefin 1(2):119–134

Jung JH et al (2012) RNAi suppression of lignin biosynthesis in sugarcane reduces recalcitrance for biofuel production from lignocellulosic biomass. Plant Biotechnol J 10(9):1067–1076. https://doi.org/10.1111/j.1467-7652.2012.00734.x

Kamm B, Kamm M (2007) Biorefineries–multi product processes. In: White biotechnology. Springer, Berlin Heidelberg, pp 175–204

Kamm B (2012) Introduction of biomass and biorefineries. In: The role of green chemistry in biomass processing and conversion. Wiley, Chapter 1, pp 1–26

Karapatsia A, Pappas I, Penloglou G et al (2017) Optimization of dilute acid pretreatment and enzymatic hydrolysis of Phalaris aquatica L. lignocellulosic biomass in batch and fed-batch processes. BioEnergy Res 10:225–236. https://doi.org/10.1007/s12155-016-9793-4

Karimi K, Taherzadeh MJ (2016) A critical review on analysis in pretreatment of lignocelluloses: degree of polymerization, adsorption/desorption, and accessibility. Bioresour Technol 203:348–356

Kenney WA, Sennerby-Forsse L, Layton P (1990) A review of biomass quality research relevant to the use of poplar and willow for energy conversion. Biomass 21(3):163–188

Kim M, Day DF (2011) Composition of sugar cane, energy cane, and sweet sorghum suitable for ethanol production at Louisiana sugar mills. J Ind Microbiol Biotechnol 38(7):803–807

Kim S, Dale BE (2004) Global potential bioethanol production from wasted crops and crop residues. Biomass Bioenergy 26(4):361–375

Kole C (2010) Genetics, genomics and breeding of sugarcane. In: Henry RJ, Kole C (eds) , 1st edn. Science Publishers, Enfield, New Hampshire

Kuhad RC, Singh A (1993) Lignocellulose biotechnology: current and future prospects. Crit Rev Biotechnol 13(2):151–172

Kumar P, Barrett DM, Delwiche MJ, Stroeve P (2009) Methods for pretreatment of lignocellulosic biomass for efficient hydrolysis and biofuel production. Ind Eng Chem Res 48(8):3713–3729

Kumar R, Tabatabaei M, Karimi K, Sárvári Horváth I (2016) Recent updates on lignocellulosic biomass derived ethanol-a review. Biofuel Res J 3(1):347–356

Kumar AK, Sharma S (2017) Recent updates on different methods of pretreatment of lignocellulosic feedstocks: a review. Bioresour Bioprocess 4:7. https://doi.org/10.1186/s40643-017-0137-9

Kumar M, Oyedun AO, Kumar A (2018) A review on the current status of various hydrothermal technologies on biomass feedstock. Renew Sust Energ Rev 81:1742–1770

Li J, Wei X, Wang Q, Chen J, Chang G, Kong L, Liu Y (2012) Homogeneous isolation of nanocellulose from sugarcane bagasse by high pressure homogenization. Carbohydr Polym 90(4):1609–1613

Limayem A, Ricke SC (2012) Lignocellulosic biomass for bioethanol production: current perspectives, potential issues and future prospects. Prog Energy Combust Sci 38(4):449–467

Loow Y-L, Wu TY, Md Jahim J et al (2016) Typical conversion of lignocellulosic biomass into reducing sugars using dilute acid hydrolysis and alkaline pretreatment. Cellulose 23:1491–1520. https://doi.org/10.1007/s10570-016-0936-8

Lü J, Sheahan C, Fu P (2011) Metabolic engineering of algae for fourth generation biofuels production. Energy Environ Sci 4(7):2451–2466

Ludueña L, Fasce D, Alvarez VA, Stefani PM (2011) Nanocellulose from rice husk following alkaline treatment to remove silica. Bioresources 6(2):1440–1453

Luo X, Zhu JY (2011) Effects of drying-induced fiber hornification on enzymatic saccharification of lignocelluloses. Enzyme Microb Technol 48(1):92–99

Macgregor AW, Fincher GB (1993) Carbohydrates of the barley grain. In: Macgregor AW, Bhatty RS (eds) Barley: chemistry and technology. American Association of Cereal Chemists

Martino DC, Colodette JL, Chandra R, Saddler J (2017) Steam explosion pretreatment used to remove hemicellulose to enhance the production of a eucalyptus organosolv dissolving pulp. Wood Sci Technol 51:557–569. https://doi.org/10.1007/s00226-016-0889-y

Matsuoka S et al (2014) Energy cane : its concept, development, characteristics, and prospects. Adv Bot 2014. https://doi.org/10.1155/2014/597275

McKendry P (2002) Energy production from biomass (part 1): overview of biomass. Bioresour Technol 83(1):37–46

Medina DP, Colorado AA (2006) Ethanol producction of banana shell and cassava starch. Dyna 73 (150):21–27

Melati RB, Shimizu FL, Oliveira G, Pagnocca FC, de Souza W, Sant'Anna C, Brienzo M (2019) Key factors affecting the recalcitrance and conversion process of biomass. Bioenergy Res 12 (1):1–20

Morais RFD, Souza BJD, Leite JM, Soares LHDB, Alves BJR, Boddey RM, Urquiaga S (2009) Elephant grass genotypes for bioenergy production by direct biomass combustion. Pesquisa Agropecuária Brasileira 44(2):133–140
Mosier N, Wyman C, Dale B, Elander R, Lee YY, Holtzapple M, Ladisch M (2005) Features of promising technologies for pretreatment of lignocellulosic biomass. Bioresour Technol 96 (6):673–686
Müller-Langer F, Kaltschmitt M (2015) Biofuels from lignocellulosic biomass – a multi-criteria approach for comparing overall concepts. Biomass Convers Bioref 5:43–61. https://doi.org/10.1007/s13399-014-0125-7
Nakamura A, Watanabe H, Ishida T, Uchihashi T, Wada M, Ando T et al (2014) Trade-off between processivity and hydrolytic velocity of cellobiohydrolases at the surface of crystalline cellulose. J Am Chem Soc 136(12):4584–4592
Nigam JN (2002) Bioconversion of water-hyacinth (*Eichhornia crassipes*) hemicellulose acid hydrolysate to motor fuel ethanol by xylose–fermenting yeast. J Biotechnol 97(2):107–116
Olsson L, Hahn-Hägerdal B (1996) Fermentation of lignocellulosic hydrolysates for ethanol production. Enzym Microb Technol 18(5):312–331
Paszner L (1988) Salt catalyzed wood bonding with hemicellulose. Holzforschung-International J Biol Chem Phys Technol Wood 42(1):11–20
Peng XUE (2009) Growth and biomass of six-year-old Eucalyptus urophylla plantation in Leizhou forestry bureau. Eucalypt Sci Technol 1(9)
Pielhop T, Amgarten J, von Rohr PR, Studer MH (2016) Steam explosion pretreatment of softwood: the effect of the explosive decompression on enzymatic digestibility. Biotechnol Biofuels 9:152. https://doi.org/10.1186/s13068-016-0567-1
Prasad S, Singh A, Joshi HC (2007) Ethanol as an alternative fuel from agricultural, industrial and urban residues. Resour Conserv Recycl 50(1):1–39
Pu Y, Zhang D, Singh PM, Ragauskas AJ (2008) The new forestry biofuels sector. Biofuels Bioprod Biorefin 2:58–73
Quiroz-Castañeda RE, Folch-Mallol JL (2011) Plant cell wall degrading and remodeling proteins: current perspectives. Biotecnol Appl 28:205–215
Ramos RLB et al (2001) Sugarcane expressed sequences tags (ESTs) encoding enzymes involved in lignin biosynthesis pathways. Genet Mol Biol 24(1–4):235–241. https://doi.org/10.1590/S1415-47572001000100031
Reddy N, Yang Y (2005) Biofibers from agricultural byproducts for industrial applications. Trends Biotechnol 23(1):22–27
REUTERS (2017) Usinas de cana da Índia devem mais do que dobrar oferta de etanol para mistura à gasoline. https://br.reuters.com/article/businessNews/idBRKBN1EE1L4-OBRBS. Accessed 10 Jan 2019
Rocha GJM, Gonçalves AR, Oliveira BR et al (2012) Steam explosion pretreatment reproduction and alkaline delignification reactions performed on a pilot scale with sugarcane bagasse for bioethanol production. Ind Crops Prod 35:274–279. https://doi.org/10.1016/j.indcrop.2011.07.010
Rocha JR d AS d C et al (2017) Bioenergetic potential and genetic diversity of elephant grass via morpho-agronomic and biomass quality traits. Ind\ Crops Prod. Elsevier B.V. 95:485–492. https://doi.org/10.1016/j.indcrop.2016.10.060
Saini JK, Saini R, Tewari L (2015) Lignocellulosic agriculture wastes as biomass feedstocks for second-generation bioethanol production: concepts and recent developments. Biotech 5(4):337–353
Sánchez C (2009) Lignocellulosic residues: biodegradation and bioconversion by fungi. Biotechnol Adv 27(2):185–194
Sannigrahi P, Ragauskas AJ (2010) Poplar as a feedstock for biofuels : A review of compositional characteristics. Biofuels Bioprod Bioref 4:209–226. https://doi.org/10.1002/bbb
Santos FA, Queiróz JH, Colodette JG, Fernandes SA, Guimarães VM, Rezende ST (2012) Potencial da palha de cana-de-açúcar para produção de etanol. Química Nova 35(5):1–7

Santos FA, Queiroz JH, Colodette JL, Manfredi M, Queiroz MELR, Caldas CS, Soares FEF (2013) Otimização do pré-tratamento hidrotérmico da palha de cana-de-açúcar visando à produção de etanol celulósico. Química Nova 37(1):56–62

Sant'Anna C, De Souza W, Brienzo M (2014) The influence of the heterogeneity, physicochemical and structural properties on the recalcitrance and conversion of sugarcane bagasse. Sugarcane: Production, Consumption and Agricultural Management Systems, Nova Science Publishers, pp 1–34

Sarkar N, Ghosh SK, Bannerjee S, Aikat K (2012) Bioethanol production from agricultural wastes: an overview. Renew Energy 37(1):19–27

Saxena RC, Adhikari DK, Goyal HB (2009) Biomass-based energy fuel through biochemical routes: a review. Renew Sust Energ Rev 13(1):167–178

Schmitt-Harsh M, Evans TP, Castellanos E, Randolph JC (2012) Carbon stocks in coffee agroforests and mixed dry tropical forests in the western highlands of Guatemala. Agrofor Syst 86(2):141–157

Sharma S, Kuila A, Sharma V (2017) Enzymatic hydrolysis of thermochemically pretreated biomass using a mixture of cellulolytic enzymes produced from different fungal sources. Clean Technol Environ Policy 19:1577–1584. https://doi.org/10.1007/s10098-017-1346-9

Shen H et al (2013) Enhanced characteristics of genetically modified switchgrass (Panicum virgatum L.) for high biofuel production. Biotechnol Biofuels 6(71):1–15

Silva TAL, Zamora HDZ, Varão LHR et al (2018) Effect of steam explosion pretreatment catalysed by organic acid and alkali on chemical and structural properties and enzymatic hydrolysis of sugarcane bagasse. Waste Biomass Valoriz 9:2191–2201. https://doi.org/10.1007/s12649-017-9989-7

Sims RE, Mabee W, Saddler JN, Taylor M (2010) An overview of second generation biofuel technologies. Bioresour Technol 101(6):1570–1580

Solomon BD, Banerjee A, Acevedo A et al (2015) Policies for the sustainable development of biofuels in the Pan American region: a review and synthesis of five countries. Environ Manage 56:1276–1294. https://doi.org/10.1007/s00267-014-0424-6

Steinbach D, Kruse A, Sauer J (2017) Pretreatment technologies of lignocellulosic biomass in water in view of furfural and 5-hydroxymethylfurfural production- a review. Biomass Convers Biorefinery 7:247–274. https://doi.org/10.1007/s13399-017-0243-0

Sticklen MB (2008) Plant genetic engineering for biofuel production : towards affordable cellulosic ethanol. Nat Publish Group 9(6):433–443. https://doi.org/10.1038/nrg2336

Sun S, Cai Y, Liu H (2001) Biomass allocation of Scirpus mariqueter along an elevational gradient in a salt marsh of the Yangtse River estuary. Acta Bot Sin 43(2):178–185

Sun G, Ranson KJ, Kharuk VI (2002) Radiometric slope correction for forest biomass estimation from SAR data in the Western Sayani Mountains, Siberia. Remote Sens Environ 79(2–3):279–287

Sun Y, Cheng J (2002) Hydrolysis of lignocellulosic materials for ethanol production: a review. Bioresour Technol 83(1):1–11

Timell TE (1964) Wood hemicelluloses: part I. In: Advances in carbohydrate chemistry. Academic press, vol 19, pp 247–302

UBRABIO, Brazilian Union of Biodiesel and Bioquerosene (2017). https://www.ubrabio.com.br/sites/1800/1891/PDFs/Apresentacoes/20171101FArumAeroespacial2017.pdf. Accessed 9 Jan 2019

Urbanowicz BR, Peña MJ, Ratnaparkhe S, Avci U, Backe J, Steet HF, Darvill AG (2012) 4-O-methylation of glucuronic acid in Arabidopsis glucuronoxylan is catalyzed by a domain of unknown function family 579 protein. Proc Natl Acad Sci 109(35):14253–14258

Wang L, Liu Y, Chen H (2018) A steam-explosion-based hydrolysis and acidification technology for cornstalk bioconversion. BioEnergy Res. https://doi.org/10.1007/s12155-018-9945-9

Werpy T, Petersen G, Aden A, Bozell J, Holladay J, White J, Jones S (2004) Top value added chemicals from biomass. In: Results of screening for potential candidates from sugars and synthesis gas, vol 1, pp 26–28

Xie G, Peng L (2011) Genetic engineering of energy crops : a strategy for biofuel production in China. J Integ Plant Biol 53(2):143–150. https://doi.org/10.1111/j.1744-7909.2010.01022.x

Ximenes E, Kim Y, Mosier N, Dien B, Ladisch M (2011) Deactivation of cellulases by phenols. Enzy Microb Technol 48(1):54–60
Yang Q, Pan X (2016) Correlation between lignin physicochemical properties and inhibition to enzymatic hydrolysis of cellulose. Biotechnol Bioeng 113(6):1213–1224
York WS, O'Neill MA (2008) Biochemical control of xylan biosynthesis - which end is up? Curr Opin Plant Biol 11(3):258–265
Yücel Y, Göycıncık S (2015) Optimization and modelling of process conditions using response surface methodology (RSM) for enzymatic saccharification of spent tea waste (STW). Waste Biomass Valoriz 6(6):1077–1084
Yung MM (2016) Catalytic conversion of biomass to fuels and chemicals. Top Catal 59(1). https://doi.org/10.1007/s11244-015-0511-9
Zakzeski J, Bruijnincx PC, Jongerius AL, Weckhuysen BM (2010) The catalytic valorization of lignin for the production of renewable chemicals. Chem Rev 110(6):3552–3599
Zheng Y, Tang Q, Wang T, Liao Y, Wang J (2013) Synthesis of a green fuel additive over cation resins. Chem Eng Technol 36(11):1951–1956
Zhao X, Zhang L, Liu D (2012) Biomass recalcitrance. Part I: the chemical compositions and physical structures affecting the enzymatic hydrolysis of lignocellulose. Biofuels Bioprod Bioref 6(4):465–482
Zhao H et al (2014) Genotypic variation of cell wall composition and its conversion efficiency in Miscanthus sinensis, a potential biomass feedstock crop in China. GCB Bioenergy 6:768–776. https://doi.org/10.1111/gcbb.12115
Zhao X, Li S, Wu R, Liu D (2017) Organosolv fractionating pre-treatment of lignocellulosic biomass for efficient enzymatic saccharification: chemistry, kinetics, and substrate structures. Biofuels Bioprod Bioref 12(5):834–845
Zhu S, Wu Y, Yu Z, Zhang X, Wang C, Yu F, Xue Y (2005) Simultaneous saccharification and fermentation of microwave/alkali pre-treated rice straw to ethanol. Biosyst Eng 92(2):229–235
Zhu JY, Pan X, Zalesny RS (2010) Pretreatment of woody biomass for biofuel production: energy efficiency, technologies, and recalcitrance. Appl Microbiol Biotechnol 87(3):847–857

Biogas: An Effective and Common Energy Tool – Part I

3

Seethalaksmi Elangovan, Sathish Babu Soundra Pandian, Geetha S. J., and Sanket J. Joshi

Abstract

Energy is a much crucial necessity for daily errands, either household or industrial. We use it as fuel (transportation or industrial commodity), to provide power, heat, electricity, etc., and we can't imagine life without it. Several kinds of fuels are available in the market, mainly non-renewables – fossil based (coal, crude oil, etc.). However, due to awareness about long-term issues related to use of fossil fuels, several other types of renewable fuels are gaining much attention. Biogas, biofuels (bioethanol, biodiesel), and biohydrogen are some of the examples for such renewables with very high future potential. However, even with those recent developments, rural areas in some of the developing countries are still using agricultural remains, cow dung, etc., for cooking and heating purposes. This kind of crude practice significantly raises environmental, economic, and public health-related worries. To achieve a worldwide sustainable progress in both developed and developing countries, clean and affordable energy could be offered by using the existing biomass resources (crop residues, agro-industrial, animal, and other type of wastes) to produce a cleaner, more efficient, and reliable energy, such as

S. Elangovan
Department of Biochemistry, Sri Sankara Arts and Science College, Affiliated to University of Madras, Chennai, India

S. B. S. Pandian
Central Analytical and Applied Research Unit, College of Science, Sultan Qaboos University, Muscat, Oman

Geetha S. J.
Department of Biology, College of Science, Sultan Qaboos University, Muscat, Oman

S. J. Joshi (✉)
Central Analytical and Applied Research Unit, College of Science, Sultan Qaboos University, Muscat, Oman

Oil & Gas Research Centre, Sultan Qaboos University, Muscat, Oman
e-mail: sanket@squ.edu.om

N. Srivastava et al. (eds.), *Biofuel Production Technologies: Critical Analysis for Sustainability*, Clean Energy Production Technologies,
https://doi.org/10.1007/978-981-13-8637-4_3

biogas. Unlike other types of renewable biofuels, biogas production is a natural non-energy intensive process, and the raw materials are mostly renewable resource and wastes – thus serving both purposes, bioremediation and energy generation. Biogas is a blend of gases, mainly methane and carbon dioxide. Over the years, several biogas plant designs are available, which are compiled in present chapter along with its advantages and disadvantages. At present several countries are already utilizing biogas for various household and industrial applications. The main applications are generating electricity, cooking, heating, and using as a fuel for transportation. The ease of operation, maintenance, and easy availability of substrate – waste materials – are some of the key selling points for biogas to be an effective and common energy tool in the near future.

Keywords
Energy · Fossil fuels · Renewables · Biofuel · Biogas

3.1 Introduction

In order to perform work on or to heat the object, energy is one of the quantitative properties that must be transferred to an object. It is a conserved quantity, where the law of conservation of energy states that energy can neither be created nor destroyed, rather only be converted in forms. Energy is an essential commodity for day-to-day human activities, either directly (as fuel) or indirectly (to provide power, electricity, mobility), which enables socioeconomic development. Villages in many developing countries, firewood, crop residues, cow dung, etc. are still used for cooking purposes. These practices are significant sources of environmental, economic, and public health-related concerns. To achieve a sustainable progress in the developing countries, clean and affordable energy could be offered by using the existing biomass resources to produce biogas – a cleaner, more efficient, and reliable energy (Surendra et al. 2014).

Biogas could be naturally produced from the decomposition of organic waste and raw materials from agronomic, municipal, sewage, or food waste, under anaerobic condition. Biogas is a blend of gases, mainly methane and carbon dioxide. Biogas production process is also known as anaerobic digestion, as the decomposition occurs under anaerobic condition. Bacteria produce roughly two-thirds of methane, as well as carbon dioxide, nitrogen, and traces of other types of gases (Armaha et al. 2017; Sibiya et al. 2017). Biogas is produced by anaerobic organisms (such as methanogen), inside a closed system. This process is non-polluting and also reduces greenhouse emissions. Since no combustion takes place in the process, there is zero emission of greenhouse gasses. It consists mainly of methane (CH_4) and carbon dioxide (CO_2). The amount of CO_2 that is produced corresponds to the amount of CO_2 captured when the biomass was created (Weiland 2010). Producing such fuel from waste is actually a great way to lessen the global warming. Biogas helps to reduce the reliance on fossil fuels (crude oil derivatives, and coal), which cause environmental pollution and global warming. Unlike other types of renewable biofuels, biogas production is a natural process, non-energy intensive, and the raw materials are mostly renewable resource like agricultural waste, and food scraps, which are easily available making it a highly viable resource.

In biogas digester, anaerobic digestion-fermentation processes take place, in which organic waste is digested to produce methane and other gases. Almost any type of organic waste can be used as initial substrate in the process; however pH and temperature of the process influence the optimum production of gas. Digester tank provides a suitable environment for anaerobic microbes to digest the biomass, thus producing digested solids, slurry, and biogas. The favourable/optimum temperature, moisture content, oxygen exclusion, pH, and a continuous food (organic waste) supply are some of the prerequisite conditions required in order to support anaerobic digestion process. Even though, in most cases, the pH is maintained by itself, the problem of pH drops by acidophilic bacteria, which grows faster than methanogens, creating an unfavourable condition (optimum pH range ~6.8–8.5) for methanogens. To maintain consistent pH, sometimes bicarbonates are also used. Like any other biogenic processes, biogas plant operators also must be aware of associated potential hazards such as explosion, suffocation, probable pathogens, or hydrogen sulphide (H_2S) poisoning and must take all sorts of preventative safety measure. As per the energy and climate policies of the European Union, the EU has set some goals for year 2020, such as at least 20% lesser greenhouse gases emissions and 20% reduction in utilization of energy by means of better energy efficacy and to cover 20% of requirement by renewable sources. To reach such set targets, Germany uses ~66% biomass as a key contributor to provide energy, and in year 2012, ~50% of electricity was produced using biogas from biomass. Overall share of biogas was ~15% to generate electricity and ~8% to provide heat supply. Biogas could play a distinct role among other renewable fuels, as it is suitable for the production of combined heat and power simultaneously and as a natural gas substitute once purified. The storage of biogas is relatively easy and it is flexible to use. Currently biogas plants in Germany are reported to be capable to generate ~420 kW, which thereby already replaced some of the coal-fired power plants, showing great potential that even small energy producers can also jointly generate such a significant magnitude of energy (Heffels et al. 2012). Brundtland Report of 1987 (https://mediathek.fnr.de/media/downloadable/files/samples/b/r/brosch.biogas-2013-en-web-pdf.pdf) states that sustainability means that "the present generation satisfies its needs without endangering the ability of future generations to meet their respective needs". All three dimensions – environmental, economic, and social – are interconnected in sustainability. Sustainable resources mean striking a balance between what is essential in terms of economics (such as consistent and high biomass produces) and practicality (for our expectation from our environs). One of the main research areas of the Federal Ministry of Food, Agriculture and Consumer Protection (BMELV) was to assess those methods in various research projects to find proper approaches for viable energy supply and natural resources management and to further develop them through appropriate and coordinated research funding. The strategies being pursued included increasing the biodiversity to cultivate different energy crops, breeding of new plant varieties, novel cultivation techniques, and reduce-reuse and recycle (reduced use of chemicals, agricultural lands with perennial vegetative cover, proficient transformation methods, renewable resources being both used as a material and as a source of energy, and the residual materials to be used as fertilizer). The sustainability principle also applies to the biogas sector, and it can't be offered as lasting energy source if those conditions are not fulfilled.

The anaerobic digestion process also reduces environmentally harmful methane emanations to quite an extent and could also reduce the nitrous oxide emissions, having an impact on the climate (Möller 2015). To a certain degree, bad smell-causing substances are also decomposed and neutralized. Other environmental advantages include the utilization and reduction of waste during electricity or heat generation, and the remaining digestate can also be used as a fertilizer. Most of non-crude-oil-producing countries are dependent on energy imports; thus with the directed development of such renewable energy systems, they could reach the goal of energy independence and stable supply. According to the Renewable Energy Sources Act (EEG), production of electricity and heat (combined heat and power production – CHP) primarily uses biogas for feeding biogas-sourced electricity into the power grids. Generally, CHP units consist of a biogas-powered combustion engine, driving a generator to produce electrical energy. To use biogas as a substitute for natural gas, it first needs to be purified of unwanted constituent elements (such that the CO_2 needs to be separated) leading to higher methane content. The upgraded biogas, called biomethane (or bio-natural gas), could then be stored in the extensive gas grid or transported to be used at any site. It can also be used in other domestic purpose like gas stove fuel and lighting (Pfeifer and Obernberger 2007; Rutz et al. 2015).

3.2 Source of Organic Waste for Biogas Production

Biogas is produced by anaerobic digestion, a natural biological process which stabilizes organic waste in the absence of oxygen. The important parameter to be considered in anaerobic digestion is the source and type of feedstock used. Almost any organic material can be utilized such as newspapers, restaurant and kitchen food waste, animal waste, agricultural waste, and waste from different types of food and beverage industries, leather processing industries, etc. Organic waste is generated from daily human activities and from the industries; thereby the amount of organic waste produced is amassed each year as the population increase. Developing countries deal with the organic waste in different ways, like soil improvement, for animal raising and also to provide a source of energy.

3.3 Materials to Be Excluded from Anaerobic Digesters

Toxic, inorganic, and poorly degradable material should be excluded from the feedstock. Toxic compounds like ammonium and sulphur, if present in feedstock at high concentration, could inhibit anaerobic organisms. Poorly degradable material requires more time for breakdown and to convert into biogas. Inorganic material like sand bed cannot contribute or be digested to biogas, which in turn can clog the pipeline or get settle in the plant, creating problem in biogas plant.

3.4 Biogas Potential of Different Feedstock

For optimized anaerobic digestion process and biogas production, feedstocks are particularly vital (Schievano et al. 2009). However, it is somewhat challenging to exactly calculate the biogas production potential of different feedstocks, due to origin of the source, organic load, chemical oxygen demand (COD), and moistness of the feedstock. Anaerobic biogasification potential (ABP) and biochemical methane potential (BMP) are some of the important criteria reported to evaluate biogas/methane potential of feedstocks (Schievano et al. 2009). BMP assay is an efficient method and an indication of the degradability of the feedstock; it helps to establish a baseline data for the performance of anaerobic digestion (Speece 1996). The combinations of a particular feedstock, microbial inoculum, and other additives are optimized during BMP assays, based on which a specific microbial inoculum is used to seed the feedstock to reduce the initial lag phase and fasten the anaerobic degradation process. BMP is a laboratory-based analysis approach to evaluate the feedstock potential by following three criteria:

1. *Characterization*: BMP assay includes different criteria such as pH, COD, and volatile and total solids (VS and TS); assay values are used to evaluate the anaerobic digestion process in terms of the obliteration of feedstock material.
2. *Total biogas production*: Gas production is analysed during the course of the BMP assays, and biogas could also be scoured of the carbon dioxide and traces (of other gases) by different chemical processes to mainly check methane content.
3. *Biogas analysis:* Generally gas chromatography (GC) is used to analyse gas composition and concentrations (such as CH_4, CO_2, N_2, and H_2S).

3.4.1 Factors Affecting BMP

Chemical composition of the feedstock such as carbohydrates, proteins, fats, cellulose, and hemicelluloses could affect the BMP. The biogas production and composition varies based on different feedstocks, which differs substantially subject to their origin and constitution (Weiland 2010; Mayer et al. 2014).

Feedstock Constituents BMP depends on the chemicals, fats, and proteins, which support better methane production than carbohydrates.

Total and Volatile Solids The TS include both organic and inorganic content of the feedstock. Organic material could be evaluated by the amount of total organic carbon present in a feedstock (Hamilton 2012). Volatile solid degradation is directly related to methane production (Moody et al. 2009) and methane yield is affected by volatile solids (Mayer et al. 2014). Three reported anaerobic digestion technologies according to the total solid content of feedstocks are wet, semi-dry, and dry methods for roughly $\leq 10\% - \geq 20\%$ TS content. It is reported that during the batch anaerobic digestion of cardboard or food waste, the biogas production decreases in the presence of high TS contents (Forster-Carneiro et al. 2008; Guendouz et al. 2012), whereas the gas yield increases with higher VS (Gao et al. 2012).

Chemical and Biological Oxygen Demand Both COD and biological oxygen demand (BOD) are used to quantify the organic matter in feedstocks and to predict its applicability in anaerobic digestion process (De Mes et al. 2003; Angelidaki and Sanders 2004). BOD is the measure of the oxygen needed by biodegrading microbes.

Carbon/Nitrogen Ratio It represents the association between the amount of carbon and nitrogen in a feedstock, and 25:1 ratio is reported to be better for optimal process (Gerardi 2003). Kwietniewska and Tys (2014) reported that 20–35:1 is an optimum C/N ratio range for anaerobic digestion. Anaerobic digestion of high protein containing substrate leads to higher ammonia content causing alkaline pH, leading to methanogens inhibition (Khalid et al. 2011). Wang et al. (2014) reported that temperature and C/N ratio directly affect anaerobic digestion performance.

Inhibitory Constituents Feedstocks may contain constituents that can inhibit anaerobic digestion, such as NH_3, H_2S, and heavy metals, where free NH_3 directly affects the process (Gerardi 2003). Among all the microbes in anaerobic digesters, methanogens are highly susceptible to NH_3 inhibition (Chen et al. 2008).

Agronomic Practices Various endeavours to improve the gas (methane) production from agro-based feedstocks are always preferred. However, the content and yield of such agro-based substrates are greatly affected by environmental and geographical conditions, land characteristics, type of plants, and overall farming practices. It was reported that Agrogas has showed better methane yield (367 NL kg/VS), mainly due to high carbohydrates and low lignin in the substrate.

3.5 Organic Waste: Types, Sources, and Uses

Depending on the feedstock source, biogas generally contains methane ($\geq 50\%$), CO_2 ($\leq 50\%$), and traces of hydrogen, water vapour, CO, N_2, and different compounds of S, $H(OSiH_2)_nOH$, $(OSiH_2)_n$, halogens, and heavy metals. The ease of availability for different feedstock for anaerobic digestion is one of the main selection criteria. Feedstock's energy production potential differs based on the type, concentration, and pretreatment of the biodegradable matter. Some of the reported common sources of organic feedstock are:

- Agro-based feedstock
- Community-based feedstock
- Industrial feedstock

3.5.1 Agro-Based Feedstock

Animal compost, energy crops, aquatic plants and farming residues are some of the examples of agro-based feedstocks. Due to the consistency of those substrates and the absence of $H(OSiH_2)_nOH$, $(OSiH_2)_n$, thus produced biogas using such substrates would be a bit easier to treat than that produced using other type of wastes.

Animal Compost Animal compost or manure is increasingly being used in anaerobic digestion for biogas production, which is one of the principal organic waste-based feedstock. Its digestion leads to increase in available nitrogen to plants and in the soil and can also be used as an efficient soil conditioner. The reported amount of gas produced by different type of animal manure-based sources is ~0.29–46 ft^3/day. In the United States, ~80% of the available energy is used in the form of power, and the rest is utilized used for heating purposes. Iran produces~573,600–6,059,600 m^3/y biogas from different manure-based sources.

Methane production potential from diverse manure sources are shown in Table 3.1. Depending on the housing pig slurry, total solids vary from 2% to 5% which makes the anaerobic digestion system uneconomic. If the pig slurry is collected by scraping, it can increase the total solids up to 5–10%, and it can generate better yields. Similarly, cow dung is usually accumulated and mixed with straw to increase the TS content. Different reports highlighted that poultry wastes are rich in organic nitrogen and relatively lower carbon source. The chicken manure contains high TS contents (~20%) and ammonia nitrogen (~8 g/l), which could inhibit the digestion process and irritating/smelly discharges during storage.

Generally, the manure is blended with other type of materials during commercial stages of biogas production. Some reports showed higher methane yield when chicken or cow manure co-digested with fruits and veggie, grass forage, or straws. Countries like the Philippines and Taiwan predominantly used pig manure as feedstock for the bioenergy production. Generally used animal wastes are cow dung, pig waste, poultry manure, horse dung, camel dung, elephant dung, fishery waste, and slaughter house wastes.

Table 3.1 Methane production potential of various types of manure with characteristics

Parameters	Cow manure	Pig manure	Chicken manure
Total solids (%)	2–12	3–8	10–30
Volatile solids (%)	75–85	70–80	70–80
Carbon-to-nitrogen ratio (C/N)	6–20	3–10	3–10
Biogas yield (M^3 kg/VS)	0.20–0.30	0.25–0.50	0.35–0.60
CH_4 content	55–75	70–80	60–80

Aquatic Plant Feedstocks such as water hyacinth, micro and macro algae, and sea weeds are suitable for bioenergy production. These feedstocks have easily hydrolysable sugars, contain lower lignin, and do not compete with land resources used as compared to arable food crop cultivation. The co-digestion of alga sludge and waste paper leading to methane production rate up to 1607 ± 17 ml/day was reported. Both water hyacinth and micro algae are mostly used as feed material because of their higher gas yield.

Farming Residues Farming activities produce plenty of biomass residues as agricultural waste which can be used as feedstock for anaerobic digestion, such as cotton, maize, and rice residues. However, high lignin content of such residues can lead to poor biodegradability and minimal biogas production. Pretreated lignocellulosic can improve biodegradability and the biogas production. Several reports showed sustainable and higher gas yield by using pretreatment, additional nutrients, and co-digestion of farming residues and other substrates (Chandra et al. 2012; Nges et al. (2012).

Energy Crops Energy crops such as C-4 plants are good candidate-substrate for biogas production. Grass is one such substrate, which is easily available throughout the year and reported to produce high methane content (~70–80%). Corn and Sudan grass are some of the most popular co-substrates in Germany and Austria. Biogas or methane production of about 0.655 to 0.72–0.77 m^3/kg of VSS from maize silage, and non-acidified or acidified maize, and ~5–181 Nm^3/tonne from sugarcane were reported during anaerobic digestion. Although energy crops have potential, lignocellulosic content is still an issue; thus pretreatment prior to anaerobic digestion is needed for better biogas production. When the sorghum was pretreated with sodium hydroxide, an increased methane production (from 8% to 19%) was reported.

3.5.2 Community-Based Feedstock

Community-based feedstock includes municipal solid waste, organic fraction of municipality waste, sewage sludge, kitchen waste (Ziana and Rajesh 2015; Apte et al. 2013; Ogur and Mbatia 2013), garden waste, institutional waste, wholesale fruit and vegetable market waste, etc.

Municipal Solid Waste (MSW) MSW comes from housing, institutions, and commercial establishments. If discarded untreated, it is both burden on treatment plants and a waste of energy and nutrients, and it could rather be used as a substrate for biogas production. MSW typically contain ~65% of food waste from the kitchen. Reports suggest that blending kitchen waste and cow manure as a co-substrate at mesophilic condition could generate ~ 60–69% methane, and ~10 kg of kitchen waste could produce ~2.292 m^3 of biogas (Reddy 2017; Singh and Sankarlal 2015; Haftu et al. 2018).

Organic Fraction of Municipality Waste (OFMSW) Based on the source of organic matter from well-sorted municipal solid waste, biogas (methane) yield of ~300–400 Nm^3/tons of VS, and ~80% volatile solids degradation is reported. As estimated, ~960,000 tons of restaurant food waste in megacities such as Beijing could generate ~300 million Nm^3 CH_4 (De Clercq et al. 2016).

Sewage Waste Worldwide sewage sludge is generated in enormous quantities, and mostly treated by anaerobic digestion process, such as in Europe and the United States. According to US Environmental Protection Agency (EPA) report in 2007, >16,000 municipal wastewater treatment facilities (WWTFs) utilized sewage sludge to generate biogas and could generate ~1 ft^3 gas from 100 gallons of wastewater/day. Municipal WWTFs also represent quite lucrative market for the utilisation of fuel cells, as roughly only 1/3 of municipal WWTFs use produced biogas for different applications, leaving a strong possibility of its use for fuel cells. The heat produced by the fuel cell during the operation is quite suitable to be utilized to heat the digester, for better microbial growth and performance, and this energy input could also be converted to useful electricity.

3.5.3 Industrial Feedstock

Industry-based feedstock includes food or beverage processing, dairy, starch industry (such as potato chips manufacturing facility), sugarcane industry, pharma and cosmetic industry, biological industry, tanneries, etc. Among industrial wastes, pulp and paper waste sludge as a feedstock offers several advantages, due to its high organic content, and it also provides economic benefit such as lower transport costs. The methane yield for different sources of pulp and paper waste sludge was reported to be ~50–200 m^3/VS added. However, some sort of pretreatment would be required for such feedstock to decrease the necessary retention time for efficient biogas production.

Generally, slaughter house waste is chemically similar to household sewage. It is entirely organic, making it the right feedstock for biogas production process. Depending on the amount of the material and used methods, the reported methane yield from slaughterhouse ranges between 160 and 500 dm^3/kg VS (Castellucci et al. 2013). Tannery industries generate huge amount of stinking wastage, during different steps (fleshing, splitting, and liming-reliming), leaving fleshy and shaving wastage. For biogas production, a mixture was reported containing fleshy tannery wastage, thioglycollate broth as anaerobic culture media, and water solution of carbohydrate and protein. Mesophilic digestion process takes 18–24 days to produce biogas, containing ~14% of methane gas.

Table 3.2 Advantages and disadvantages of anaerobic and aerobic digestion processes

Anaerobic digestion	No aeration required	Can generate energy from biogas
	Required low energy	Low surplus sludge
Aerobic digestion	Required aeration	Completely treated water
	High energy cost	High surplus sludge

3.6 Biogas Production

To reduce the change in global climate patterns, specifically the production of increased level of atmospheric carbon dioxide by the utilization of fossil fuels is a difficult task for the society. The replacement of fossil fuels with one of the renewable sources such as biogas is an important step to reduce the greenhouse gas emission thus preventing the climate change. Aerobic digestion could also effectively degrade the organic material. However, an aerobic process produces huge amounts of sludge along with fully treated wastewater, rather than converting feedstock entirely to methane. Different types of advantages and disadvantages of both processes are listed in Table 3.2.

3.6.1 Anaerobic Digestion

Anaerobic digestion is an indigenous process for natural anaerobic ecosystem. It is a microbiological process by the consortia of microbes which are partly syntrophic and decompose the organic material, specifically in the absence of oxygen. As previously mentioned, it is the naturally occurring process in the sediments at the bottom of lakes and ponds, in bogs, in neighbourhoods of wet sites like swamps and slurry pits, and in the intestines of ruminant animals such as cows. The anaerobic digestion is a more efficient and environmentally promising technology for biogas production. The different stages of decomposition process are interlinked to each other, as mentioned below and as shown in Fig. 3.1:

1. Disintegration and hydrolysis
2. Acidogenesis
3. Acetogenesis
4. Methanogenesis

3.6.1.1 Disintegration and Hydrolysis

Disintegration involves lysis, non-enzymatic degeneration, phase separation, and breakdown of biomass into polymers (Batstone et al. 2002). In hydrolysis, water reacts with long-chain insoluble organic polymers to form soluble short-chain polymers (Fig. 3.2). The hydrolytic bacteria involved in hydrolysis excretes

Fig. 3.1 The different stages of decomposition process during biogas production

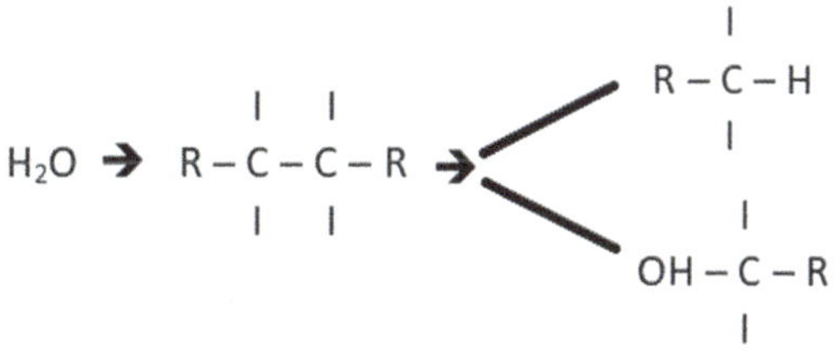

Fig. 3.2 The enzymatic hydrolysis of organic polymers

hydrolytic enzymes (exoenzymes of obligatory and facultative anaerobic bacteria) to convert the insoluble biopolymers into soluble smaller compounds (Gerardi et al. 2003). The organic waste produced by this process is further degraded into soluble form which is utilized by the microorganisms for their own metabolic process.

The carbohydrates are hydrolysed within few hours, proteins and lipids after few days. But the lignocelluloses and lignins are hydrolysed very slowly and incompletely (Deublein and Steinhauser 2008). Different types of enzymes are responsible for the breakdown of various substrates, and the breakdown products are listed in Table 3.3.

3.6.1.1.1 Hydrolysis of Polysaccharides

Polysaccharides are compounds comprised of long-chain monosaccharides connected by glycosidic bonds. Common polysaccharides are cellulose, hemicelluloses, starch, glycogen, and pectin. Polysaccharides are classified into:

1. Linear-chain polysaccharides (cellulose, starch)
2. Branched-chain polysaccharides (hemicellulose, starch, glycogen, pectin) (Peter 2009)

Cellulose is hydrolysed into cellobiose and glucose by the enzyme cellulases. Starch and glycogen get converted into glucose units. Other sugars are also formed by the conversion of hemicelluloses and pectin. In the case of recalcitrant structure, like cellulose, the degradation takes too long. To improve the hydrolysis, the pretreatment is needed to convert the raw material into the form that is more susceptible to microbes and the enzymatic attack.

3.6.1.1.2 Hydrolysis of Proteins

Proteins are polymers of amino acids linked together by peptide bonds. In hydrolysis, these bonds are degraded by proteases, producing building blocks – amino acids. Some proteins present in cell membranes like glycoproteins (contains carbohydrates) are harder to hydrolyse, which needs pretreatment before digestion (Salminen et al. 2003).

Table 3.3 Different types of enzymes responsible for breakdown of various substrates

Enzymes	Proteinase	Cellulase	Hemicellulase	Amylase	Lipase	Pectinase
Substrates	Proteins	Cellulose	Hemicellulose	Starch	Fats	Pectin
Breakdown products	Amino acids	Cellobiose and glucose	Glucose, xylose, mannose, and arabinose	Glucose	Fatty acids and glycerol	Galactose, arabinose, and polygalacturonic acid

3.6.1.1.3 Hydrolysis of Lipids

Lipids, commonly present in the form of triglycerides and in digestion process, is taken as fat, oils, and greases (FOG). The lipids are degraded by the enzyme lipases (Alves et al. 2009).

3.6.1.2 Acidogenesis or Fermentation

This is the second stage of AD process, where the products of the hydrolysis stage are used as a substrate by acidogenic bacteria. The substrates such as soluble monomers are converted to short-chain organic acids (volatile fatty acids – acetic, butyric, lactic, propionic, and succinic acid). The other products from hydrolysis are converted to alcohols, ammonia, hydrogen, and carbon dioxide. The formation of compound depends on the organism present, substrate and environmental conditions (Bischofsberger et al. 2005).

3.6.1.3 Acetogenesis

The products produced in the acidogenesis are used as substrates for the acetogenic microorganisms, which are active acetogenes. So the volatile fatty acids (VFA) and alcohols are oxidized into methanogenic substrates like acetate, hydrogen, and carbon dioxide. This intermediary translation is crucial for successful biogas production (Bischofsberger et al. 2005).

3.6.1.4 Methanogenesis

The final step in the AD process is methanogenesis. This is critical and also the slowest step in whole process. The methanogenic bacteria utilize the products of previous steps as substrate. Two different microbes produce methane by two well-balanced processes. In the first process, the acetic acid produces methane and carbon dioxide by acetoclastic microorganisms. In the second process, hydrogen is reduced to form carbon dioxide and methane by hydrogenophilic methanogens, and based on thermodynamics, scientists identified a new reaction (Fig. 3.3).

3.7 Biogas Plant

"Biogas plant" is a physical structure which is designed to perform the anaerobic digestion process. The principal part of the biogas plant is a digester (Fig. 3.4). The digester is an airtight tank where the bacteria break down the organic waste material in an anaerobic digestion process (Samah 2016) (Fig. 3.5). During the last century, different types of digesters were reported. Based on the nature of feeding material (organic waste), the biogas plant can be divided into several types as mentioned below.

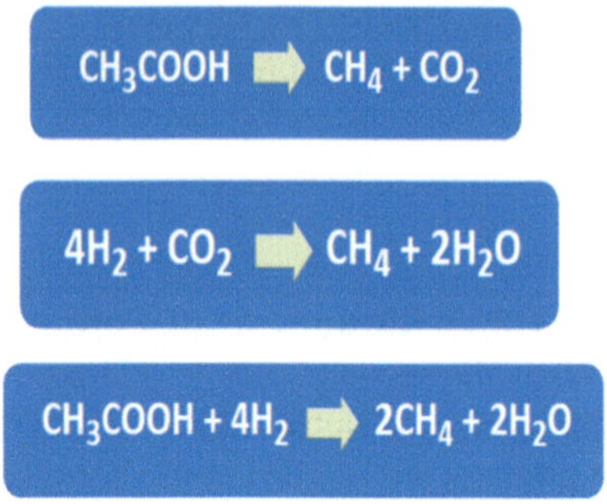

Fig. 3.3 Overall chemical process of three different stages of biogas production

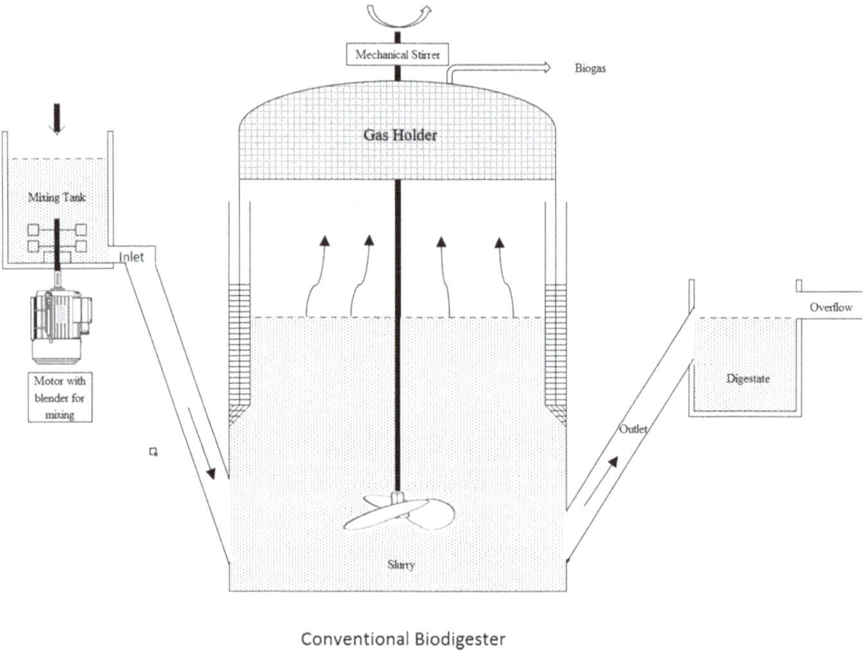

Fig. 3.4 Conventional design of a biogas plant. (Modified from Bhardwaj and Das 2017)

3.7.1 Classification of Anaerobic Digestion Technologies

3.7.1.1 Total Solid Content (Wet/Dry)

Based on the total solid content of the feed, the digester designs are classified into wet and dry systems. TS content of wet reactors is 16%, and the TS content of dry reactors ranges between 22% and 40%. Dry reactors are more advantageous than wet reactors since it requires a smaller reactor volume and minimal energy. Because of the dry nature, the digestate can be easily used as a fertilizer after digestion.

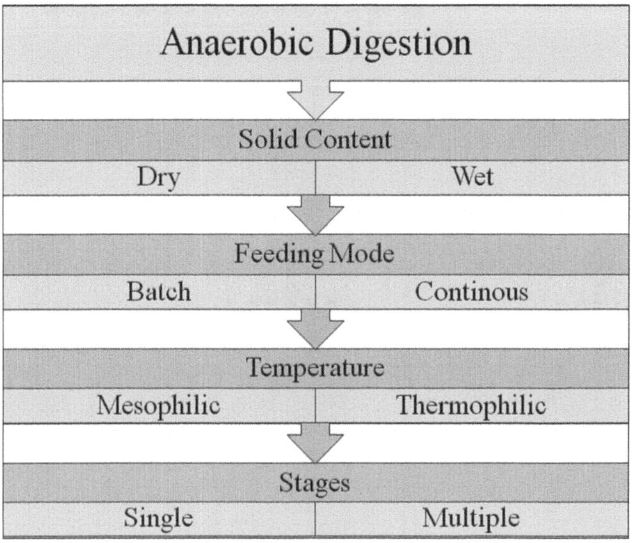

Fig. 3.5 Schematics of an anaerobic digestion process based on the nature of feeding material (organic waste)

3.7.1.2 Feeding Mode

Based on the feeding type, the digesters are classified into:

1. Batch flow digester
2. Continuous flow digester

3.7.1.2.1 Batch Flow Digester

The organic materials are filled at once, maintained closed for a specific period, and then unfed after a provided retention time (Khalid et al. 2011). The retention time varies from 50 to 60 days. Batch flow digester is designed to digest dry organic vegetable waste (Florentino 2003). The vegetable wastes (dry solid organic wastes) have higher C/N ratio than the dung, so this type of plants needs some addition of fermented slurry like organic nitrogen producers to begin the digestion process (Fig. 3.6).

Advantages of Batch Flow Digester

1. Sporadic raw material or limited wastes.
2. Depending on the waste material and operating temperature, this digester slowly starts the biogas production and also increases the production with time and after the 4–8 weeks which can end up slowly. So this is one of the best digesters to always produce the biogas (Florentino 2003).
3. It doesn't require that much of daily attention.

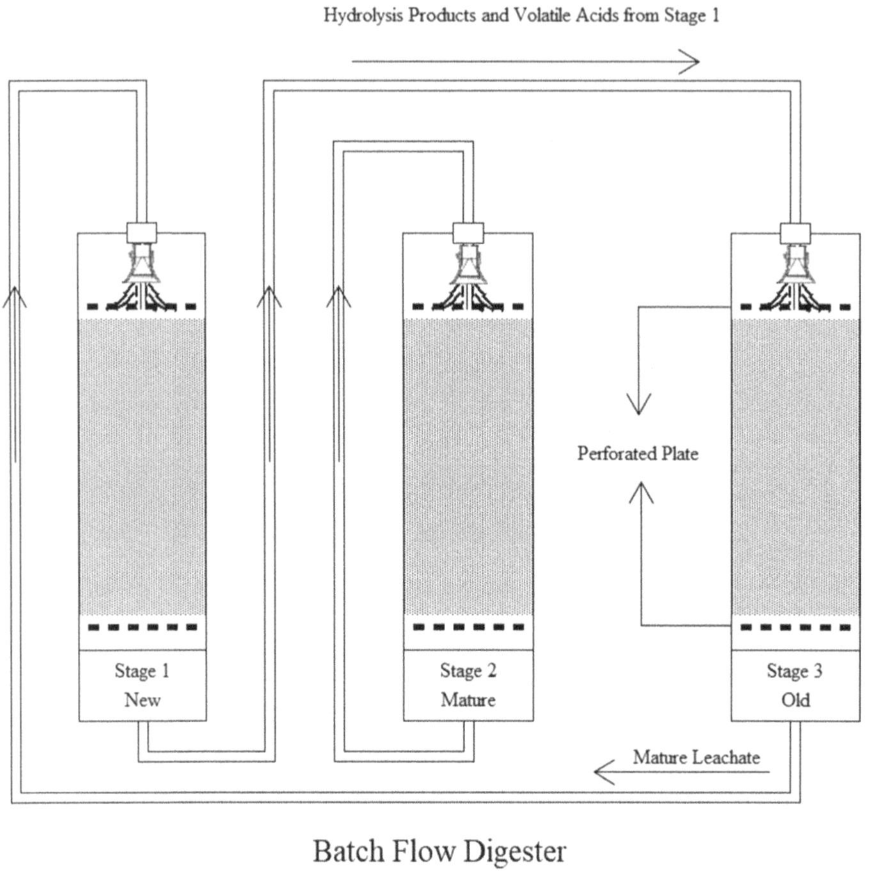

Fig. 3.6 Schematics and workflow of a batch flow digester. (Modified from Prakash et al. 2015)

Disadvantages of Batch Flow Digester

1. The irregular retention time leads to the intermittent gas production. So bigger gasholders or the simultaneous/parallel digesters are required for continuous gas production.
2. The addition of fermented slurry is a somewhat expensive process.
3. The battery of the digesters is sometimes emptied and recharged. So it is expensive and labour intensive (Khan 2009).
4. Non-economical for rural areas.

3.7.1.2.2 Continuous Flow Digester

Continuous flow digesters require daily loading of biomass. Inside the digester, the biomass digested by the hydraulic heat difference, between the substrate and

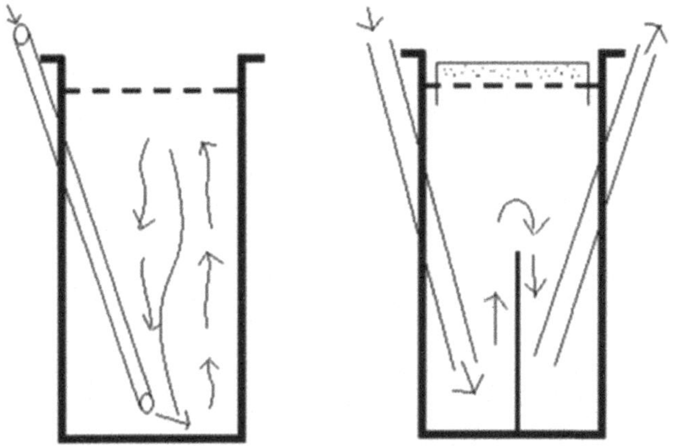

Fig. 3.7 Schematics and work flow for vertical mixing in a continuous flow digester. (Modified from Al-Sadi 2010; Fry 1974; Patel 1951)

digestate. So the digester completely digests the raw material and the digested slurry is collected for further uses. The time required when the biomass stayed inside the digester is known as retention time.

The retention time depends on raw material and the temperature. This type of digester can be stopped in case of sludge removal (undigested biomass residue); otherwise it is operated continuously. The continuous flow digesters are also divided into two basic models (Khan 2009).

- Vertical mixing – contains vertical chambers where the raw materials are added. The digested slurry overflows at the top of the chamber (Fig. 3.7).
- Displacement – contains long-cylinder chambers which are parallel to the ground (Fig. 3.8). The digested slurry can be displaced towards the end, where the maximum fermentation carries on.

Advantages of Continuous Flow Digester

1. The most important advantage of this type of system is the constant and automatic adjustment of the speed to guarantee the complete digestion of organic waste.
2. It has great investment potential.
3. Most convenient for rural households.
4. Constant gas production.

Disadvantages of Continuous Flow Digester

1. Raw materials need to be diluted.

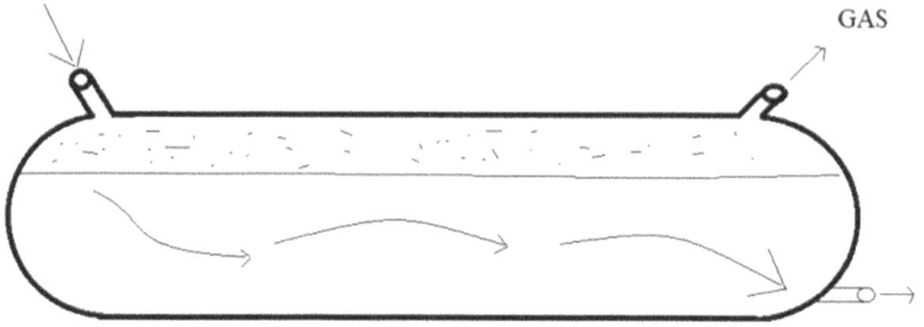

Fig. 3.8 Schematics and work flow for displacement in a continuous flow digester. (Modified from Al-Sadi 2010; Fry 1974; Patel 1951)

3.7.1.3 Floating Drum Digester

The floating drum plant was first developed in India. It is generally known as "Gobar Gas Plant". It was developed by Jashu Bhai J. Patel in 1956 (Singh et al. 1987). It contains a brick-lined pit which is partly underground and the mild steel floating drum (also acting as a gas holder) which is above the ground, coated with paints to avoid erosion (Fig. 3.9). The popular floating drum design is Khadi and Village Industries Commission (KVIA), India (Dana 2009). Nowadays, the steel drum is replaced by fibreglass-reinforced plastic or galvanized sheet metal (Nzila et al. 2012). The drum is the most expensive part in the plant covered by concrete work (burnt clay brick and cement) with the partition wall that provides an optimum growth condition to anaerobic microbes. The central guide frame is used to hold the gasholder and also guides it to move in a vertical position during gas production (Khan 2009).

Operating Procedure

1. The digester is feed with organic waste through inlet pipe.
2. The produced gas collected through a valve which is presented in the top of the gasholder.
3. The production of gas increases the drum starts to rise and the stored gas is collected; the drum sinks.
4. The rotation of the drum helps to break the scum.
5. After completion of digestion, the slurry is further collected.
6. In case of any maintenance, the digester may be emptied by another outlet at the bottom of the digester (Kossmann et al. 1999).

Advantages of Floating Drum Digester
1. The operation procedure is quite easy.
2. The drum weight assists the plant to discharge the gas at constant pressure.

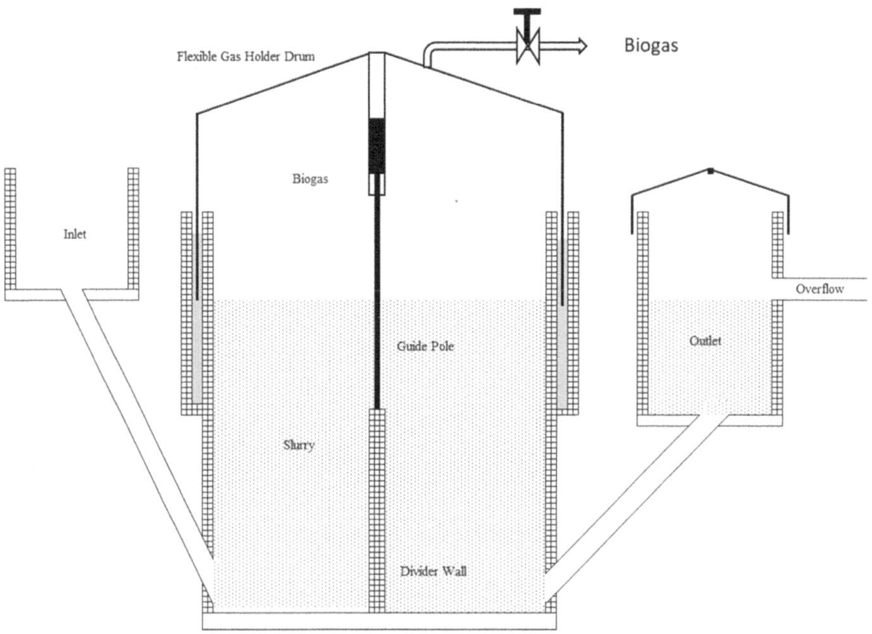

Fig. 3.9 Schematics and work flow of a floating drum type digester (or Gobar gas plant). (Modified from Florentino 2003; Samer 2012; Sasse 1988; Vögeli et al. 2014)

3. The drum level indicates of gas volume. So the volume of stored gas is easily visualized.
4. No gas leakage.
5. The well-maintained metal gas holder has a lifespan of 3–12 years depending on the humidity.

Disadvantages of Floating Drum Digester

1. Highly expensive.
2. Heat loss through metal gasholder.
3. Regular painting of gasholder.
4. UV rays can damage the main gas pipe, so regular maintenance is important.
5. Sometimes the steel drum can get stuck (Sasse 1988; Kossmann et al. 1999).

3.7.1.4 Fixed-Dome Digesters

The archetype of fixed-dome digester was first experimented in China. The plant contained a closed-dome shape digester with a fixed fermentation chamber and gasholder, a feedstock inlet, and a compensation tank. To reduce the cost of plant construction, researchers have developed a fixed-dome digester (Fig. 3.10). Instead of high-value drum, the dome acts as a gasholder. So the fixed-dome digesters are otherwise called as drum less digesters. Generally, the complete plant is constructed beneath the ground to maintain the absolute environment for anaerobic fermentation

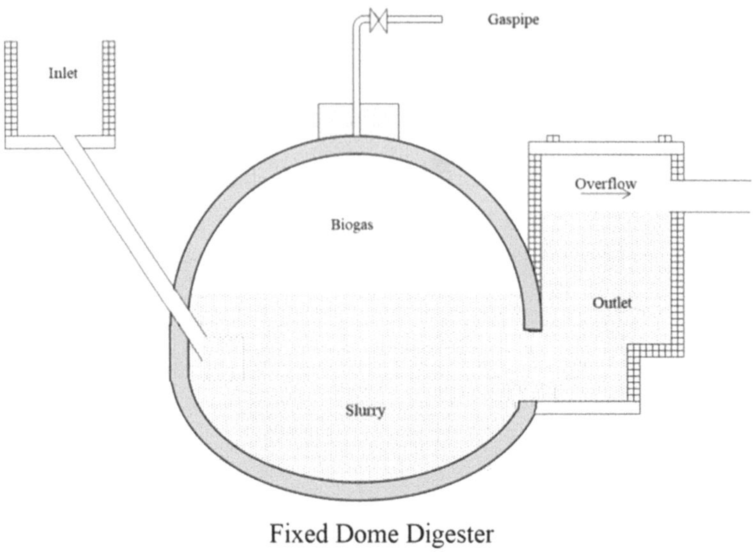

Fig. 3.10 Schematics and workflow of fixed- dome digester. (Modified from Florentino 2003; Samer 2012; Sasse 1988; Vögeli et al. 2014)

to take place. Moreover, the cracking of the dome is also avoided by the temperature and moisture change (Singh et al. 1987).

Operating Procedure

1. Feed the organic wastes like manure, dung, human excreta, etc.
2. The methanogenic bacteria decompose the raw material.
3. Biogas and the digestate are produced.
4. Gas is stored in the dome, and the slurry is displaced into the compensation tank.
5. When the gas is utilized, the pressure decreases and the proportional quantity of the slurry pushed back into the digester. So the pressure of gas varies depending on the gas production and the usage (Kudaravelli 2013; Stalin 2007).

Advantages of Fixed-Dome Digester

1. Inexpensive.
2. Longer lifespan if well constructed (20 years or more).
3. Less susceptible to corrosions.
4. Sub-surface assembly saves space and reduces variation in temperature (Werner et al. 1989).
5. Creates local employment.

Disadvantages of Fixed-Dome Digester

1. To construct such gas-tight structure and bedrock, it requires capital and expertise/technical skills.

2. The special sealant is required to plaster the gasholder and to prevent damage to the structure.
3. The fluctuated gas pressure complicates the gas utilization.
4. The construction should be with excellent structural strength (Sharma and Giuseppe 1991).
5. Repair and maintenance of such underground plant is bit difficult.

3.7.1.5 Balloon Plants

The balloon digester was developed in Taiwan at 1960, which is mainly discovered to solve the difficulties created with concrete and metal digesters (Karki 2005). The balloon plant contains a long-cylinder weather-resistant PVC or red mud plastic or rubber bag (balloon) that functions both as a digester and also as a gasholder (Fig. 3.11). Wherever the skin of the balloon is not damaged and also has an even and high temperature, the balloon plants are recommended. The inlet (lower part) and outlet (upper part) are connected to the balloon. This is a type of plug flow reactor.

Operating Procedure

The digestion of biomass is occurred within the bag.

- The plant is fed with inlet pipe and it is placed in a slightly deep trench to avoid damage.
- The bag inflates, if the gas produced.
- The gas is removed through the outlet pipe in the top.
- To maintain the inside pressure, small bags of sand are placed on the edge of the plastic.
- Gas pressure increased if placed any weight on the top of the balloon (Vögeli et al. 2014).

Advantages of Bag Digester/Balloon Plants

1. Inexpensive.
2. Easy for transportation.
3. Simple technology.
4. Easy to clean.

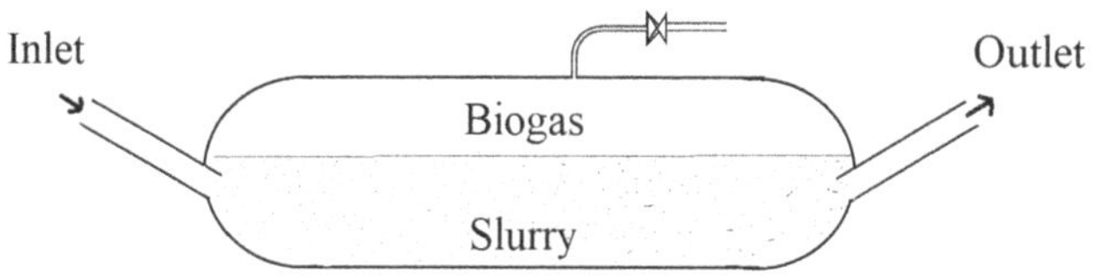

Fig. 3.11 Schematics and workflow of balloon- type digester. (Modified from Sasse 1988; Vögeli et al. 2014)

5. Easy to empty and maintain.
6. Difficult materials, like water hyacinths, can also be digested in this type of plant.
7. High temperature even in warm climates.

Disadvantages of Bag Digester/Balloon Plants
1. Plastic balloons are easily damaged and susceptible to physical and mechanical damage leading to shorter lifespan (2–5 years).
2. Difficult to repair.
3. High-grade plastic material is required to construct plant.
4. Insulation is quite difficult.
5. No local employment.
6. No scum removal technology.

3.7.1.6 Plug Flow Digester

The first report for plug flow digester was reported from South Africa in 1957 (Fry 1974). The design of the plug flow digester is very similar to bag digester. It contains a concrete-lined trough, which is five times longer than its width and also insulated with flexible gasholder (steel or fibreglass), reinforced concrete, or galvanized iron sheet (Singh et al. 1987).

1. The digester is feed with manure at a constant rate.
2. The feed moves along the digester as a plug.
3. Gas outlet is presented on the top of the digester connected with gas collector.

Advantages of Plug Flow Digester
1. Simple to understand
2. Economically good and easy to operate

Disadvantages of Plug Flow Digester
1. Because of plug flow type, sometimes the parts of the waste material travel faster than others.
2. Limited in applications with substances which floats or settles on the digester.
3. Difficult to clean if particles are settled or float.

Operating Temperature
Based on the temperature, the AD systems are divided into two types such as mesophilic and thermophilic systems. The psychrophilic system is not suitable for AD process (Vögeli et al. 2014).

Stages
The single and multistage systems are available in AD process. Multistage system is comparatively advantageous than single stage, since it does not require the optimal environmental conditions for both. In single-stage processes, all three stages of anaerobic process occur in single digester (Tucker 2008), whereas in multistage

Table 3.4 Role of different type of bacteria in the biogas production process

Bacteria	Electron acceptor	Electron donor	Final product	Reaction
Fermentative	Organic carbon	Organic carbon	CO_2	Fermentation
Syntropic	Organic carbon	Organic carbon	H_2	Acetogenesis
Acetogenic	Organic carbon/H_2	CO_2	CH_3COOH	Acetogenesis
Methanogenic	Organic carbon/H_2	CO_2	CH_4	Methanogenesis

processes, more than one digesters separate different stages (Lozano et al. 2009; Klocke et al. 2008).

3.8 Microbiology of Anaerobic Digestion

Microbiology of anaerobic digestion is a process which involves decomposition of several types of complex organic wastes, through consortia of metabolically active microorganisms such as hydrolysing, acidifying, acetogenic, and methanogenic, into fuel and manure (Demirel and Scherer 2008; Nealson 1997) (Table 3.4). These microbes' works as consortia are mutually beneficial, and slight disruption of this cooperation leads to reduced efficiency and also the breakdown of the process (Wijekoon et al. 2011).

The metabolic activities of microbes involved in anaerobic process rely on the chemical constitution of the biomass, ecological factors, and the working conditions of the digesters (Cha et al. 2001). The microbes involved in this digestion can be grouped into:

A. Acidogens
B. Acetogens
C. Methanogens

3.8.1 Acidogens

The bacterial species involved in hydrolysis are also active in acidogenesis. So both the hydrolytic and acidogenic bacteria are collectively known as fermentative bacteria. They can be either facultative or strict anaerobes. Most common genera of bacteria in hydrolysis are *Bacteroides, Lactobacillus, Propionibacterium, Sphingomonas, Sporobacterium,* and *Megasphaera.* The *Streptococcus* and the family of *Enterobacteriaceae or enteric bacteria* are also responsible for the hydrolysis. In acidogenic process, anaerobic facultative and the obligatory microbes are *Clostridium* spp., *Peptococcus anaerobius, Bifidobacterium* spp., *Desulfovibrio* spp., *Corynebacterium* spp., *Lactobacillus* spp., *Staphylococcus* spp., *Micrococcus or Flavobacterium,* and *E. coli* (Sharma 2008; Metcalf 2004). The commonly involved acidogens are *Bacillus cereus, B. megaterium, C. carnofeetidium, Pseudomonas formacans*, etc. (Karki 2005). Other organisms like *Caldicellulosiruptor saccharolyticus*, *Thermotoga maritima*, *C. thermocellum*, *Anaerocellum thermophilum*, *E. coli*, *C. kluyveri*, and *Ruminococcus albus* (Blumer et al. 2008; Wirth et al. 2012) are also involved. The most common cellulose fermenter in nature is *Clostridium*. The other cellulose-degrading bacteria involved in fermentation process are *Aminobacterium*, *Psychrobacter*, *Anaerococcus*, *Bacteroides*, *Acetivibrio*, *Butyrivibrio*, *Halocella*, *Spirochaeta*, *Caldicellulosiruptor*, and *Cellulomonas* (Burrell et al. 2004; Li et al. 2013).

3.8.2 Acetogens

Acetogenic bacteria (acetate and hydrogen producing bacteria) – acetogens such as *Syntrophobacter wolinii*, *Syntrophomonas wolfei*, *Syntrobacter wolinii*, and *Smithella* sp. – are involved in alcohol and short-chain fatty acid metabolism (Vavilin et al. 2008). Some of the amino acids and aromatic compounds produce by methanogens are also utilized as substrate by this group of bacteria (McInerney et al. 2008), (Fig. 3.12).

Syntrophic acetogens converts intermediary metabolites to acetate and other substrates with syntrophic partner (hydrogenotrophs), which oxidize acetate to methane (Hori et al. 2011). The syntrophic acetogens are both mesophilic and thermophilic in nature (Table 3.5). The *Acetobacterium woodii, Clostridium aceticum, Methanobacterium suboxydans*, and *Methanobacterium propionicum* are also responsible for acetogenesis process (Weiland 2010).

3.8.3 Methanogens (Archaea)

Methanogens are unique in nature and also strict anaerobes. The presence of oxygen is lethal for their activities. Methanogens are known by several names and found in anaerobic habitats including freshwater and marine water sediments, sewage digesters, waterlogged soils, the intestine of animals, wood-eating insects, etc.

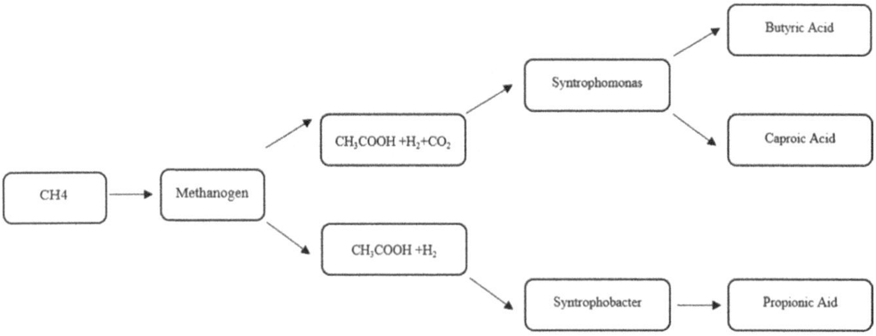

Fig. 3.12 Alcohol, aromatics, and short- chain fatty acid (produced by methanogens) metabolism by acetogens

Table 3.5 Temperature optima for syntrophic acetogens

Acetate oxidizing bacteria	Optimum temperature	Hydrogenotrophic methanogens	References
Thermacetogeniumphaeum	Thermophilic (55–58 °C)	*Methanothermobacter thermoautotrophicus TM*	Hattori et al. (2005)
Thermotogalettingae	Thermophilic (65 °C)	*Methanothermobacter thermoautotrophicus* or *Thermodesulfovibrio yellowstonii*	Balk et al. (2002)
Syntrophaceticusschinkii	Mesophilic (25–40 °C)	*Methanoculleus* sp. strain MAB1	Westerholm et al. (2010)

Table 3.6 Different types of carbon substrates oxidized by methanogenic bacteria

Genus	Substrate used	End products
Methanobacterium	Formate	$CH_4 + HCO_3$
Methanobacillus	Formate	$CH_4 + HCO_3$
Methanococcus	Formate	$CH_4 + HCO_3$
Methanosarcina	Acetate, methanol	$CH_4 + HCO_3$

(Zhu et al. 2004; Attwood et al. 2007; Ver Eecke et al. 2012; Brune, 2010). Morphologically they have several shapes such as cocci, bacilli, spirilla, and sarcina. Different types of carbon substrates utilized by methanogenic bacteria and their end products are listed in Table 3.6.

Methanogens Are Classified into Six Orders Macario (2008)

1. *Methanobacteriales*
2. *Methanococcales*
3. *Methanomicrobiales*
4. *Methanosarcinales* (acetoclastic having two families *Methanosarcinaceae* and *Methanosaetaceae*)
5. *Methanocellales*
6. *Methanopyrales*

The methanogens are the oldest bacteria and belong to the domain Archaea (ancient). These have slower metabolism and growth rate, responsible for rate limiting process in anaerobic degradation. Methanogens are important for the breakdown of substrate into gaseous form. They play important part in breakdown of substrate into gas form. In the absence of exogenic electron acceptors, they convert acetate and hydrogen to gaseous products. Based on the substrates, methanogens are divided into hydrogenotrophic and acetotrophic methanogens. Substrate for hydrogenotrophic methanogens are hydrogen and carbon dioxide and the substrate for acetotrophic methanogens is acetate. Some other substrates like methyl amine, alcohols, and formates can also be degraded (Table 3.7).

Table 3.7 The methanogenesis reaction

Acetate	$CH_3COOH \rightarrow CH_4 + CO_2$
Carbon monoxide	$4CO + 2H_20 \rightarrow CH_4 + 3H_2CO_3$
Dimethylamine	$2(CH_3)_2NH + 2H_2O \rightarrow 3CH_4 + CO_2 + 2NH_3$
Formate	$4HCOOH \rightarrow CH_4 + CO_2 + 2H_2O$
Hydrogen	$4H_2 + CO_2 \rightarrow CH_4 + 2H_2O$
Methanol	$4CH_3OH \rightarrow 3CH_4 + CO_2 + 2H_2O$
Methylamine	$4(CH_3)NH_2 + 2H_2O \rightarrow 3CH_4 + CO_2 + 4NH_3$
Methyl mercaptans	$2(CH3)_2S + 3H_2O \rightarrow 3CH_4 + CO_2 + H_2S$
Metals	$4Me^{\circ} + 8H^{+} + CO_2 \rightarrow 4Me^{++} + CH_4 + 2H_20$
Trimethylamine	$4(CH_3)NH + 6H_2O \rightarrow 9CH_4 + 3CO_2 + 4NH_3$

Modified from Demirel and Scherer (2008)

Table 3.8 Different steps and important parameters to carry out a successful anaerobic digestion process

	pH	Redox potential	Temperature	C/N ratio	C/N/P/S ratio	Trace elements
Hydrolysis/acidogenesis	5.2–6.3	+400 to −300 mV	25–35 °C	10–45	500:15:5:3	–
Acetogenesis/methanogenesis	6.7–7.5	< −250 mV	Mesophilic 30–40 °C Thermophilic 50–60 °C	20–30	600:15:5:3	Ni, Co, Mo, Se

3.9 Factors Affecting Biogas Production Operational Parameters of Biogas Production

Anaerobic digestion is a convenient technology to generate energy from wastes. For the successful process, number of factors needs to be optimized. Every bacterium has different living environment. The acidogenic bacteria requires different environment than methanogenic archaea. In the case of mixed culture, the need of the methanogenic bacteria should be considered with priority because of its nature (strict anaerobes, longer RT, and slow growth). To avoid the imbalance of AD, the two-stage process has developed. Here, each group of microorganisms have the optimum environmental condition (Table 3.8).

Factors Which Affect the Production of Biogas Are Follows

1. pH
2. Temperature
3. Organic loading rate
4. Hydraulic retention time
5. Ratio of carbon to nitrogen (C/N)
6. Toxicants
7. Mixing/agitation

3.9.1 pH

The optimum pH for AD process is 6.5–7.5 (Mata 2003; Khalid et al. 2011). During the process, the hydrolysis and acidogenesis take place at lower pH (~5.5–6.5) as compared to methanogenesis (~6.5–8.2). Generally, in every digester there are two buffering systems which ensure the optimum pH range: carbon dioxide–hydrogen carbonate and the ammonia–ammonium. Alternatively, lime, sodium carbonate, soda ash, and sodium hydroxide are also used for pH adjustments. In the case of immediate response, the sodium salts are used (Igoni et al. 2007). Addition of alkali making the enlargement of particulate organics leads to more susceptible enzymatic attack on the cellular substances.

3.9.2 Temperature

Digester temperature is a vital factor which influences the speed and total amount of biogas produced (Samah 2016). Normally, three widely used temperature ranges are psychrophilic (<20 °C), mesophilic (15–45 °C), and thermophilic (40–65 °C). Microbes are somewhat susceptible to alteration in temperature. The mesophilic are the most common bacteria used in AD process with varying temperature up to 3 °C (Chen et al. 2014); thermophilic consume additional power due to its increased temperature requirement (Chen et al. 2008). The psychrophilic are the slowest than others. The biogas production is lower in mesophilic digestion when compared to the thermophilic digestion.

3.9.3 Organic Loading Rate

The determination of biological transformation in the digestion process is known as organic loading rate (OLR). In other words, volume of substrate is introduced per digester volume in a given time. This is another controlling parameter of AD system, as overloading leads to increased volatile fatty acid concentration can result in bacterial degradation and system failure. This is the index parameter to declare the stress enforced on the microbes which also affects the total biogas production, COD stabilization, and alkalinity. The correct OLR is determined by the content of biomass and its biodegradability, where higher TS leads to higher OLR. In industrialized countries, OLRs are in the range of 4–8 kg VS/m^3 reactor per day for biowaste treatment. OLR is important for continuously stirred digesters. In non-stirred AD systems, the recommended OLR is below 2 kg VS/m^3 reactor (Vandevivere et al. 2003).

3.9.4 Retention Time

Hydraulic retention time is the time which depicts the number of days the liquid fraction stayed in the digester before they pushed through the outlet. The HRT is determined by the ratio of active slurry amount to biomass input flow rate.

$$\text{High HRT------------------more biodegradation.}$$

3.9.5 Solid Retention Time

This is the number of days solids remain in the digester. Solids are retained due to sedimentation in the case of unstirred digesters.

$$\text{High SRT------------------decrease HRT.}$$

So the manual removal of solids is important for desired HRT and gas production (Verma 2002).

3.9.6 Carbon-to-Nitrogen Ratio

The C/N ratio is the measure of relative amount of carbon and nitrogen in organic materials. It is important to estimate the nutrient (C) insufficiency and process inhibition by ammonia by adding extra “C” source. The reported optimum C/N is 16:25.

$$\text{High C/N-------lower gas production (quick consumption of nitrogen by methanogens).}$$
$$\text{Low C/N--------increased pH (lethal to methanogens).}$$

3.9.6.1 Sample Raw Materials C/N Ratio

To maintain the desirable level of C/N ratio, different materials with high C/N ratio are mixed with those with low ratio, such as municipal waste, biosolids, animal manure, etc. (Karki et al. 1994). The C/N ratio of different types of frequently available raw materials is shown in Table 3.9.

3.9.7 Stirring

Stirring is the process which increases the biogas production. Mixing combines the fresh feed with digestate and the microbes. So the stirring increases the fermentation. Mixing should be done in such a manner that air (oxygen) doesn’t enter the digester. Violent or mechanical agitation stops the digestion. Stirring is advantageous in some types of digesters. If not agitated properly, scum will be formed negatively affecting

Table 3.9 The C/N ratio of different types of raw materials

Sample raw materials	C/N ratio
Bird excreta	8–10
Human excreta	8
Animal dung	12–43
Water hyacinth	25
Agricultural waste	60–90

Table 3.10 The inhibitory and toxic concentration of heavy metals

Metal	Inhibition start (mg/l)	Toxicity to adopted microorganism (mg/l)
Cr^{3+}	130	260
Cr^{6+}	110	420
Cu	40	170
Ni	10	30
Cd	70	600
Pb	340	340
Zn	400	600

the release of biogas. To avoid such problem, continuous feeding is advisable, since the fresh feed will disrupt the scum and will also provide continuous mixing. According to the type of digester and TS content, the performance of the stirring equipment varies. The most promising AD technologies in developing countries contain no stirring equipment. Recirculation of removed digestate into the inlet of the digester achieves a passive mixing process, which tends to flush the inlet pipe and helps blending of new feed with digestate rich in microbes (Deublein and Steinhauser 2008).

3.9.8 Toxicants

Antibiotics and the disinfection agents are also the important factors affecting the production process. These are used in cattle to treat the sick animals and also to keep the farm and the milk parlours clean. Low concentration doesn't have any negative impact. Continuous feeding leads to inhibition. Other substances like heavy metals, salts, oxygen, hydrogen, sulphide, organic acids, free ammonia, tannins, herbicides, insecticides, and micronutrients are also inhibiting the biogas production process (Table 3.10). Ammonia nitrogen is the most common inhibiting agent of AD process. In optimum pH condition, the total inorganic nitrogen share is in the form of ammonium, but at higher pH and temperature, ammonia concentration increases (Chen et al. 2008). The increased ammonia diffuses the cell membrane and disrupts the proton and potassium balance leading to cell function damage. So, the intermediate digestion products are accumulated which results in acidification process, and the biogas production could be ceased (Kayhanian 1999). Low concentrations of heavy metals, minerals, ions, and detergents are important for bacterial growth but in higher rate inhibit the bacterial growth.

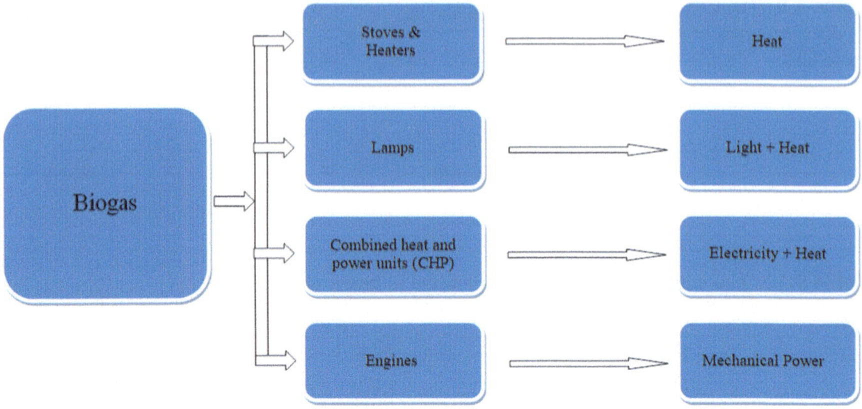

Fig. 3.13 Various applications of biogas in day-to-day life

3.10 Biogas Applications in Day-to-Day Life and Industries

Biogas can be used for various domestic and industrial purposes (Fig. 3.13), while major household applications are in cooking (using specially designed burners) or accordingly customized devices, and water and room heaters and for electricity production.

3.10.1 Domestic Applications

3.10.1.1 Biogas Stove

Biogas can be used in similar ways as natural gas in gas stoves. It is a colourless, odourless blue burning gas that can be obtained from various feedstocks with heating value of 2 MJ/m^3 and can be used for domestic purposes like cooking, heating, and illumination. Pipeline is used to transport biogas from plant to the cooking place (Sasse 1988). It is a clean fuel and produces fewer pollutants during cooking as compared to other unrefined raw fuel sources. Commonly used stoves and ovens in the kitchen can be adapted for the use with biogas by modification in burners for proper biogas combustion and efficient use of energy. The factors that influence the use of biogas as a combustible gas are the ratio of gas/air mixing, gas pressure, flame speed, and ignition temperature. When compared to LPG, biogas needs less air/m^3 for combustion (almost 5–6 times less air/l of biogas, as compared to butane or propane) (Sasse et al. 1991). Although it is a renewable and carbon-negative clean fuel, it requires high initial capital for construction and set-up of the digester, accessories, and increase workload of the end user, physical burden to feed the substrate and water to the digester.

Different types of stoves with gas consumption of 0.22–0.44 m^3 per hour are commonly used (Surendra et al. 2014). Those stoves are primarily manufactured to work with at least 75% air, and if biogas/air mixture is not proper or too little air is

Table 3.11 Reported calorific value (CV) of biogas

S.I #	CV	Source and year
1	19	Fulford (1988)
2	22	Itodo (2007)
3	20	Pathak et al. (2009)
4	21	Surendra et al. (2014)

available, the burning efficiency will be drastically reduced and considerable amount of the unused gas will be wasted. Biogas stoves with cyclone burner are more efficient than the usual burner of biogas stove. The biggest power generated is equal to 1.52488 kW for cyclone burner. The highest efficiency is equal to 58.42% for cyclone-shaped burner (Syamsuri and Yustia 2015). Thermal proficiency for the biogas stoves ranges between 50% and 60% (Clean Energy Solutions Vienna). Different reports showed calorific value (CV) of biogas in the range of 19–22 (Table 3.11).

By using biogas as fuel for cooking, we can prevent faecal-borne and parasitic diseases and reduction in household air pollution due to openly defecated dung. Biogas as a fuel substitution for firewood in the rural areas of the developing countries could reduce fuel collection time, smoke, and uncontrollable fire. Biogas slurry from plant is a more potent fertilizer, which can be used in gardening and farming practices (Bond and Templeton 2011; Brown 2006; Chen et al. 2010; De Alwis 2002; Jian 2009; Ovueni 2014; Van and Weber 1994).

3.10.1.2 Biogas Light

Biogas can be used for illuminating the house in the developing countries where electricity is not available or is still irregular. Biogas lighting system uses a special type of gauze mantle lamps consuming 0.07–0.14 m^3 of gas/h. Biogas lamps can work with a proficiency rate of 3–5% (Everson and Smith 2016). Several commercially single- or double-mantle biogas lamps are available in the market, which could be used for both indoors and outdoors. Biogas lamp mantle can produce clear and bright light equivalent to 40–100 candle powers. In china different types of biogas lamp are in use, being economical and having ease of operation. Clay lamp is also used by Chinese farmers which do not need many skills to manufacture. Porous burners (PB) are reported to be the hi-tech alternatives which could consistently supply illumination and warmth, with much lower greenhouse gases. Takeno and Sato (1979) conceptualized the idea by using a porous component to maintain the flame with lower fuel consumption rates. Such porous components have some advantages such as light weight and higher efficiency.

3.10.1.2.1 Biogas as a Renewable Energy for both Public and Industrial Use

Electricity generation by biogas is one of the main applications. In addition to power, many applications require heat, like hospitals, schools, office, industries with heat-intensive manufacturing units, shopping centres, etc. They generate heat from the public grid. Current biogas plants are equipped with cogeneration unit CHP (combined heat and power) in which it can be simultaneously used for both heat and

Table 3.12 Comparison of total resource efficiency between a power plant and boiler vs CHP system

Separate production of electricity and heat	Fuel (100) → Power plant → Electricity (36)	Total efficiency (η) = 0.58
	Fuel (100) → Boiler → Heat (80)	
Cogeneration of heat and electricity	Fuel (100) → CHP plant → Electricity (30) + Heat (55)	Total efficiency (η) = 0.85

electricity generation. CHP enhances gas engine fuel economy; overall efficiency of 75–80% and efficiencies up to 90% are achievable (Table 3.12). The longer the annual operating hours, the greater will be the potential for profitable-P.

Cogenerated power and heat can be used for internal demands, and the excesses can be fed into the public grid, while the thermal energy can be utilized for heating purposes or deployed as process heat. By preheating the substrate, it creates an ideal condition for the microorganisms processing the organic matter.

The largest heating demands (over 90%) in the digester operation are heating the substrate (Zupancic and Roš 2003). CHP can deliver the heat demands for the common digester temperature range from 38 to 44 °C for typical mesophilic digestion, but for thermophilic digestion, it requires additional heating. Different technologies are used to heat the digester by heating pipes along the fermenter walls, by pumping the digestate through a heat exchanger, or by heat exchange between substrate outflow to substrate inflow (Fig. 3.14). Fifteen percent of the heat produced by the CHP unit is used to heat the digester, and on an average the consumption of heat energy for the system is between 70 and 120 kWh for the different months of the year.

CHP system can be used as cooling energy, like air conditioning in an office building by using absorption or an adsorption chiller. This plays an important role in building temperature control (office and data centres) and process cooling in industrial manufacturing unit. The cogeneration of heat and power from organic waste, a regenerative resource, is a carbon-neutral means of energy production. CHP systems accrue heat during the combustion process via heat exchangers. Using heat by this process reduces energy usage up to 40% compared to conventional power systems (Pfeifer and Obernberger 2007). The CHP process includes internal combustion engines, combustion gas turbines, micro turbines, fuel cells, steam turbines, and Stirling engine.

3.10.1.3 Internal Combustion Engines

The internal combustion engines could be easily operated as it can operate using both liquid and gaseous fuels for the generation of heat and energy. Internal combustion diesel engines could be modified to use biogas as a fuel: by dual-fuel operation with ignition by pilot fuel injection and/or biogas alone with spark ignition. In dual-fuel engine, the normal fuel injection system still supplies a certain amount of diesel (between 10% and 20% of the original amount needed). A compressed mixture of air and biogas together with the diesel fuel is sprayed in for ignition. When biogas is not available or has less supply, operation on diesel fuel alone or substituting a corresponding part with diesel for continuous operation is

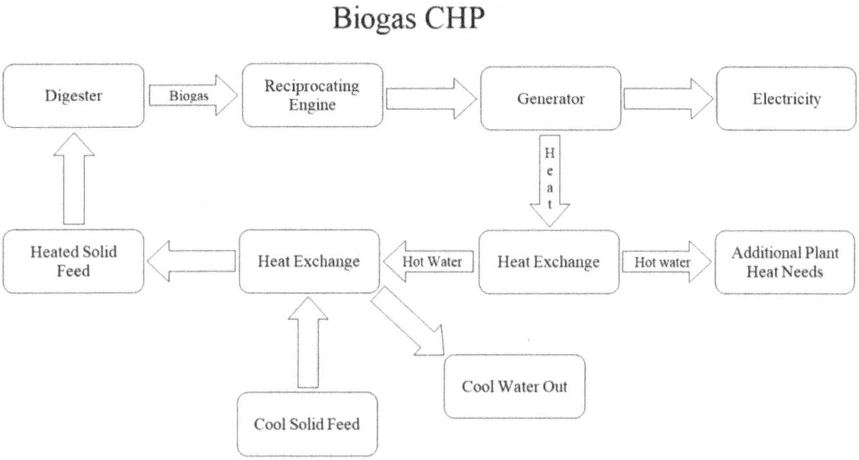

Fig. 3.14 Schematics and flow diagram of a biogas-CHP system

possible. However diesel for initial ignition is a must for dual-fuel engine. In gas-alone spark engines, basic modification is the air/gas mixer set-up, instead of a carburettor. Rich-burn and lean-burn types of spark-ignition internal combustion engines are mainly used for low-BTU (British thermal unit) gas CHP applications. However, as the engine modification is permanent, original fuel cannot be used (Pruthviraj 2016; Stefan 2004).

3.10.1.4 Combustion Gas Turbines

In combustion gas turbines, heat and energy are produced by both compression and ignition of atmospheric air and fuel mixture within the turbine. The compressed intake air and biogas is ignited in the combustion chamber. The gases then enter the turbine at high pressure and drive the generator. Energy is then utilized from the expanded, high-temperature gases to move turbine blades that produce electricity. The gas turbines consist of turbine assembly with gas compressor, the combustion chamber, and monitoring equipment.

The turbine exhaust gas leaves at temperature 400–600 °C, which can be utilized with a heat exchanger for producing hot water or it can be utilized to run a steam turbine or to preheat the air used in the turbine. Gas turbines can generate100 kW of electricity and a heating output of 165 kW, with very low emission values (Deublein and Steinhauser 2008). Over 50% of efficacy can be improved by combining a gas turbine with a steam turbine. Since it requires high-pressure feed gas supply, gas turbines require gas compressor, which in turn will increase the initial investment costs. Generally higher capital is required to build it, as compared to other types of engines but the overall operating and maintenance costs are much lower.

3.10.1.5 Micro Turbines

Micro gas turbines are small high-speed turbine, an established technology in sizes above 500 kW. Micro turbines in biogas application, with range of 25–100 kW, have been introduced. Micro turbines are high-speed, integrated power plants that include a turbine of radial design, compressor, generator, heat exchanger, and power electronics to produce power. The compressor, the turbine, and the generator are fixed on a single shaft. The biogas mixed with the air is supplied to the combustion air in the combustion chamber, and the turbine sucks in the combustion air. Overall efficiency of current generation micro turbines is still low (Deublein and Steinhauser 2008).

3.10.1.6 Fuel Cells

The fuel cell converts the chemical energy to electric current and heat. Biogas-powered fuel cells hold great potential to convert biogas directly into electricity with zero emission. Major multinational companies and many residential buildings are currently using biogas-powered fuel cells to generate energy for various purposes. It simply utilizes chemical reactions to produce the energy. Fuel cells boost net output of electricity by a minimum of 60%, with an average 0.6–0.9 V/single cell. Single cells could be further arranged in a stack to get the desired voltage. However, biogas has to be purified before using as a fuel in the fuel cell by mainly removing the CO and H_2S. Proton exchange membrane (PEM) and solid oxide electrolytes are the currently used fuel cells (Chambers and Potter 2002).

3.10.1.7 Steam Turbine

High-pressure and superheated steam is injected into shaft power that drives the generator to produce electric power in steam turbine. Generally steam turbine CHP systems produce electricity as a by-product of steam generation. Biogas can be used as a fuel to generate high-pressure steam from the boiler to power the turbine for electricity generation. Low-pressure steam could be directly extracted from the turbine and used as a thermal energy for other needs. The steam is condensed and pumped back to the boiler, thus completing Rankine cycle (thermodynamic cycle). Radial and axial flows are the two types of turbine in action. If the exhaust steam contains >10% water, it can erode nozzle and blades. Special design in the turbine to remove the moisture can be used when the superheated steam temperature is limited. Reheating of the superheated steam after partial expansion will increase the cycle efficiency. Three types of steam turbines are in operation: condensing steam turbine, extraction turbines, and back pressure turbine. Steam turbines are one of the oldest (100 years) technologies still in commercial production, with 50 kW–>100 MW capacity (Ion and Popescu 2016).

3.10.1.8 Stirling Engines

A Stirling engine produces heat and electricity by external combustion. One of the advantages of a Stirling engine is its external combustion; it can utilize a multitude of fuels including biogas where other engines cannot. The engine uses nitrogen as a "working fluid". Heat exchanger from the combustion chamber is utilized to produce thermal energy. Since the engine has external combustion chamber, it is not required to refine the fuel as it does for other types of engines. The complete combustion of fuel in the external combustion chamber also provides less emission of unburned hydrocarbons in the exhaust. The Stirling engine can consume renewable fuels and produces minimal emissions (less CO_2 and other emissions). The combustion happens in the external combustion chamber, and there is no fuel or oil contamination within the engine. It also contains 50% less moving parts than internal combustion engines. The CO_2 and CO emissions for biogas were less than that for natural gas (Pourmovahed et al. 2011).

3.11 Conclusion and Future Outlook

The fossil fuels are non-renewable source of energy and are also the prime cause of environmental pollution. Sooner or later we need to find an alternative – continuous sources of energy, which are both energy efficient and environmental friendly at the same time. Currently several types of renewable energy sources are either at trial or some sort of application stage – solar power, wind energy, hydrothermal, nuclear, and others. One major issue with such energy sources are either lack of efficiency or efficient storage and transport from the source of generation to usage. Other types of energy sources are biofuels, such as bioethanol, biodiesel, biohydrogen, etc., which are also in their third generation of research and applications (Ingale et al. 2014, 2019). However production of such fuels is energy intensive at the moment, and it lags behind with respect to economy due to competition with food crops and other issues. Biogas production is a natural non-energy intensive, and the raw materials are mostly renewable resource and wastes – thus serving both purposes (bioremediation and energy generation). The ease of biogas plant operation, maintenance, and easy availability of substrate – waste materials – is the key selling point which makes it an effective and common energy tool in the near future.

References

Al-Sadi M (2010) Design and building of biogas digester for organic materials gained from solid waste. M. Sc. thesis, Faculty of Graduate Studies, An-Najah National University, Nablus, Palestine

Alves MM, Pereira MA, Sousa DZ, Cavaleiro AJ, Picavet M, Smidt H, Stams AJM (2009) Waste lipids to energy: how to optimize methane production from long-chain fatty acids (LCFA). Microbial Biotechnol 2:538–550

Alwis AD (2002) Biogas – a review of Sri Lanka's performance with a renewable energy technology. Energy Sustain Dev 6(1):30–37

Angelidaki I, Sanders W (2004) Assessment of the anaerobic biodegradability of macropollutants. Rev Environ Sci Biotechnol 3(2):117–129

Apte A, Cheernam V, Kamat M, Kamat S, Kashikar P, Jeswani H (2013) Potential of using kitchen waste in a biogas plant. Int J Environ Sci Dev 4:370

Armaha EK, Tetteha EK, Boamah BB (2017) Overview of biogas production from different feedstocks. Int J Sci Res Publ 7(12):158

Attwood GT, Kelly WJ, Altermann EH, Leahy SC (2007) Analysis of the *Methanobrevibacter ruminantium* draft genome: understanding methanogen biology to inhibit their action in the rumen. Aust J Exp Agric 48:83–88
Balk M, Weijma J, Stams AJM (2002) *Thermotogalettingae sp.* nov., a novel thermophilic, methanol-degrading bacterium isolated from a thermophilic anaerobic reactor. Int J Syst Evol Microbiol 52:1361–1368
Batstone DJ, Keller J, Angelidaki I, Kalyuzhnyi SV, Pavlostathis SG, Rozzi A, Sanders WTM, Siegrist HA, Vavilin VA (2002) Anaerobic digestion model no. 1 (ADM1), task group for mathematical modelling of anaerobic digestion processes. IWA Publishing, London
Bhardwaj S, Das P (2017) A review: advantages and disadvantages of biogas. Int Res J Eng Technol 04(10)
Bischofsberger W, Dichtl N, Rosnwinkel KH, Bohnke CFSB (2005) Anaerob technik. 2. Auflage. Heidelberg. Germany. ISBN:978-3-540-06850-1
Blumer SSE, Kataeva I, Westpheling J, Adams MWW, Kelly RM (2008) Extremely thermophilic microorganisms for biomass conversion: status and prospects. Curr Opin Biotechnol 19:210–217
Bond T, Templeton MR (2011) History and future of domestic biogas plants in the developing world. Energy Sustain Dev 15:347–354
Brown VJ (2006) Biogas: a bright idea for Africa. Environ Health Perspect 114(5):A301–A303
Brune A (2010) Methanogenesis in the digestive tracts of insects. In: Timmis KW (ed) Handbook of hydrocarbon and lipid microbiology. Springer, Berlin/Herdelberg, pp 707–728
Burrell PC, O'Sullivan C, Song H, Clarke WP, Black-all LL (2004) The identification, detection and spatial resolution of Clostridium populations responsible for cellulose degradation in a methanogenic landfill leachate bioreactor. Appl Environ Microbiol 70:2414–2419
Castellucci S, Cocchi S, Allegrini E, Vecchione L (2013) Anaerobic digestion and co-digestion of slaughterhouse wastes. J Agric Eng Res 44(s2)
Cha GC, Chung HK, Kim DJ (2001) Characteristics of temperature change on the substrate degradation and bacterial population in one-phase and two-phase anaerobic digestion. Environ Eng Res 6:99–108
Chambers AK, Potter I (2002) Gas utilization from sewage waste. Alberta Research Council, Alberta, pp 1–13
Chandra R, Takeuchi H, Hasegawa T, Kumar R (2012) Improving biodegradability and biogas production of wheat straw substrates using sodium hydroxide and hydrothermal pretreatments. Energy 43(1):273–282
Chen Y, Cheng JJ, Creamer KS (2008) Inhibition of anaerobic digestion process: a review. Bioresour Technol 99(10):4044–4064
Chen Y, Yang G, Sweeney S, Feng Y (2010) Household biogas use in rural China: a study of opportunities and constraints. Renew Sust Energ Rev 14(1):545–549
Chen JL, Ortiz R, Steele TWJ, Stuckey DC (2014) Toxicants inhibiting anaerobic digestion: a review. Biotechnol Adv 32(8):1523–1534
Dana B (2009) Build manual: ARTI floating dome biodigester, Appropriate Infrastructure Development Group (AIDG)
De Clercq D, Zongguo W, Fei F, Luis C (2016) Biomethane production potential from restaurant food waste in megacities and project level-bottlenecks: a case study in Beijing. Renew Sust Energ Rev 59:1676–1685
De Mes TZD, Stams AJM, Reith JH, Zeeman G (2003) Methane production by anaerobic digestion of wastewater and solid wastes. In: Reith JH, Wijffels RH, Barten H (eds) Bio-methane and bio-hydrogen: status and perspectives of biological methane and hydrogen production. Dutch Biological Hydrogen Foundation, Petten, pp 58–102
Demirel B, Scherer P (2008) The roles of acetotrophic and hydrogenotrophic methanogens during anaerobic conversion of biomass to methane: a review. Rev Environ Sci Biotechnol 7 (2):173–190
Deublein D, Steinhauser A (2008) Biogas from waste and renewable resources. Wiley. ISBN:978-3-527-31841-4

Everson TM, Smith MT (2016) Improving rural livelihoods through biogas generation using livestock manure and rainwater harvesting. Volume 2: guideline report. Report to the water research commission

Florentino H (2003) Mathematical tool to size rural digesters. Sci Agric 60(1):185–190. ISSN:0103-9016

Forster-Carneiro T, Pérez M, Romero LI (2008) Influence of total solid and inoculum contents on performance of anaerobic reactors treating food waste. Bioresour Technol 99(15):6994–7002

Fry LJ (1974) Practical building of methane power plants for rural energy independence. The EPA National Library Catalog, Andover. ISBN:10: 0960098410

Fulford D (1988) Running a biogas programme: a handbook. Intermediate Technology Publications, London

Gao R, Yuan X, Zhu W, Wang X, Chen S, Cheng X, Cui Z (2012) Methane yield through anaerobic digestion for various maize varieties in China. Bioresour Technol 118:611–614

Gerardi MH (2003) The microbiology of anaerobic digester. Wiley. ISBN:978-0-471-20693-4

Guendouz AA, Brockmann D, Trably E, Dumas C, Delgenès JP, Steyer JP, Escudie R (2012) Total solids content drives high solid anaerobic digestion via mass transfer limitation. Bioresour Technol 111:55–61

Haftu G, Solomon M, Giday G (2018) Qualitative and quantitative feasibility of biogas production from kitchen waste. Am J Energy Eng 6(1):1–5

Hamilton DW (2012) Organic matter content of wastewater and manure. BAE 1760, Oklahoma Cooperative Extension Service, Stillwater, Oklahoma

Hattori S, Galushko AS, Kamagata Y, Schink B (2005) Operation of the CO dehydrogenase/acetyl coenzyme A pathway in both acetate oxidation and formation by the syntrophically acetate oxidizing bacterium *Thermacetogeniumphaeum*. J Bacteriol 187:3471–3476

Heffels T, McKenna R, Fichtner W (2012) Direct marketing of electricity from biogas and biomethane: an economic analysis of several business models in Germany. J Manag Control 23:53–70

Hori T, Sasaki D, Haruta S, Shigematsu T, Ueno Y, Ishii M, Igarashi Y (2011) Detection of active, potentially acetate-oxidizing syntrophs in an anaerobic digester by flux measurement and formyltetrahydrofolate synthetase expression profiling. Microbiology 157:1980–1989

Igoni AH, Ayotamuno MJ, Eze CL, Ogaji SOT, Probert SD (2007) Designs of anaerobic digesters for producing biogas from municipal solid waste. Appl Energy 85:430–438

Ingale S, Joshi SJ, Gupte A (2014) Production of bioethanol using agricultural waste: banana pseudo stem. Braz J Microbiol 45(3):885–892

Ingale S, Parnandi VA, Joshi SJ (2019) Bioethanol production using *Saccharomyces cerevisiae* immobilized in calcium alginate–magnetite beads and application of response surface methodology to optimize bioethanol yield. In: Srivastava N, Srivastava M, Mishra P, Upadhyay S, Ramteke P, Gupta V (eds) Sustainable approaches for biofuels production technologies, Biofuel and biorefinery technologies, vol 7. Springer, Cham, pp 147–181

Ion VI, Popescu F (2016) Efficiency improvement of a biogas engine-driven CHP plant. Sci Work Univ Food Technol 63(1)

Itodo IN, Agyo GE, Yusuf P (2007) Performance evaluation of a biogas stove for cooking in Nigeria. J Energy S Afr 18(4):14–18

Jian L (2009) Socioeconomic barriers to biogas development in rural Southwest China: an ethnographic case study. Hum Organ 68(4):415–430

Karki AB (2005) Biogas, as renewable source of energy in Nepal theory and Development, BSP-Nepal

Karki AB, Gautam KM, Karki A (1994) Biogas installation from elephant dung at Machan Wildlife Resort, Chitwan, Nepal. Biogas Newsletter, Issue No 45

Kayhanian M (1999) Ammonia inhibition in high-solids biogasification: an overview and practical solutions. Environ Technol 20:355–265

Khalid A, Arshad M, Anjum M, Mahmood T, Dawson L (2011) The anaerobic digestion of solid organic waste – review. Waste Manag 31(8):1737–1744

Khan BH (2009) Non-conventional energy resources, Mechanical engineering series. Mc Graw Hill Education, New Delhi

Klocke M, Nettmann E, Bergmann I, Mundt K, Souidi K, Mumme J, Link B (2008) Characterization of the methanogenic archaea within two-phase biogas reactor systems operated with plant biomass. Syst Appl Microbiol 31:190–205

Kossmann W, Pönitz U, Habermehl S, Hoerz T, Krämer P, Klingler B, Kellner C, Wittur T, von Klopotek F, Krieg A, Euler H (1999) Biogas digest volume I – IV. German Agency for Technical Cooperation (GTZ), Eschborn

Kudaravelli K (2013) Biogas plant construction manual – fixed dome Deenbandhu model digester: 2 to 6 cubic meter size. Egyptian Environmental Affairs Agency, Cairo

Kwietniewska E, Tys J (2014) Process characteristics, inhibition factors and methane yields of anaerobic digestion process, with particular focus on microalgal biomass fermentation. Renew Sust Energy Rev 34:491–500

Li A, Chu Y, Wang X, Ren L, Yu J, Liu X, Yan J, Zhang L, Wu S, Li S (2013) A pyrosequencing-based metagenomic study of methane-producing microbial community in solid-state biogas reactor. Biotechnol Biofuels 6:3

Lozano CJS, Mendoza MV, de Arango MC, Monroy EFC (2009) Microbiological characterization and specific methanogenic activity of anaerobe sludges used in urban solid waste treatment. Waste Manag 29:704–711

Macario DEC (2008) Taxonomy of methanogens. In: Bergey's manual of systematic bacteriology, 2nd edn. Springer, New York

Mata AJ (2003) Biomethanization of the organic fraction of municipal solid wastes. IWA Publishing, London

Mayer F, Gerin PA, Noo A, Foucart G, Flammang J, Lemaigre S, Sinnaeve G, Dardenne P, Delfosse P (2014) Assessment of factors influencing the biomethane yield of maize silages. Bioresour Technol 153:260–268

McInerney MJ, Struchtemeyer CG, Sieber J, Mouttaki H, Stams AJM, Schnink B, Rohlin L, Gunsalus RP (2008) Physiology, ecology, phylogeny, and genomics of microorganisms capable of syntrophic metabolism. Ann N Y Acad Sci 1125:58–72

Metcalf E (2004) Wastewater engineering: treatment and reuse. In: Franklin L, Burton H, Stensel D (eds) Revised by George tchobanoglous, 4th edn. McGraw-Hill, New York

Möller K (2015) Effects of anaerobic digestion on soil carbon and nitrogen turnover, N emissions, and soil biological activity. A review. Agron Sustain Dev 35:1021–1041

Moody LR, Burns R, Wu-Haan W, Spajić R (2009) Use of biochemical methane potential (BMP) assays for predicting and enhancing anaerobic digester performance. In: Proceedings of the 4th international and 44th Croatian symposium of agriculture, Optija

Nealson KH (1997) Sediment bacteria: who's there, what are they doing, and what's new? Annu Rev Earth Planet Sci 25:403–434

Nges IA, Escobar F, Fu X, Björnsson L (2012) Benefits of supplementing an industrial waste anaerobic digester with energy crops for increased biogas production. Waste Manag 32 (1):53–59

Nzila C, Dewulf J, Spanjers H, Tuigong D, Kiriamiti H, van Langenhove H (2012) Multi criteria sustainability assessment of biogas production in Kenya. Appl Energy 93:496–506

Ogur EO, Mbatia S (2013) Conversion of kitchen waste into biogas. Int J Eng Sci 2(11):70–76

Ovueni UJ (2014) Comparative study of the heating capacity of biogas and conventional cooking gas. Int J Eng Sci 3(1):7–10

Patel J (1951) Digestion of waste organic matter and organic fertilizer and a new economic apparatus for small scale digestion. Poona Agri Coll Mag (India) 42(3):150–159

Pathak H, Jain N, Mohanty S, Gupta N (2009) Global warming mitigation potential of biogas plants in India. Environ Monit Assess 157:407–418

Peter JJ (2009) Biogas –green energy. Digisource Danmark A/S. ISBN:978-87-992243-2-1

Pfeifer J, Obernberger I (2007) Technology evaluation of an agricultural biogas CHP plant as well as definition of guiding values for the improved design and operation. In: 15th European biomass conference & exhibition, 7–11 May 2007, Berlin, Germany, pp 1864–1868

Pourmovahed A, Opperman T, Lemke B (2011) Performance and efficiency of a biogas CHP system utilizing a stirling engine. In: Proceedings of international conference on renewable energies and power quality, Las Palmas de Gran Canaria, Spain (Vol. 1315)

Prakash O, Anil K, Pandey A, Kumara A, Laguria V (2015) A review on biogas plant. Int J New Technol Sci Eng 2(4). ISSN:2349-0780

Pruthviraj NB (2016) Introduction to biogas & applications. Int J Adv Res Mech Eng Technol (IJARMET) 2(4)

Reddy SN, Satyanarayana SV, Sudha G (2017) Bio gas generation from biodegradable kitchen waste. Int J Environ Agric Biotechnol 2(2):0689–0694

Rutz D, Mergner R, Janssen R (2015) Sustainable heat use of biogas plants. A handbook, biogas heat. WIP Renewable Energies, Munich

Salminen E, Einola J, Rintala J (2003) The methane production of poultry slaughtering residues and effects of pre-treatments on the methane production of poultry feather. Environ Technol 24:1079–1086

Samah E (2016) Measuring small-scale biogas capacity and production. International Renewable Energy Agency (IRENA), Abu Dhabi. ISBN:978-92-95111-12-7

Samer M (2012). Chapter 17: Biogas plant constructions, biogas. In: Kumar S (ed). ISBN:978-953-51-0204-5. InTech

Sasse L (1988) Biogas Plants by A Publication of the Deutsches Zentrum für Entwicklungstechnologien – GATE in: Deutsche Gesellschaft für Technische Zusammenarbeit (GTZ) GmbH

Sasse L, Kellner C, Kimaro A (1991) Improved biogas unit for developing countries. Vieweg and Sohn, Eschborn

Schievano A, Scaglia B, D'Imporzano G, Malagutti L, Gozzi A, Adani F (2009) Prediction of biogas potentials using quick laboratory analyses: upgrading previous models for application to heterogeneous organic matrices. Bioresour Technol 100(23):5777–5782

Sharma KR (2008) In: Khanal SK (ed) Kinetics and modeling in anaerobic processes in anaerobic technology for bioenergy production: principles and applications. Wiley-Blackwell, Ames

Sharma N, Giuseppe P (1991) Anaerobic biotechnology and developing countries – technical status. Energy Conversion 32:447–469

Sibiya NT, Muzenda E, Mbohwa C (2017) Evaluation of potential substrates for biogas production via anaerobic digestion: a review. In: Proceedings of the world congress on engineering and computer science WCECS 2017, vol II, October 25–27, San Francisco, USA

Singh TS, Sankarlal P (2015) Production of biogas from kitchen waste using cow manure as co-substrate. In: Proceedings of the conference on "Energy conversion and conservation", 27/03/2015

Singh JB, Myles R, Dhussa A (1987) Manual on Deenbandhu biogas plant. Tata McGraw Hill Publishing Company Limited, New Delhi

Speece RE (1996) Anaerobic biotechnology for industrial wastewaters. Archae Press, Nashville

Stalin N (2007) Performance evaluation of partial mixing anaerobic digester. ARPN J Appl Sci 2:1–6

Stefan M (2004) Biogas fuel for internal combustion engines. Annals of the Faculty of Engineering Hunedoara – 2004 Tome II. Fascicole 3

Surendra KC, Takara D, Hashimoto AG, Khanal SK (2014) Biogas as a sustainable energy source for developing countries: opportunities and challenges. Renew Sust Energ Rev 31:846–859

Syamsuri S, Yustia WM (2015) Performance analysis of biogas stoves with variations of flame burner for the capacity of biogas 1 m^3/day. ARPN J Eng Appl Sci 10:22

Takeno T, Sato K (1979) An excess enthalpy theory. Combust Sci Technol 23:73–84

Tucker MF (2008) Farm digesters for small dairies in Vermont. Bio Cycle 49:44

Van DDL, Weber JA (1994) Biogas production from animal manures: what is the potential?' Industrial uses of agricultural materials, USDA/ERS outlook report IUS-4

Vandevivere P, Baere LD, Verstraete W (2003) In: Mata-Alvarez J (ed) Types of anaerobic digesters for solid wastes, in biomethanization of the organic fraction of municipal solid wastes. IWA Publishing, Barcelona, pp 111–140

Vavilin VA, Qu X, Mazéas L, Lemunier M, Duquennoi C, He P, Bouchez T (2008) Methanosarcina as the dominant aceticlastic methanogens during mesophilic anaerobic digestion of putrescible waste. Antonie Van Leeuwenhoek 94:593–605

Ver Eecke HC, Butterfield DA, Huber JA, Lilley MD, Olson EJ, Roe KK, Evans LJ, Merkel AY, Cantin HV, Holden JF (2012) Hydrogen-limited growth of hyperthermophilic methanogens at deep-sea hydrothermal vents. Proc Natl Acad Sci 109(34):13674–13679

Verma S (2002) Anaerobic digestion of biodegradable organics in municipal solid wastes. Department of Earth & Environmental Engineering Fu Foundation School of Engineering and Applied Science, Columbia University

Vögeli Y, Lohri CR, Gallardo A, Diener S, Zurbrügg C (2014) Anaerobic digestion of biowaste in developing countries: practical information and case studies. Swiss Federal Institute of Aquatic Science and Technology (Eawag), Dübendorf. ISBN:978-3-906484-58-7

Wang X, Lu X, Li F, Yang G (2014) Effects of temperature and carbon-nitrogen (C/N) ratio on the performance of anaerobic co-digestion of dairy manure, chicken manure and rice straw: focusing on ammonia inhibition. PLoS One 9(5):e97265

Weiland P (2010) Biogas production: current state and perspectives. Appl Microbiol Biotechnol 85 (4):849–860

Werner U, Stöhr U, Hees N (1989) Biogas plants in animal husbandry. A publication of the Deutsches Zentrum für Entwicklungstechnologien – GATE, a division of the Deutsche Gesellschaft für Technische Zusammenarbeit (GTZ) GmbH

Westerholm M, Roos S, Schnürer A (2010) Syntrophaceticusschinkii gen. nov., *sp.* nov., an anaerobic syntrophic acetate-oxidizing bacterium isolated from a mesophilic anaerobic filter. FEMS Microbiol Lett 309:100–104

Wijekoon KC, Visvanathan C, Abeynayaka A (2011) Effect of organic loading rate on VFA production, organic matter removal and microbial activity of a two stage thermophilic anaerobic membrane bioreactor. Bioresour Technol 102(9):5353–5360

Wirth R, Kovács E, Maròti G, Bagi Z, Rakhely G, Kovács KL (2012) Characterization of a biogas—Producing microbial community by short-read next generation DNA sequencing. Biotechnol Biofuels 5:41

Zhu W, Reich CI, Olsen GJ, Giometti CS, Yates JR (2004) Shotgun proteomics of *Methanococcus jannaschii* and insights into methanogenesis. J Proteome Res 3:538–548

Ziana Z, Rajesh P (2015) Production and Analysis of Biogas from Kitchen Waste. Int Res J Eng Technol 02:04

Zupančič GD, Roš M (2003) Heat and energy requirements in thermophilic anaerobic sludge digestion. Renew Energy 28(14):2255–2267

Biogas: An Effective and Common Energy Tool – Part II

4

Gupta Monika, Panpatte Deepak, Jhala Yogeshvari, and Vyas Rajababu

Abstract

Combustion of nonrenewable energy sources brings about emission of greenhouse gases which lead to global warming. A large number of renewable energy sources are available as an alternative for mitigation of climate change, among which biogas seems to be more popular and attractive option. Biogas is presently still typically used for heating and electricity generation, but in the future, it may find its way as vehicle fuel. Biogas production technology has potential to utilize a large number and variety of lignocellulosic biomass such as vegetable wastes, crop residues, food waste, cattle dung, and other organic fractions. Anaerobic degradation of waste to yield biogas is a widely adopted cost-effective strategy for generation of renewable energy. In addition to energy generation, biogas technology provides additional benefits, such as reduction of odor, improved sanitation, and removal of organic waste, thereby solving a majority of modern-day problems. The slurry left after biogas production can be utilized as manure and thereby aids in nutrient recycling to the soil. Besides a large number of applications, full potential of biogas technology cannot be harnessed due to various limitations associated with it. A large number of technological improvements are done during recent years to increase conversion rates of biomass to biogas. The present article provides an insight regarding recent research for sustainable biogas production.

Keywords

Biogas · Renewable energy · Lignocellulosic biomass · Slurry · Manure

G. Monika (✉)
Amity Institute of Biotechnology, Amity University Madhya Pradesh, Gwalior, India

P. Deepak · J. Yogeshvari · V. Rajababu
Department of Agricultural Microbiology, BACA, AAU, Anand, India
e-mail:

N. Srivastava et al. (eds.), *Biofuel Production Technologies: Critical Analysis for Sustainability*, Clean Energy Production Technologies,
https://doi.org/10.1007/978-981-13-8637-4_4

4.1 Introduction

Biogas production through anaerobic digestion (AD) is an environment-friendly method which gets accelerated due to a growing amount of natural waste formed around the world. A broad range of waste including municipal waste and industrial waste, food industrial waste, and agricultural as well as plant residues may be used with this technique. The most important manufactured goods of this treatment, i.e., biogas, is a renewable energy resource, whereas the derivative left in the digester may be used as manure due to the high nutrient offered for plants (Ward et al. 2008). The performance of anaerobic digestion process highly depends on the activity of feedstock, along with the activity of microorganisms involved in various phases (Batstone et al. 2002). The change of organic materials during the course of biogas production may be divided into three phases: methane production, acid formation, and hydrolysis. In these dissimilar steps, which are made parallel, different groups of bacteria make anaerobic food chain, where the product of one reaction will be the substrate of the second reaction. If the degradation rate of different steps is in balance, then this process progresses efficiently (Yong et al. 2015; Sárvári et al. 2016).

This review presents a summary of the biogas industry throughout the world and converses several novel techniques that are aimed at using new substrates and increasing efficiency of the process.

4.2 Evolution of Biogas Policies in India and Their Current Status

India was a net petroleum importer. The combination of a worldwide energy crisis and local energy shortages increased the national energy security risk from increasing energy import expenses as well as the pressure on the national budget to satisfy increasing energy subsidies for domestic fuels, mainly kerosene, utilized by the rural and urban poor for very fundamental cooking and lighting requirements.

By the end of the 1970s, it was clear to Indian policymakers that the usual restricted energy resources, for example, animal waste, fuel wood, and agricultural wastes, were not accessible in several pastoral areas and there was a requirement for conservation of restricted property. Many pastoral plains such as manure management program, national biogas program, as well as grid biogas power generation were initiated. In 1981, the biogas development program was part of many approaches for reducing rural energy crisis (Shukla 2007). Growing concerns about solid waste management and climate change are the main drivers behind these policy initiatives to boost the growth of biogas in metropolitan regions. Table 4.1 represents a policy timeline setting out numerous public measures adopted over the last three decades to increase waste in the energy and biogas sectors (Mittal et al. 2018). Programs and projects to boost the waste-to-energy industry from municipal solid waste and industrial waste are of latest creation; therefore it is still hard to determine the impact of new policies on the implementation of biogas technology in urban areas.

Table 4.1 Policy timeline

Timeline	Development
1981	The first national program for biogas production was launched
	Capital subsidy is provided for installing small biogas plant
	Ownership of the two to three cattle is one of the criteria to gain subsidy provided under the program
1995	National program to recover energy from municipal solid, industrial, and agricultural waste was launched
2006	The government renamed the NPBD program by the Biogas and Manure Management Programme (NBMMP) in 2006 auspices of Ministry of New and Renewable Energy (MNRE) retaining the same objective
	An off-grid biogas power generation program was also initiated by the NMRE in 2006 to promote decentralized power option. Financial incentives are offered to private/public players for setting up biogas power or cogeneration plants as well as production of bio-CNG using bio-methanation technology
2016	Rules regarding management and handling of solid waste, first notified in 2000 by the Ministry of Environment and Forest, were revised in 2016 extending its area of jurisdiction
	The central government released a new tariff policy that made it mandatory for the electricity distribution companies to procure 100% electricity generated from waste to energy plants

Source: Mittal et al. (2018)

The level of biogas dissemination in urban regions is small, and the proportion of biogas in private homes in the petrol blend is small. Around five million household biogas plants (40%) were built under the biogas development program against the complete capacity of 12 million national biogas plants assessed by the MNRE (CSO 2014). Besides, biogas plants for family, four hundred biogas off-grid authority plants have been set up with about 5.5 MW power generation capacities (MNRE 2015). In urban areas, the part of anaerobic digestion in organic wastewater treatment is currently very low compared to other competitive waste treatment techniques due to low revenue growth and capital cost prospects. At present in India, there are only 56 equipped biogas-based power plants; most of them are situated in three states, i.e., Maharashtra, Karnataka, and Kerala (CPCB 2013).

4.3 Overview of Successful Commercial Biogas Tool

4.3.1 India

It is said that the first anaerobic digestion (AD) plant in Bombay, India, in 1859 produced biogas from a therapy facility and was used for street lighting. India began programs to encourage biogas in the nation in the 1970s (Ahammad and Sreekrishnan 2016). Sensing the need to value agricultural disposal, animal manure, and other forms of biomass, India has launched programs such as the National Biogas and Manure Management Program and the Off-Grid Biogas Power

Generation Program to provide renewable energy for cooking and heating. About 300 MW of viable biogas production plants in India has been established at the end of 2017 (Shukla 2019).

Although less efficient than the Chinese biogas program, the Indian government has given much assistance to India's biogas sector. The Ministry of New and Renewable Energy has been formed to supervise biogas and additional renewable energy venture programs (Sehgal 2018). The Ministry had introduced several programs including plan and subsidy programs. The fertilizer management and national biogas plan were also started to deal with subprograms like grid power production program, revival of energy from urban wastes, and revival of energy from industrial wastes (Renewable Energy 2018; Kemausuor et al. 2018). Promotion of society scale biogas plants was encouraged. The railway company in India is presently working on its networks with bio-based engines, including biogas.

4.3.2 Other Countries

Other significant nations promoting business of biogas are the United States, the United Kingdom, and Italy. Italy is the second biggest biogas producer in Europe with more than 1300 biogas plants and 8 GWh power generations (Valenti et al. 2018). The production of biogas plants in Italy was helped by a high feed-in tariff of 0.28/kWh for plants with less capacity of less than 1 MW for a period of 20 years (Torrijos 2016). Like a novel feed-in tariff substrate installed in 2012, the installed power differs depending on the use of technology and the use of heat in which there is an overall objective for using small-sized plants to use agriculture and animal waste (Torrijos 2016).

The United Kingdom is a biogas user in Europe. Bioelectricity production facilities established in the United Kingdom have reached 6 GW in 2017, with a grouping of frozen biomass fuel, MSW, and (AD) anaerobic digestion (Renewable Energy 2018), with an energy invention of 31.8 TWh. Countries with moderate biogas production at the lower end contain Denmark, Austria, the Czech Republic, and Bulgaria in Europe. Denmark is extremely knowledgeable to utilize viable biogas services and has been encouraging technologies for the treatment of co-digested compost, source-separate municipal solid waste (MSW), and clean biological industrial wastes for decades. Denmark utilizes "green pricing," a strategy instrument that offers opportunities for companies using biogas to produce electricity to encourage the use of biogas (Aryal et al. 2018). There is a subsidy system to assist in the creation of an agricultural biogas plant that processes purposeful yield in the Czech Republic (Vochozka et al. 2018). Bulgaria has promoted biogas plants by utilizing the praise line up for "energy efficiency and renewable energy sources for Bulgaria" in which plan may accept up to 20% grants (Kolchakov et al. 2017).

In the beginning of 2002, Austria had increased the installed ability of biogas from 15 MW to about 80 MW by the end of 2015 due to feed-in tariff and green electricity act (Stürmer 2017). Feed-in tariff differs depending on plant's technology and capacity and the source of biogas such as landfill gas or sewage gas (Pablo-Romero et al. 2017).

Swedish biogas production reached 1.7 TW in 2013, in which production source co-digestion plant (34%), waste treatment plant (40%), landfill gas valorization (14%), and farm-scale plants (5%) as well as industrial waste treatment plants (7%) are used in more than 50% of advanced and used vehicles, making Sweden the world organizer in the use of biogas for transport (Larsson et al. 2016).

Apart from India and China, other Asian countries encouraging biogas system include Thailand, Vietnam, Nepal, and Bangladesh. Vietnam had created above 1 million biogas digestion plants, including 150,000 industrial plants and 500,000 medium-scale plants by 2013, which treated industrial wastewater (Sehgal 2018). A strategy introduced by the Government of Thailand led to enhancement in generation of local electrical energy from 120 MW to 600 MW from biogas in 2011 (Chaiyapong and Chavalparit 2016). In Japan, efforts have been made to promote industrial biogas plants (Takeuchi et al. 2018).

The United States and Brazil are among the prominent biogas developers. As of 2016, the American Biogas Council reports that there are more than 2100 operational biogas systems in the United States with more than 11,000 prospective sites (American Biogas Available online 2018). Biomethane production capacity alone is estimated to be 5,128,334.6 million gallons per year, which is equivalent to 4360.41 million gallons of diesel and 4883.29 million gallons of gasoline (Pasqual et al. 2018).

State governments of the United States provide federal taxation for promotion of biogas projects by providing performance-based incentives, like tax incentives, soft loans, etc. Some examples include tax credits value of $ 0.015 per kWh for 10 years in Iowa; Pennsylvania and Oregon grants 25–50% of the total project cost to agricultural plants; Massachusetts offers 75% for phase cost and 25% for construction-phase expenses, and six conditions give $ 0.1 per kWh incentives from $ 0.015 (Sam et al. 2017).

Approximately 127 profitable biogas plants are functioning in Brazil by 2017, which are mostly running with farm animals dung and municipal waste. It is estimated that around 100 million m^3 of methane could be produced annually from Brazil's livestock manure, agricultural scrap, wastewater, and municipal waste (Langeveld and Peterson 2018). Biomethane's ability from livestock alone in Brazil is projected at 1.961.171.9 million gallons per annum, which could substitute 1667.42 million gallons of diesel or 1848.34 million gallons of gasoline (Pasqual et al. 2018). Brazil is the main producer of biofuel in South America, whose expected production is 49 TWh (Renewable Energy 2018).

4.4 Current Biogas Process Technologies

Production of biogas during anaerobic digestion provides most important benefits more than other types of bioenergy assembly. It has been cleared as the most environmentally friendly and energy-efficient technology for bioenergy production in fact (Deublein and Steinhauser 2011). The process of degradation may be separated into four phases, methanogenesis, acetogenesis, acidogenesis, and

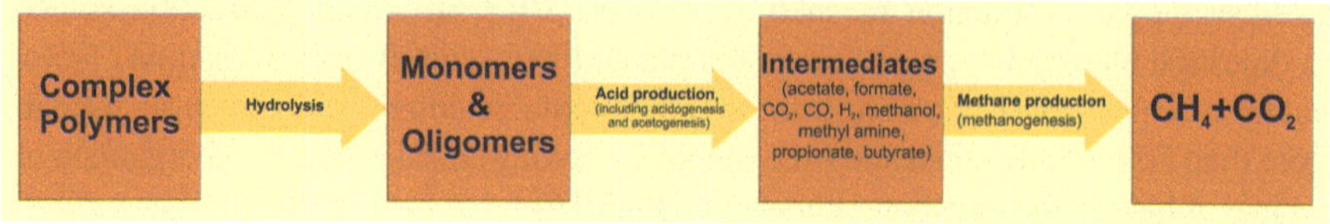

Fig. 4.1 Degradation process taking place during AD, i.e., hydrolysis, acidogenesis, acetogenesis, and methanogenesis. (Source: Merlin Christy et al. 2014; Chasnyk et al. 2015; Abdeshahian et al. 2016)

hydrolysis, and in each individual stage, various types of faculty or compulsory anaerobic microorganisms are included as demonstrated in Fig. 4.1 (Merlin Christy et al. 2014; Chasnyk et al. 2015; Abdeshahian et al. 2016).

4.4.1 Steps Involved in Biogas Production Process

4.4.1.1 Hydrolysis

Hydrolysis is the first phase in biogas manufacturing in which the main elements or complicated organic molecules of scrap natural matter/biomass (carbohydrates, lipids, and proteins) break down into lower units (monomer sugars, amino acids, alcohols, and fatty acids) by cellulolytic, lipolytic, and proteolytic organisms, respectively. Most commonly bacteria of genera *Megasphaera*, *Sporobacterium*, *Sphingomonas*, *Bifidobacterium*, *Bacteroides*, *Lactobacillus*, and *Propionibacterium* are found to be involved in hydrolysis including both facultative and obligatory anaerobes. The rate of hydrolysis is typically represented by using first-order kinetics (Sharma 2008).

4.4.1.2 Acidogenesis or Acid Production

After hydrolysis of organic matter present in biomass, acidogenic bacteria convert the products of hydrolysis into short-chain organic acids like volatile fatty acids, i.e., propionic, lactic and butyric acids, NH_3, H_2S, and H_2. Carbon dioxide (CO_2) and alcohols, for instance, ethanol, are also formed through this method. Various obligate and facultative anaerobic fermentative bacteria, viz., *Desulfovibrio* spp., *Corynebacterium* spp., *Clostridium* spp., *Bifidobacterium* spp., *Peptostreptococcus anaerobius*, *Escherichia coli*, *Lactobacillus* spp., and *Staphylococcus* spp., were found to be involved in the process (Sharma 2008; Metcalf and Eddy 2004).

4.4.1.3 Acetogenesis or Acetic Acid Production

In this step volatile fatty acids and ethanol produced by acidogenic microbes are converted into acetic acid (CH_3COOH)/acetate (CH_3COO^-), H_2, and CO_2 by acetogenic bacteria. Acetogenic bacteria like *Syntrobacter wolinii* and *Syntrophomonas wolfei* change alcohol and fatty acids into carbon dioxide, acetate, and hydrogen.

4.4.1.4 Methanogenesis or Methane Production

Methanogenic bacteria belonging to the genus *Methanosarcina* and *Methanosaeta* carry out acetoclastic methanogenesis wherein acetic acid is changed into carbon dioxide and methane which ultimately reduce oxygen in the system. Some microbial species, called hydrogenotrophic methanogens, may produce methane from the CO_2 and H_2 produced as products in prior stages (Alexander 1961).

In addition to energy generation, degradation of organic wastes also suggests some other benefits, including decrease in odor discharge and decrease in pathogens levels. Apart from all these, nutrient-rich digested residues can be used as organic fertilizers rather than mineral fertilizers and a biological substrate for the cultivation in greenhouse (de Vries et al. 2012, Abdeshahian et al. 2016). Among the raw materials for biogas production, waste streams of the farm and animals and biological materials obtained from domestic and industrial activities are important sources for biogas production.

4.4.2 Substrates Traditionally Used

Generally lignocellulosic waste obtained from municipal, agricultural, and other routine may be used as feedstock for biogas production. Usually, slurry and animal manure, sewage sludge, food waste, and municipal solid waste are used as feedstock comprising of hemicelluloses, proteins, carbohydrates, fats, and celluloses. Recent trend for enhancing biogas production involves addition of co-substrates like organic wastes from collected municipal biowaste and/or agriculture-related industries and food waste from households. The yield and composition of biogas are determined by the type of co-substrate composition and feedstock. Even though proteins and carbohydrates demonstrate earlier conversion rates than fats, it is reported that the latter provide a higher biogas yield (Braun 1982, Braun 2007; Zubr 1986).

Large quantities of organic solid waste are produced through human activity, which can be discussed in the form of feedstock for the production of biogas as before. On the basis of utility, agricultural waste, various waste streams, and municipal solid waste (MSW) can be classified as waste derived from industrial activities and urban activities. Three billion urban residents were generated by 1.3 billion tons of MSW per year according to a 2012 World Bank report, which would enhance to 2.2 billion tons by 2025 (Hoornweg and Bhada-Tata 2012). MSW mainly consists of paper, glass, plastic, wood, metal, yard trimmings, paper, paperboard, and food waste. However, its structure varies depending on those countries and regions in which it is collected. In order to be able to use this fraction for biogas production, inert material including all metal, glass, and plastic should be removed before marketing. In addition, about 15 billion tons of waste, animal manure, and crop residues are generated annually from the agricultural sector worldwide (Donkin et al. 2013).

The food processing industry also produces waste, although its estimation is extremely difficult, because it relies greatly on the industry and useful technology. As an example, in the juice-producing industry, about 50% of processed fruit will finish up as waste. In addition, 30% of the weight of chicken is not appropriate for human utilization, and therefore it is removed as waste through slaughtering and other processing steps (Salminen and Rintala 2002; Forgács et al. 2012).

As all these waste components are dissimilar, their biogas production capacity is quite different. Biogas production mainly depends on biodegradability of waste and structure. Theoretically, highest production of biogas could be obtained from lipid (1.01 Nm^3 CH_4/kg VS) followed by protein (0.50 Nm^3 CH_4/kg VS), and carbohydrate (0.42 Nm^3 CH_4/kg VS) (Møller et al. 2004). Conversely, biodegradability describes how much of a particular material is truly used throughout the method. Various compounds such as sugars get spoiled rapidly and totally, whereas some other ingredients lead to corrosion.

4.4.3 Pretreatment for Enhanced Biogas Production

It is essential to identify novel substrates to be used for anaerobic digestion (AD) to fulfill the ever-increasing needs for biogas production. Throughout the world, along with the abundance and availability of lignocellulosic biomass, their high carbohydrate content makes these materials a valuable feedstock for biofuel production. About 50% of the biomass in the world has been computed for lignocellulose, and simultaneously production of lignocellulose can be up to 200 billion tons per year (Claassen et al. 1999; Zhang 2008). Presently, the use of lignocelluloses as a feedstock for methane production is not extensive due to its recalcitrant structure, which is the main challenge (Lehtomäki 2006; Seppälä et al. 2007; Hendriks and Zeeman 2009).

Hydrolytic bacteria change insoluble complex organic matter into monomers and soluble oligomers into amino acids, sugars, and fatty acids during the first phase of AD, i.e., in hydrolysis phase (Fig. 4.1). In this process, enzymes like lipase, cellulase, protease, hemicellulase, and amylase are included (Taherzadeh and Karimi 2008). Consequently, almost all types of substrates may be hydrolyzed in biogas processes. On the other hand, the hydrolysis step is very much reliant on the characteristics of a given substrate. Hydrolysis could progress earlier if the essential enzymes are produced by microorganisms and have suitable surface area for physical contact between substrates and enzymes (Taherzadeh and Karimi 2008). However, substrate with complex structure, like cellulose, requires long periods to be degraded, and the degradation is generally not completed (Deublein and Steinhauser 2011). Therefore, while using these types of substrates, the hydrolysis step is often considered a rate-limited step (Vavilin et al. 1996; Taherzadeh and Karimi 2008).

The pretreatment steps convert the recalcitrant raw material into forms which can be easily degraded by enzymatic and microbial processes. With the disruption of the secondary cell wall structure, lignocelluloses reduce its complexity and thus facilitate downstream procedures (Zhang 2008). Alternatively, a pretreatment should be expensive, and the polysaccharide-rich substrate should be obtained with limited amount of inhibitory products.

Several types of manifestation have been suggested to enhance biogas production from lignocellulosic biomass, which may be classified as biological, chemical, and physical pretreatment (Chandra et al. 2007; Yang and Wyman 2008; Hendriks and Zeeman 2009; Taherzadeh and Karimi 2008). Milling is proved to be effective by reducing the degree of polymerization in the specific surface area by shear and polymerization degree between the physical structures. Thus, hydrolysis yield improved from 5% to 25% (Jin and Chen 2006; Zeng et al. 2007). These types of development depend on the type of biomass in addition to the duration and type of milling (Jin and Chen 2006; Monavari et al. 2009; Lennartsson et al. 2011; Teghammar et al. 2012). Overall, it has been repeatedly seen that small particles get more sugar yields. This is the reason that physical exposure is often done in conjunction with other pretreatment technique. On the other hand, the element representative intended for pretreatment may perform as a possible blocker for the microbial community concerned in AD in some cases. It was established that remaining residues negatively affected the digestive process when forest residues were mixed with organic solvent, *n*-methylmorpholine-*n*-oxide, even in concentrations up to 0.008% (Kabir et al. 2013). Apart from this, in spite of the optimization of the pretreatment conditions, a few inhibitors still get produced in the slurry (Ahring et al. 1996; Hendriks and Zeeman 2009).

It was recently shown that use of alcohol or weak organic acids seems to be an interesting way to digest lignocelluloses (Kabir et al. 2015).

4.5 Recent Advances in Biogas Production Technology

To overcome the problems associated with poor utilization of lignocellulosic waste for biogas production and poor methane yield, a number of alteration in the existing technology have been done like pretreatment of lignocellulosic waste, addition of substrate, and use of microbial consortia and additive incorporation to accelerate the biogas production process and enhance gas yield.

4.5.1 Pretreatment

Due to complex structure of lignocellulosic waste, it became less economic feedstock for biogas production process. Pretreatment of lignocellulosic waste is an attractive option for accelerating anaerobic digestion process and increasing biogas yield. Pecorini et al. (2016) reported that recalcitrant compounds of municipal waste can be hydrolyzed by autoclaving and microwave oven treatment. Model biomass pretreatment decreases crystallinity of the cellulosic structure which makes substrate easily accessible to microbes and enables them to completely or partially digest substrate into fermentable sugars. Pretreatment of lignocellulosic waste by milling increases specific surface area, thereby improving hydrolysis yield by 5–25%. The degree of such progress depends on the type of biomass and the time and type of milling (Jin and Chen 2006; Zeng et al. 2007). Many of the chemical agents are suggested for pretreatment, but in one or other case, they may serve as inhibitors of microbial community involved in biogas production process. Chemical pretreatment involving alcohols or weak organic acid seems to be an attractive means as both are intermediary metabolic processes during biogas formation. Kabir et al. (2015) pretreated forest residues with ethanol, methanol, or acetic acid prior to anaerobic digestion and showed higher methane production and also suggested that methanol can be a cost-effective chemical agent utilized for pretreatment due to its lower cost and easy recovery after completion of biogas production process.

4.5.2 Use of Microbial Consortia

The conversion of all the biowaste hydrolysis products such as pentoses, hexoses, volatile products, and soluble lignin to methane is practical using a mixture of microbes and a very good way to improve anaerobic digestion process (Fox et al. 2003). The quantities of the microbial groups during each step of biogas production affect the rate of the whole reaction (Griffin et al. 1998). Among all the groups of microorganisms involved in biogas production process, methanogens are very sensitive to fluctuations in environmental conditions, temperature, pH, redox potential, and inhibitors and hence are considered to be a rate-limiting factor in biogas production process (Chen et al. 2008). One obvious strategy proposed by researchers working in the field of biogas process improvement is genetic modification of microorganisms involved in the process of biogas formation so as to get higher metabolic efficiency which ultimately leads to production of energy-rich biofuels (Xu and Koffas 2010). Besides, substitute strategies suggest the blocking of undesired metabolic pathways to divert energy flow toward target-based metabolism of microorganisms present in anaerobic digestion system (Weng et al. 2008).

4.5.3 Additives

Biogas yield can be enhanced by accelerating microbial activities in the biogas digester plant. Generally additives are used as nutrients for microbes, and proper monitoring of its concentration is needed (Chen et al. 2008; Demirel and Scherer 2011). Incorporation of additive calcium salts and magnesium improved methane production and reduced foaming of slurry (Yadvika et al. 2004). Moreover, incorporation of additives for stabilization of pH and reducing concentration of hydrogen sulfide and ammonia are also recommended (Kuttner et al. 2015). Enrichment of crop residues like water hyacinth, wheat straw, onion storage waste, maize stalks, rice straw, cotton stalks, etc. with moderately digested cattle manure enhanced gas production to the tune of 10–80%. Additives like zeolite enhance biogas production by 15%, and calcium carbonate improved output by 8%. Iron salts such as iron chloride decrease hydrogen sulfide concentration in biogas with no side effects when added at the rate of 0.03 and 0.06 g l^{-1} (Kuttner et al. 2015).

4.5.4 Biosensors

Success of biogas production process involves monitoring of volatile fatty acids and organics present during the fermentation process. Presently the available monitoring methods are gas chromatography (Diamantis et al. 2006), spectroscopy (Falk et al. 2015; Stockl and Lichti 2018), and HPLC (high-performance liquid chromatography) (Zumbusch et al. 1994; Schiffels et al. 2011), but these methods do not provide real-time monitoring; thus, indecision among the plant operators takes place. For monitoring of accumulation of biogas intermediates, organic acid biosensors were developed to manage the association between these process stabilities and intermediates which have resulted in numerous studies being carried out for the optimization and expansion of an organic acid biosensor, including enzyme assembly for exact discovery of formate, ethanol, and D/L lactate, contrary to the partial concentration of the VFA (volatile fatty acid) biosensors (Crable et al. 2011; Pilas et al. 2017; Kaur et al. 2013). These analytes are identified during microbial fuel cells (Kaur et al. 2013), microbial electrolysis cells (Jin et al. 2017), or soften oxygen probes with a powerless biofilm (Sweeney et al. 2018), while enzyme-based sensors were intended for the irregular purpose of individual substrates, like propionate and acetate (Mizutani et al. 2001; Mieliauskiene et al. 2006; Sode et al. 2008).

4.5.5 Nanotechnology

Nanoparticles can enhance degradability of organic matter present in the municipal waste and thereby increase rate of biogas production. Use of iron nanoparticles for biogas production is newer aspect wherein concentration of CH_4 and biogas production is improved using nanoparticles in an anaerobic digester. Traditional biogas production process can convert only 30–40% of the biomass into gas, and the energy potential of generated biogas is also low. Addition of iron into the digester can enhance biogas production, but there might be toxicity to the functional groups of bacteria present in the rector. To overcome this issue, biodegradable nanoparticle-based delivery system is being utilized, so that problems like inhibition of bacterial activity can be minimized by production of ions into the reaction medium.

4.6 Biogas Production from Municipal Solid Waste

Municipal solid waste contains heterogeneous mass of organic matter and composed of kitchen scraps, food residue, food processing wastes, grass cuttings, etc. which can be degraded at the faster rate, whereas organic matter such as coarser wood, paper, and cardboard is degraded at slower rate. Moreover, municipal solid waste also comprises of inert fraction like stones, glass, sand, metal, etc. Metals from the municipal solid waste can be recycled using metal re-claimers, whereas other materials like stone, sand, etc. can be utilized as building material.

4.6.1 Pretreatment of Municipal Solid Waste

To upgrade and homogenize the feedstock for digestion and to remove inert and non-biodegradable material pretreatments are essential. They can be manual, mechanical, thermochemical, biological, etc. There are several ways in which this can be accomplished.

First of all bulky materials and specific hazardous waste materials are removed prior to mechanical processing from municipal solid waste. The best pretreatment process is source separation which provides clean waste with some fraction of plastic. Metal, glass, plastics, paper, etc. are recovered regularly at waste collection beans and disposal sites by waste pickers. Though this process of scavenging reduces the total volume of the waste and enriches the waste with high organic concentrations, this waste contains a number of items such as dust, foils, some plastics, metals, papers, discarded construction materials, grits, ash, broken ceramics, etc.

To further process manually cleaned municipal solid waste, it is being passed through trommel screens where oversized materials and other foreign materials are separated followed by hammer crushing to break down larger raw materials to small pieces and thereby making it more accessible to bacteria which in turn reduces retention time. The grinder also acts as a mixer. Then the municipal solid waste passes through a drum magnetic separation mechanism, where a strong magnet separates the ferrous metals. A hydro-pulper then sorts incoming solid waste into heavy and light fractions of nonorganic material as well as creates mixed organic waste. For thorough mixing of the waste and water (slurry), a slurry mixture machine should be fitted in the inlet of a digester. It is also necessary to remove inert materials such as stones from the inlet before feeding the slurry into the digester. Otherwise, the effective volume of the digester will decrease.

Chemical pretreatment has been tried in a variety of temperature regions and over a variety of time periods, from 15 to 120 min. These strategies particularly help with the degradation of fats, which is troublesome because of their insolubility in water and their semi-solidification. For fats hydrolysis, they must be emulsified to enhance their bioavailability in water. Pretreatment with sodium hydroxide, lithium hydroxide, or potassium hydroxide increases the hydrolysis rate. Lime, sodium hydroxide, and ammonia are the least expensive of these chemicals. For biological method, bacterial growth (anaerobic microorganisms) is stimulated by the addition of some organic compounds (e.g., amino acids, cofactors, cell content) in the inlet tank of the digester.

Anaerobic digestion of solid waste has been demonstrated as a technically feasible process, duly deserving further consideration in any integrated waste management concept addressing municipal solid waste. Anaerobic digestion provides an important opportunity to generate 100% renewable energy from biodegradable waste. The conversion of a sizeable part of organic waste into a convenient source of energy, i.e., biogas, is a precious asset, not in the least in times of oil scarcity and of economic support for renewable forms of energy. The simultaneous generation of digest, which can be turned into a soil amendment, may be an added advantage

4.7 Risk Factors of Biogas Technology

The risks associated with biogas technology are as follows which should be taken care while working for biogas production unit:

1. Leakage in the storage tank and/or in the biogas distribution network
2. Formation of flammable mixtures during the digester maintenance operations
3. Accidental release of H_2S and effluent discharge
4. Overflow of sewage systems or stormwater control due to exceptional downpours, presence of dangerous products in the raw material used to produce biogas, overflow, freezing of valves, and high pressure inside the digester

4.8 Conclusion

Biogas is a potential green energy resource that can be produced from organic waste material as a feedstock. Biogas production process is simple, clean, and easy. Biogas production utilizes agricultural biomass, thereby reducing greenhouse gas emission due to unsuitable dumping of organic wastes. It also improves energy safety and decreases fossil fuel depletion. Biomass energy will progressively substitute fossil fuel, and the community can even rely on biogas energy for satisfying local needs especially in rural areas; full-fledged utilization of biogas for energy generation should be explored and needs concentrated efforts, and this technology has found its utility in domestic, farming, and small-scale industries. Presently reduced efficiency of biogas production plants can be overcome by utilization of modern technologies to enhance microbial activity within the biogas production units to get efficient yield of the energy supply. Policy and government producer should have sympathetic strategies such as subsidization of biogas plan to maintain the growth of technology in the remotest places. Additional research must be done on superior bioreactor plan and feedstock presentation improvement to get better cost-effectiveness of anaerobic digestion process.

References

Abdeshahian P, Lim JS, Ho WS, Hashim H, Lee CT (2016) Potential of biogas production from farm animal waste in Malaysia. Renew Sustain Energy Rev 60:714–723

Ahammad SZ, Sreekrishnan TR (2016) Biogas: an evolutionary perspective in the Indian context. In: Green fuels technology; green energy and technology. Springer, Cham, pp 431–443

Ahring BK, Jensen K, Nielsen P, Bjerre AB, Schmidt AS (1996) Pretreatment of wheat straw and conversion of xylose and xylan to ethanol by thermophilic anaerobic bacteria. Bioresour Technol 58(2):107–113

Alexander M (1961) Introduction to soil microbiology. Wiley, New York

American Biogas Council American Biogas Council Announces 2016 Biogas Industry Awardees.. Available online: http://americanbiogascouncil.org/pdf/ABC%20Press%20Release2016%20Biogas%20Award% 20Winners.pdf. Accessed 4 July 2018

Aryal N, Kvist T, Aryal N, Kvist T (2018) Alternative of biogas injection into the Danish gas grid system – a study from demand perspective. Chem Eng 2:43
Batstone DJ, Keller J, Angelidaki I, Kalyuzhnyi SV, Pavlostathis SG, Rozzi A, Sanders WTM, Siegrist H, Vavilin VA (2002) The IWA anaerobic digestion model no 1 (ADM 1). Water Sci Technol 45(10):65–73
Braun R (1982) Biogas – methane treatment of organic waste. Springer, Wien
Braun R (2007) Anaerobic digestion: A multi-faceted process for energy, environmental management and rural development. In: Ranalli P (ed) Improvement of crop plants for industrial end uses. Springer, Dordrecht, pp 335–415
Chandra RP, Bura R, Mabee WE, Berlin A, Pan X, Saddler JN (2007) Substrate pretreatment: the key to effective enzymatic hydrolysis of lignocellulosics? In: Olsson L (ed) Biofuels. Springer, Berlin/Heidelberg, pp 67–93
Chasnyk O, Sołowski G, Shkarupa O (2015) Historical, technical and economic aspects of biogas development: case of Poland and Ukraine. Renew Sustain Energy Rev 52:227–239
Chaiyapong P, Chavalparit O (2016) Enhancement of biogas production potential from Acacia leaf waste using alkaline pre-treatment and co-digestion. J Mater Cycles Waste Manag 18:427–436
Chen Y, Cheng J, Creamer KS (2008) Inhibition of anaerobic digestion process: a review. Bioresour Technol 99:4044–4064
Claassen PAM, Van Lier JB, Contreras AML, Van Niel EWJ, Sijtsma L et al (1999) Utilisation of biomass for the supply of energy carriers. Appl Microbiol Biotechnol 52(6):741–755
CPCB (2013) Consolidated annual review report on implementation of municipal solid wastes (management and handling) rules, 2000. In: BOARD, C.P.C. (ed) Ministry of Environment Forests & Climate Change, New Delhi
Crable BR, Plugge CM, McInerney MJ, Stams AJM (2011) Formate formation and formate conversion in biological fuels production. Enzyme Res:1–8
CSO (2014) Energy statistics. In: Office, C.S. (ed) Ministry of Statistics and Programme Implementation, New Delhi. https://doi.org/10.1016/j.enpol.2017.10.027
de Vries JW, Vinken TMWJ, Hamelin L, De Boer IJM (2012) Comparing environmental consequences of anaerobic mono- and codigestion of pig manure to produce bio-energy – a life cycle perspective. Bioresour Technol 125:239–248
Demirel B, Scherer P (2011) Trace element requirements of agricultural biogas digesters during biological conversion of renewable biomass to methane. Biomass Bioenergy 35:992–998
Deublein D, Steinhauser A (2011) Biogas from waste and renewable resources: an introduction. Wiley-VCH Verlag GmbH & Co. KGaA, Mörlenbach, p 578
Deo P, Modak S, Shukla PR (1991) Decentralized energy planning. Oxford and IBH Publishing Co. Pvt. Ltd, New Delhi, India
Diamantis V, Melidis P, Aivasidis A (2006) Continuous determination of volatile products in anaerobic fermenters by on-line capillary gas chromatography. Anal Chim Acta 573–574:189–194
Donkin SS, Doane PH, Cecava MJ (2013) Expanding the role of crop residues and biofuel co-products as ruminant feedstuffs. Anim Front 3(2):54–60
Falk HM, Reichling P, Andersen C, Benz R (2015) Online monitoring of concentration and dynamics of volatile fatty acids in anaerobic digestion processes with mid-infrared spectroscopy. Bioprocess Biosyst Eng 38:237–249
Forgács G, Pourbafrani M, Niklasson C, Taherzadeh MJ, Sárvári Horváth I (2012) Methane production from citrus wastes: process development and cost estimation. J Chem Technol Biotechnol 87(2):250–255
Fox MH, Noike T, Ohki T (2003) Alkaline subcritical-water treatment and alkaline heat treatment for the increase in biodegradability of newsprint waste. Water Sci Technol 48(4):77–84
Griffin ME, McMahon KD, Mackie RI, Raskin L (1998) Methanogenic population dynamics during start-up of anaerobic digesters treating municipal solid waste and biosolids. Biotechnol Bioeng 57(3):342–355

Hendriks ATWM, Zeeman G (2009) Pretreatments to enhance the digestibility of lignocellulosic biomass. Bioresour Technol 100(1):10–18

Hoornweg D, Bhada-Tata P (2012) What a waste: a global review of solid waste management. World Bank, Washington, DC

Horváth SI, Tabatabaei M, Karimi K, Kumar R (2016) Recent updates on biogas production – a review. Biofuel Res J 10:394–402

Jin S, Chen H (2006) Superfine grinding of steam-exploded rice straw and its enzymatic hydrolysis. Biochem Eng J 30(3):225–230

Jin X, Li X, Zhao N, Angelidaki I, Zhang Y (2017) Bio-electrolytic sensor for rapid monitoring of volatile fatty acids in anaerobic digestion process. Water Res 111:74–80

Kabir MM, del Pilar Castillo M, Taherzadeh MJ, Sárvári Horváth I (2013) Enhanced methane production from forest residues by N-methylmorpholine-N-oxide (NMMO) pretreatment. Bio Resour 8(4):5409–5423

Kabir MM, Rajendran K, Taherzadeh MJ, Sarvari Horvath I (2015) Experimental and economical evaluation of bioconversion of forest residues to biogas using organosolv pretreatment. Bioresour Technol 178:201–208

Kaur A, Kim JR, Michie I, Dinsdale RM, Guwy AJ, Premier GC (2013) Microbial fuel cell type biosensor for specific volatile fatty acids using acclimated bacterial communities. Biosens Bioelectron 47:50–55

Kemausuor F, Adaramola MS, Morken J (2018) A review of commercial biogas systems and lessons for Africa. Energies 11:2984

Kolchakov V, Petrova V, Mitova T, Ivanov P, Marinova S (2017) Possibilities for biogas production from waste—potential, barriers, and legal notices. In: Energy solutions to combat global warming; lecture notes in energy. Springer, Cham, pp 181–191

Kuttner P, Weißböck AD, Leitner V, Jäger A (2015) Examination of commercial additives for biogas production. Agron Res 13:337–347

Langeveld JWA, Peterson EC (2018) Feedstocks for biogas production: biogas and electricity generation potentials. In: Biogas; biofuel and biorefinery technologies. Springer, Cham, pp 35–49

Larsson M, Grönkvist S, Alvfors P (2016) Upgraded biogas for transport in Sweden—effects of policy instruments on production, infrastructure deployment and vehicle sales. J Clean Prod 112:3774–3784

Lehtomäki A (2006) Biogas production from energy crops and crop residues. University of Jyväskylä, Jyväskylä

Lennartsson PR, Niklasson C, Taherzadeh MJ (2011) A pilot study on lignocelluloses to ethanol and fish feed using NMMO pretreatment and cultivation with zygomycetes in an air-lift reactor. Bioresour Technol 102(6):4425–4432

Merlin Christy P, Gopinath LR, Divya D (2014) A review on anaerobic decomposition and enhancement of biogas production through enzymes and microorganisms. Renew Sustain Energy Rev 34:167–173

Metcalf and Eddy (2004) Wastewater engineering: treatment and reuse. Revised by Tchobanoglous G, Franklin LB, Stensel DH. 4th edn. McGraw-Hill

Mieliauskiene R, Nistor M, Laurinavicius V, Csöregi E (2006) Amperometric determination of acetate with a tri-enzyme based sensor. Sens Actuat Biochem 113:671–676

Mittal S, Ahlgren EO, Shukla PR (2018) Barriers to biogas dissemination in India: a review. Energy Policy 112:361–370

Mizutani F, Sawaguchi T, Yabuki S, Iljima S (2001) Amperometric determination of acetic acid with a tri enzyme/poly (dimethyl siloxane)-bilayer based sensor. Anal Chem 73:5738–5742

MNRE (2015) Annual report, 2015–16. In: Energy, M.O.N.A.R. (ed) New Delhi

Møller HB, Sommer SG, Ahring BK (2004) Methane productivity of manure, straw and solid fractions of manure. Biomass Bioenergy 26(5):485–495

Monavari S, Galbe M, Zacchi G (2009) Impact of impregnation time and chip size on sugar yield in pretreatment of softwood for ethanol production. Bioresour Technol 100(24):6312–6316

Pablo-Romero MDP, Sánchez-Braza A, Salvador-Ponce J, Sánchez-Labrador N (2017) An overview of feed-in tariffs, premiums and tenders to promote electricity from biogas in the EU-28. Renew Sust Energ Rev 73:1366–1379

Pasqual J, Bollmann H, Scott C, Edwiges T, Baptista T, Pasqual JC, Bollmann HA, Scott CA, Edwiges T, Baptista TC (2018) Assessment of collective production of biomethane from livestock waste for urban transportation mobility in Brazil and the United States. Energies 11:997

Pecorini I, Baldi F, Carnevale EA, Corti A (2016) Biochemical methane potential tests of different autoclaved and microwaved lignocellulosic organic fractions of municipal solid waste. Waste Manag 56:143–150

Pilas J, Yazici Y, Selmer T, Keusgen M, Schöning MJ (2017) Optimization of an amperometric biosensor array for simultaneous measurement of ethanol, formate, D- and L-lactate. Electrochim Acta 251:256–262

Renewable Energy Network of the 21st Century. Renewables 2019 Global Status Report. 2018. Available online: http://www.ren21.net/status-of-renewables/global-status-report/. Accessed 2 Feb 2019

Salminen E, Rintala J (2002) Anaerobic digestion of organic solid poultry slaughterhouse waste-a review. Bioresour Technol 83(1):13–26

Sam A, Bi X, Farnsworth D (2017) How incentives affect the adoption of anaerobic digesters in the United States. Sustainability 9:1221

Schiffels J, Baumann MEM, Selmer T (2011) Facile analysis of shortchain fatty acids as 4-nitrophenyl esters in complex anaerobic fermentation samples by high performance liquid chromatography. J Chromatogr 1218:5848–5851

Sehgal K (2018) Current state and future prospects of global biogas industry. In: Biogas; biofuel and biorefinery technologies. Springer, Cham, pp 449–472

Seppälä M, Paavola T, Rintala J (2007) Methane yields of different grass species on the second and third harvest in boreal conditions. In: Proceedings of 11th IWA world congress on anaerobic digestion, Brisbane, Australia, 23–27

Sharma KR (2008) Kinetics and modeling in anaerobic processes. In: Khanal SK (ed) Anaerobic technology for bioenergy production: principles and applications. Wiley-Blackwell, Ames

Shukla PR (2007) Biomass energy strategies for aligning development and climate goals in India. Netherlands Environmental Assessment Agency

Shukla PR (2019) Biomass energy strategies for aligning development and climate goals in India. Available online: https://www.rivm.nl/bibliotheek/rapporten/500101002.pdf. Accessed 2 Feb 2019

Sode K, Tsugawa W, Aoyagi M, Rajashekhara E, Watanabe K (2008) Propionate sensor using coenzyme-A transferase and Acyl-CoA oxidase. Protein Pept Lett 15:779–781

Stockl A, Lichti F (2018) Near-infrared spectroscopy (NIRS) for a real time monitoring of the biogas process. Bioresour Technol 247:1249–1252

Stürmer B (2017) Biogas—part of Austria's future energy supply or political experiment? Renew Sust Energ Rev 79:525–532

Sweeney JB, Murphy CD, McDonnell K (2018) Development of a bacterial propionate-biosensor for anaerobic digestion monitoring. Enzym Microb Technol 109:51–57

Taherzadeh M, Karimi K (2008) Pretreatment of lignocellulosic wastes to improve ethanol and biogas production: a review. Int J Mol Sci 9(9):1621–1651

Takeuchi Y, Andriamanohiarisoamanana FJ, Yasui S, Iwasaki M, Nishida T, Ihara I, Umetsu K (2018) Feasibility study of a centralized biogas plant performance in a dairy farming area. J Mater Cycles Waste Manag 20:314–322

Teghammar A, Karimi K, Sárvári Horváth I, Taherzadeh MJ (2012) Enhanced biogas production from rice straw, triticale straw and softwood spruce by NMMO pretreatment. Biomass Bioenergy 36:116–120

Torrijos M (2016) State of development of biogas production in Europe. Procedia Environ Sci 35:881–889

Valenti F, Zhong Y, Sun M, Porto SMC, Toscano A, Dale BE, Sibilla F, Liao W (2018) Anaerobic co-digestion of multiple agricultural residues to enhance biogas production in southern Italy. Waste Manag 78:151–157
Vavilin VA, Rytov SV, Lokshina LY (1996) A description of hydrolysis kinetics in anaerobic degradation of particulate organic matter. Bioresour Technol 56(2):229–237
Vochozka M, Maroušková A, Šuler P (2018) Economic, environmental and moral acceptance of renewable energy: a case study-the agricultural biogas plant at Pěčín. Sci Eng Ethics 24:299–305
Ward AJ, Hobbs PJ, Holliman PJ, Jones DL (2008) Optimisation of the anaerobic digestion of agricultural resources. Bioresour Technol 99(17):7928–7940
Weng JK, Li X, Bonawitz ND, Chapple C (2008) Emerging strategies of lignin engineering and degradation for cellulosic biofuel production. Curr Opin Biotechnol 19(2):166–172
Xu P, Koffas MAG (2010) Metabolic engineering of *Escherichia coli* for biofuel production. Biofuels 1(3):493–504
Yadvika S, Sreekrishnan TR, Kohli S, Rana V et al (2004) Enhancement of biogas production from solid substrates using different techniques – a review. Bioresour Technol 95:1–10
Yang B, Wyman CE (2008) Pretreatment: the key to unlocking low-cost cellulosic ethanol. Biofuels Bioprod Biorefin 2(1):26–40
Yong Z, Dong Y, Zhang X, Tan T (2015) Anaerobic codigestion of food waste and straw for biogas production. Renew Energy 78:527–530
Zeng M, Mosier NS, Huang CP, Sherman DM, Ladisch MR (2007) Microscopic examination of changes of plant cell structure in corn stover due to hot water pretreatment and enzymatic hydrolysis. Biotechnol Bioeng 97(2):265–278
Zhang YHP (2008) Reviving the carbohydrate economy via multiproduct lignocellulose biorefineries. J Ind Microbiol Biotechnol 35(5):367–375
Zubr J (1986) Methanogenic fermentation of fresh and ensiled plant materials. Biomass 11 (3):159–171
Zumbusch PV, Meyer-Jens T, Brunner G, Märkl H (1994) On-line monitoring of organic substances with high-pressure liquid chromatography (HPLC) during the anaerobic fermentation of waste-water. Appl Microbiol Biotechnol 42:140–146

Biogas: An Effective and Common Energy Tool – Part III

5

Sheelendra Mangal Bhatt and Shilpa Bhat

Abstract

Biogas is one of the best future alternatives against depleting fossil fuel. Current Indian production of biogas is very low. There are many challenges for BioCNG production which is suitable for vehicle use but needs to adapt various technologies to enhance the content of biomethane.

Therefore, in current article, technological improvement in Biogas production intended for high production has been discussed in detail.

For use in vehicle, enhanced methane is required. Current article had focused on concise presentation of accumulated knowledge in current past. Bio-CNG can be produced from various biomass biowaste, kitchen waste, algae, and other biowastes which may be a very good option for Bio-CNG production. We have discussed socioeconomic challenges, suitable sources, barrier in production of biogas, and biochemical steps in production of biogas in normal verses reactor conditions, and also application of nanotechnology for green energy applications have been discussed.

Keywords

Biogas · Fermentative microorganism · CNG · Fossil fuels · Biofuels · Renewable energy

S. M. Bhatt (✉)
Amritsar Group of Colleges (ACET), Amritsar, Punjab, India

S. Bhat
C.G.C. Landran, Mohali, Punjab, India

N. Srivastava et al. (eds.), *Biofuel Production Technologies: Critical Analysis for Sustainability*, Clean Energy Production Technologies,
https://doi.org/10.1007/978-981-13-8637-4_5

5.1 Introduction

Depleting fossil fuels is the main concern today. Alternative to fossil fuels is the availability of biogas. Biogas primarily consists of 50–70% methane and 30–50% carbon dioxide, with trace amounts of other particulates and contaminants. It can be produced from various types of waste sources, including landfill materials, animal manure, wastewater, industrial, institutional and commercial organic wastes, kitchen waste, and various agricultural wastes.

Biogas is one of the viable alternatives to the burning energy question. In 2014–2015, about 20,700 lakh cubic meters of biogas is produced in India which is equivalent to 5% of the total LPG consumption in India. The government is also extending substantial subsidy for setting up of new biogas plants. At a time when the viability and safety of energy alternatives is being debated, it is pertinent to look at one of the oldest renewable energy alternatives, Biogas.

Biogas is made up of methane and carbon dioxide. It may have small amounts of hydrogen sulfide, moisture, and siloxanes. The gases methane, hydrogen, and carbon monoxide can be combusted or oxidized with oxygen. This energy release allows biogas to be used as a fuel; it can be used for any heating purpose, such as cooking. It can also be used in a gas engine to convert the energy in the gas into electricity and heat.

There is unique method of conversion of substrate into biogas under anaerobic oxidation using methanogenic bacteria under closed system by a process called as fermentation. Due to enhanced nitrogen content after biogas formation, agrowaste mixed with dung often results in good biofertilizers. Biogas production shall be a suitable option for reducing greenhouse gases. Methane gas production can be a suitable versatile source for vehicle fuel and can be produced in the wet or dry fermentation system. Most commonly used one is wet fermentation system where vertical stirred tank is used(Weiland 2010). Biogas is highly combustible due to presence of CH_4, H_2, and CO, so it is proved to be a better biofuel. Additional features are as follows: it can be used in engines to convert electricity and heat and thus also can be used as fuel for cars and other vehicles. Biogas can be compressed well similar to other gases like CNG or natural gas and is very useful in terms of technoeconomical aspects. Caterpillar Inc. and biogas turbine is used to convert biogas into electricity and heat.

5.2 Socioeconomic Status and Biogas Production in India

Currently biogas production in India is around 2.07 billion m^3/year which could be increased up to 29–48 billion m^3/year. Currently, biogas has less content of methane (40–70%) which needs to be enhanced for its use in transportation fuel. Small country like Nepal is currently producing around 1 million m^3 biogas (Zwolinski et al. 2013). Out of 215 large biogas plants in the world, India has only 15 biogas plants till 2013 (Kadam and Panwar 2017), due to low availability of methane (only 45% mostly) which poses problem as compared to natural gas which contains mostly

90% methane. These gases can be compressed and liquefied; therefore, biogas can be purified by using various techniques. Mostly target is to remove H_2S which is present as impurity, and in impure form methane cannot be used as vehicle fuel and also presence of water can be harmful for the compressor. Mostly technique used for purification of biogas is water scrubbing, membrane separation and biofilters, chemical absorption, and pressure swing absorption. Sweden is a world leader in the production of transportation fuel in 2012. *India* is trying hard to achieve the goal and trying to upgrade the production of biogas. The most common technology to purify methane is water scrubbing.

Simple anaerobic procedure leads to conversion of biomass into biofuel energy and manure, which enhances the financial and social improvement of farmers. Socially, methane production from cow dung or other kinds of dungs and waste rice plants is harmful for the environment, since it is a greenhouse gas and by properly using it for cooking, electricity production, or vehicle fuel, we can reduce and can manage air pollution (Mittal et al. 2018).

A family type biogas plant generates biogas from organic substances such as cattle, dung, and other biodegradable materials such as biomass from farms, gardens, kitchens and night soil wastes, etc. The process of biogas generation is called anaerobic digestion (AD). The following are the benefits of the Biogas technology. It provides clean gaseous fuel for cooking and lighting. Chemical fertilizers can be done away with since the digested slurry obtained from the biogas plants can be used as enriched biomanure. It is good for the climate and for sanitation problems since toilets can be linked directly with biogas plants.

Biogas can be used as clean energy fuel, since it finds applications in running vehicles and in cooking, heating, and electricity production. According to a report from cattle dung of 304 million cattle, around 18,240 million cubic meter of biogas can be obtained per annum (Kadam and Panwar 2017). Now, India is focusing on clean energy production due to which various plans have been implemented in the past few years.

5.3 Barriers in Biogas Production

After extensive literature review, the following barriers may be accounted for biogas production:

1. Low percentage of methane and its production
2. Costly
3. Laborious

In addition to these above factors, in India, various other problems exist, which have been discussed in detail in Mittal et al. (2018) (Table 5.1).

Table 5.1 Composition of biogas and natural gas

Component	Biogas	Natural gas
Methane (%)	40–75	87–97
Carbon dioxide (%)	25–55	0.1–1
Hydrogen sulfide (ppm)	50–5000	NA
Ammonia (%)	0–1	NA
Water (%)	0–10	NA
Nitrogen (%)	0–5	0.2–5.5
Oxygen (%)	0–2	.01–.1
Hydrogen (%)	0–1	Trace– 0.02

Adapted from Kadam and Panwar (2017) and Mittal et al. (2018)

5.4 Sources for Biogas Production

There are various reports of biogas production from various sources, e.g., slaughterhouse (Granada et al. 2018), farm animal waste (Abdeshahian et al. 2016), animal manure (Recebli et al. 2015), solid organic waste (Nasir et al. 2012), kitchen waste (Agrahari and Tiwari 2013), municipal wastewater (Appels et al. 2008), food and green wastes (Liu et al. 2009), coffee waste (Battista et al. 2016), rice straw, and pig manure (Ye et al. 2013).

Codigestion of manure and organic wastes (Angelidaki and Ellegaard 2003) such as lignocellulosic waste for biogas production has been reported by some workers (Ziemiński et al. 2012).

Some work has been done to produce economical biogas, for example 0.41 g ethanol/g glucose and 178 ML hydrogen per gram sugar obtained from agrowaste produces (Kaparaju et al. 2009), while methane was produced from 0.381 m^3/kg volatile solid. These workers emphasized that while using lignocellulosic biomass, multiple route of biofuel production must be adapted in order to obtain the economical biofuel production.

Use of rice straw is also recommended for production of biogas after proper milling and pretreatment (Mustafa et al. 2017)

Kitchen waste produces around 60% of biogas, but other types of sources produce around 40% of biogas; from cotton wastes, the production of biogas (CH_4) was approximately 65%, while 55% methane production was reported from waste disposal. Kitchen waste was proved to produce more biogas if cow dung was mixed with waste of water hyacinth (Tasnim et al. 2017). The calorific value of biogas is very good (around 4700 kcal or 20 MJ at around 55% methane content).

There are few developments of nano-based technology for rapid digestion of biomass and thus production of biogas via “nano-clean” technology where nanoparticles (iron oxide nanoparticles) are used to enhance production of biogas. Application of Nannotechnology is certainly going to double the production of biogas using same amount of agrowaste (e.g. wheat straw).

Life cycle assessment (LCA) studies of biogas systems from around Europe were done by some workers, and they reported that (Hijazi et al. 2016) most suitable substrate for high biogas (methane) production is animal dung and maize.

As compared to other biomass such as cellulose municipal solid wastes, some workers have compared various microalgae and cyanobacteria for production of biogas (Mussgnug et al. 2010). It was reported that green algae *Chlamydomonas reinhardtii* is more beneficial in terms of biogas methane production as compared to other microalgae.

In biogas production, mostly carbon is used in the production of methane while nitrogen left can be used as manure which enriches soil fertility by replacing chemical fertilizers depending on the kind of solid agrowaste used. Rice straw is posing huge problems of disposal instantly and its recalcitrant nature makes it difficult for its conversion to methane. Some workers stressed on changing the pattern of pretreatment which can get rid of extra silica and can make it a suitable biomass for conversion for biogas production (Gurung et al. 2013).

5.5 Lignocellulosic Biomass for Biogas Production

Lignocellulosic biomass is present in abundance in the nature, which can be utilized for conversion to biogas. Various lignocellulosic feedstocks with their cellulose, hemicellulose, and lignin content have been summarized in Table 5.2 and Fig. 5.1.

5.6 Biogas Production Mechanism

Hydrolysis of biomass is a must for large organic biomass in order to get rid of recalcitrant structure such as silica and lignin which impregnate cellulosic and hemicellulosic biomass. Anaerobic bacteria can be used in anaerobic digesters to access the energy potential of the material, which can be broken down into their smaller constituent parts. Sugars released in this way are readily used by other acetogenic and further by methanogenic bacteria. Acetate and hydrogen produced in the first stages can be used directly by methanogens.

The third stage of anaerobic digestion is acetogenesis where microbial acetogens are added to produce largely acetic acid as well as carbon dioxide and hydrogen. The terminal stage of anaerobic digestion is the methanogenesis. Here methanogens utilize the intermediate products of the preceding stages and convert them into methane, carbon dioxide, and water. It is these components that makes up the majority of the biogas emitted from the system. Methanogenesis is sensitive to both high and low pH and occurs between pH 6.5 and pH 8. The remaining, nondigestible material which the microbes cannot feed upon, along with any dead bacterial remains, constitutes the digestate.

Table 5.2 Lignocellulosic biomass fraction of some feedstocks (Isikgor and Becer 2015)

Lignocellulosic feedstocks	Cellulose (%)	Hemicellulose (%)	Lignin (%)
Sugar cane bagasse	42	25	20
Softwood	45–50	25–35	25–35
Hardwood	40–55	24–40	18–25
Corn stover	38	26	19
Corn cobs	45	35	15
Nut shells	25–30	25–30	30–40
Rice Straw	32	24	18
Grasses	25–40	25–50	10–30
Newspaper	40–55	25–40	18–30
Banana waste	13.2	14.8	14
Wheat straw	29–35	26–32	16–21
Bagasse	54.87	16.52	23.33
Sponge gourd fibers	66.59	17.44	15.46
Agricultural residues	5–15	37–50	25–50
Hardwood	20–25	45–47	25–40
Softwood	30–60	40–45	25–29
Grasses	0	25–40	35–50
Waste papers from chemical pulps	6–10	50–70	12–20
Newspaper	12	40–55	25–40
Sorted refuse	60	20	20
Leaves	15–20	80–85	0
Cotton seed hairs	80–95	5–20	0
Paper	85–99	0	0–15
Switch grass	45	31.4	12
Sweet sorghum	45	27	21

A simplified generic chemical equation for the overall processes outlined above is as follows:

$$C_6H_{12}O_6 \rightarrow 3CO_2 + 3CH_4$$

5.6.1 Anaerobic Digestion [AD]

Biogas can be produced in four simple steps as shown in Figs. 5.2 and 5.3 flowchart. There are four key biological and chemical stages of anaerobic digestion:

1. Hydrolysis
2. Acidogenesis
3. Acetogenesis
4. Methanogenesis

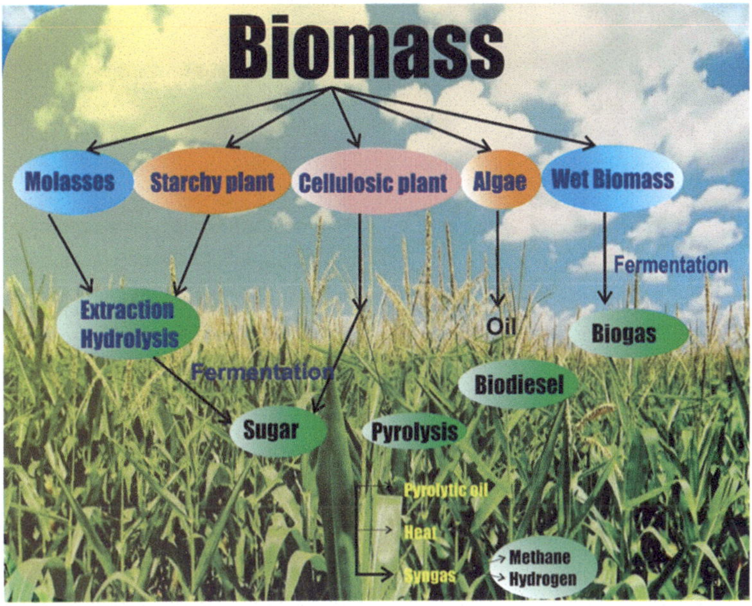

Fig. 5.1 Different biomass for biofuel production (redrawn for illustration)

Fig. 5.2 Biogas production mechanism

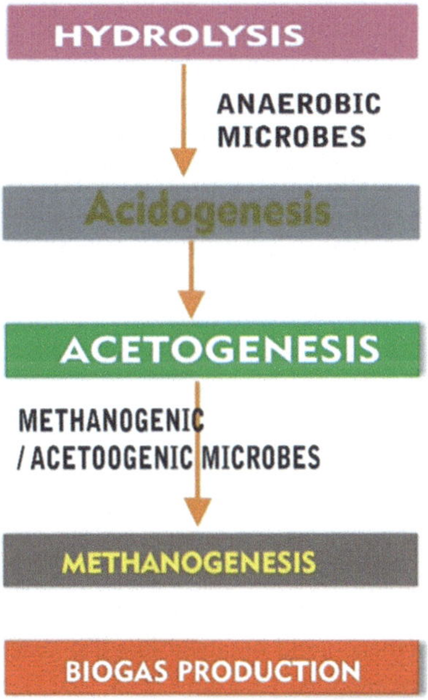

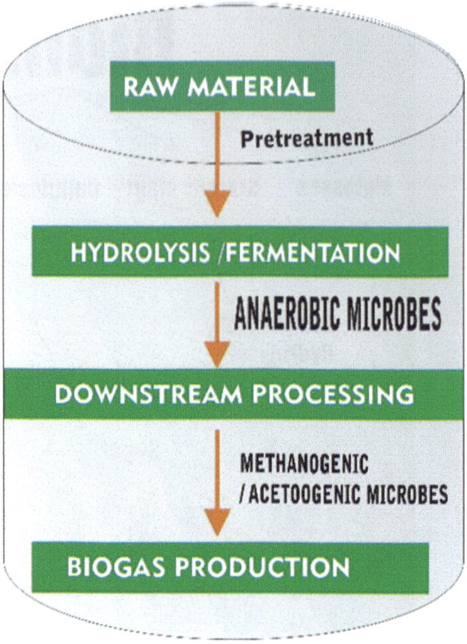

Fig. 5.3 Steps for biogas production

Anaerobic digestion involves basically four steps as discussed earlier and involves three microbes such as pshychrophilc, mesophilic, and thermophilic. As shown in Table 5.2 and Fig 5.3, mesophilic seems to be most suitable one as its temperature range lies in 30 to 42°C range and also has optimum retention time of 30 to 40 days. First step is complex and requires 7 days to 15 days minimum to digest the complex material into simplest one where lignin silica and other complex materials of plant cell wall are released. Now comes second step which is acidogenesis, which is fastest and involves formation of organic acids from simple carbohydrate with the help of acidogenic fermentative bacteria microbes.

Third stage is acetogenesis where organic acid is converted into acetic acid, CO_2, and hydrogen with the help of acetogenic microbes. Last stage is conversion of CO_2 and hydrogen into methane by methanogenic microbes leaving behind organic residue by process of hydrogen trophic bacteria.

Therefore, biogas is mainly affected by microbial growth and substrate utilization rate.

There are two types of anaerobic digestion (see Fig. 5.3): 1) solid state and 2) liquid state applied as per need and composition of waste or biomass (Li et al. 2011). Highest methane yield obtained in solid state anaerobic digestion of rice 258 L/kg of biogas was reported (Mustafa et al. 2017). It was reported that in absence of technology, large-scale burning of rice straw has been in practice by farmers and that results in release of open combustion of one ton of rice straw release 3 kg particulate matter, 60 kg CO, 1460 kg CO_2, and 2 kg SO_2 to the atmosphere (Sanchis et al. 2014).

5.7 Strategies to Enhance Biogas Production

There are various strategies to enhance biogas production (Keanoi et al. 2014; Gu et al. 2015; Amnuaycheewa et al. 2016; Soam et al. 2017). To enhance biogas production, cow dung addition and effect of other factors have been studied (Keanoi et al. 2014). The benefit of addition of cow dung and codigestion along with rice straw was attempted in different mode; as a result, 1650 mL/day methane was obtained when natural water, rice straw, cow dung, and water hyacinth is used in the ratio of 2:1:1:1. The recommended C:N ratio was 32: 25. Mustafa et al. emphasize on biological pretreatment strategies to improve the digestibility of substrate, and worker used *Pleurotus ostreatus*. According to a report, addition of increasing amount of goat manure, dairy manure, with other agricultural waste, such as straw material, for example, wheat straw (WS), corn stalks (CS), and rice straw (RS) (at C/N ratio of 35:61), is helpful in increasing biogas production. A similar report was given by Li et al. (2014) where biogas was enhanced by adding dairy manure with a mixture of three straw rice straw, corn stalks, and wheat straw. In another attempt C/N ratio was improved by adding urea, in optimized condition 1% NaOH-pretreated rice, and biogas was obtained after 15 days, 514 L/kg VS/day.

In another attempt, biogas production was enhanced by optimizing inoculum to substrate ratio (Candia-García et al. 2018). The optimum condition was temp. 5 °C to 27 °C time 60 days and biogas produced was 410 L/kgVS at 0.8 of I/S ratio, and methane obtained 70%.

5.7.1 Enhancing Biogas Production via Application of Microbial Consortia in Pretreatment

Recently, some workers reported use of thermophilic microbial consortium (MC1) (Yuan et al. 2014). This pretreatment method is novel in the sense that it has opportunity to digest lignocellulosic biomass with high efficiency to produce biogas and methane because of high mass of ethanol, acetic acid, propionic acid, and butyric acid.

A novel microbial consortia was developed named as BYND-5 which was based on mixture of microbes from mesophilic microbes that were mixed in order to digest the rice straw. Digestion efficiency of BYND-5 was checked for rice straw which was around 49% after 7 days (Yan et al. 2012). The species were identified by molecular techniques called as ARDRA (Amplified ribosomal DNA restriction analysis).

5.7.2 Improved Biogas Production by Bioaugmentation Technique

Another method to improve biogas production reported was bioaugmentation technique by use of two fungi in anaerobic two-stage system for digesting two substrates, corn silage and cattail (Table 5.3).

Table 5.3 AD feedstock and biogas yield (Seadi et al. 2008)

Substrates	Biogas yield (m^3/tFF[a])	Methane percent
Liquid pig manure	28	65
Liquid cattle manure	25	60
Distillers grains with soluble	40	61
Pig manure	60	60
Cattle manure	45	60
Chicken manure	80	60
Organic waste	100	61
Beet	88	53
Sweet sorghum	108	54
Grass silage	172	54
Corn silage	202	52
Forage beet	111	51

[a]Tons fresh feed

5.8 Biogas Plant

As an estimate of biogas production in India in 2014–2015, biogas produced was around 20,757 lakh cubic meters which is equivalent to 6.6 crore domestic LPG cylinders. This is equivalent to 5% of the total LPG consumption in the country today and state-wise Maharashtra tops in the production with 3578 lakh cubic meters, while second highest biogas production was at Andhra Pradesh that comes next with 2165 lakh cubic meters.https://factly.in/biogas-production-in-india-is-about-5-percent-of-the-total-lpg-consumption/

5.9 Application of Nanotechnology in Enhancing Biogas Production

There is now plethora of reports using nanocatalyst for enhancing biogas production. Three types of catalyst are used: (1) metal oxide, e.g., Cu, Ti, Zn, and mainly Al, Fe, and Mg, (2) zero valance metals, and (3) carbon nanomaterials.

Cu nanomaterials are reported to have negative effect, while ZnO have positive effect. Ti, Al, and Si oxide have no effect on methane production (Otero-González et al. 2014). Fe_2O_3 has a positive effect on methane production and around 180% increase was reported after 60 days (Ganzoury and Allam 2015).

5.10 Biogas Global Production Scenario

Leaders in global biogas productions are well-developed biogas industry Germany, Denmark, Austria, and Sweden followed by the Netherlands, France, Spain, Italy, the United Kingdom, and Belgium (Horváth et al. 2016). And biogas plants are classified based on (1) digestion methodology or codigestion (2) farm technology.

Table 5.4 Microbes used in AD (Kadam and Panwar 2017)

Facultative and obligate anaerobic fermentative bacteria	Acetogenesis or acetic acid production	Methanogenic bacteria
Clostridium spp., *Peptococcusanaerobius*, *Bifidobacterium* spp., *Desulphovibrio* spp., *Corynebacterium* spp., *Lactobacillus* spp., *Staphylococcus* spp., and *Escherichia coli*	*Syntrobacterwolinii and Syntrophomonaswolfei*	*Methanosarcina and Methanosaeta*, *Methanobacteriumruminantium*, *Methanosaeta* sp., *Methanosarcina* sp.

European countries are using biogas for two purposes either for heat production or for electricity. Basic technology is totally different from the industrial production technology (Bauer et al. 2013) (Tables 5.4, 5.5 and 5.6).

5.11 Bio-CNG

The need for liquid and gaseous fuel for transportation application is growing very fast. There are various development in this sector, and various countries such as Bangladesh are adapting this technology to improve the production (Shah et al. 2017). Bio-CNG is a very good alternative form of gasoline. CNG is actually found over landfill gases or above oil as a separate layer and is compressed below 1% volume of atmospheric pressure or around 20–25 MPa. Biogas produced is actually compressed and is rich in methane and filled in cylinder after removing impurities such as water, H_2S, etc. Bio-CNG has been tested for various vehicle operations (Ryan and Caulfield 2010). Compared to diesel, bus fleet gave equivalent performance and has benefit of reducing greenhouse gas emission (Fig. 5.4).

A case study has been done for packaging transportation fuel in Ireland whose main focus was to study policy barriers for Bio-CNG (Goulding et al. 2017)

Bio-CNG production has been done from municipal sewage sludge (MSS) waste (Bharathiraja et al. 2014), and due to heavy availability of municipal sewage sludge, this has the opportunity to establish Bio-CNG plant in every city. MSS is rich in organic and inorganic media which can be easily converted to Bio-CNG.

Bio-CNG plant has been installed in India in various city to cope with the pollution problems also with and also to facilitate its heavy production to meet the required demands. The sewage has been utilized in United Kingdom for production of electricity which produces around 14,000 m^3/day and is used in dual fuel engines. It has been reported that using dual fuel engine is more beneficial as compared to using biogas alone and as a result 70% of diesel or petrol can be saved.

Table 5.5 Additives used for biogas production

Additives category	Used compound/ material	Feedstock	Result	References
Biological additives	Plant extracted Aquasan and teresan	Cattle dung and kitchen waste	Increased gas yield	Singh et al. (2001)
	Lantana residue, wheat straw, apple leaf litter, and peach leaf litter	Cattle dung	1–6% increased gas yield	Hassan Dar and Tandon (1987)
	Mustard meal/cake	Cattle dung		Satyanarayan et al. (2008)
	Homo- and hetero-fermentative bacteria	Maize	22.5% increased gas yield	Vervaeren et al. (2010)
	Auricularia Auricula-judae (Fungi)	Sweet chestnut leaves and hay	15% increase in biogas production	Mackuľak et al. (2012)
	Ceriporiopsis Subvermispora ATCC 96608(Fungi)	Yard Trimmings	154% increased methane yield	Zhao (2013)
	Enzymes	Lignocellulosic Biomass	Up to 34% increased gas yield	Zheng et al. (2014)
Chemical Additives	NaOH	Fallen leaves	24 times higher biogas yield than control	Liew et al. (2011)
	NaOH	Wheat Straw	112% increase in methane yield	Chandra et al. (2012)
	NaOH	Wheat Straw	47% increase of biochemical methane potential	Sambusiti et al. (2012)
	NaOH	Rice Straw	30% higher biogas yield	Chen et al. (2010)
	NaOH	Lignocellulosic feedstocks	Positive effect of methane yield	Zheng et al. (2014)
	Lime	Rice straw	Improved biogas production	Song et al. (2013)
	$Ca(OH)_2$	OFMSW	172% higher methane yield	Llore et al. (2008)
	Ammonium hydroxide	switch grass	65% increased methane yield	Himmelsbach et al. (2010)
	Diluted $H_2\ SO_4$	Sugarcane Bagasse	166% increase in methane yield	Badshah et al. (2012)
	4% HCl or 10% $FeCl_3$	Sunflower stalks	21–29% increase methane potential	Monlau et al. (2012)
	H_3PO_4	OPEFB	40% improved methane yield	Nieves et al. (2011)
	$FeSO_4$	Cow dung and poultry litter	40% and 42% increased methane production respectively	Preeti Rao and Seenayya (1994)
	$FeCl_3$	Swine excreta	60% increased bio gas production	Hansen et al. (1999)
	20 mM sulfate		Twofold increase in biogas production	Sai Ram et al. (1993)

Table 5.6 AD temperature ranges with the corresponding retention time

Process	Temperature (°C)	Minimum retention time (days)
Psychrophilic	<20	70–80
Mesophilic	30–42	30–40
Thermophilic	43–55	15–20

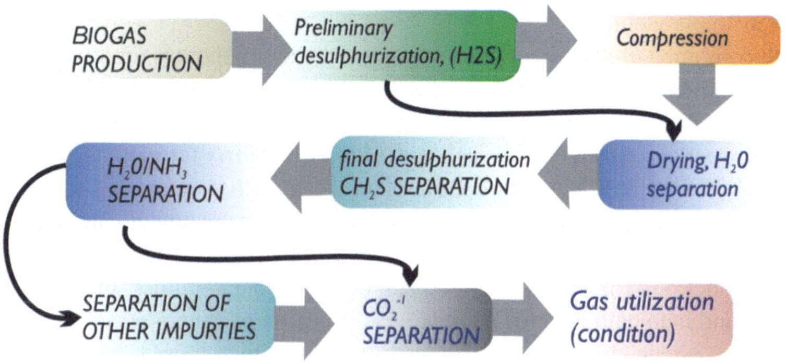

Fig. 5.4 Operation flowchart Bio-CNG production. (Redrawn and adapted from Kadam and Panwar 2017)

5.12 Methods to Improve Anaerobic Digestion of Algae

For biogas production, a suitable digester design is essential which can be applied to:

1. Municipal solid waste treatment
2. Treatment of microalgae residue for getting high value production such as methane and fertilizers

The main problem lies in anaerobic digestion for algae treatment, as digestion of its tough cell wall since many components of cell wall mainly consists of cellulose and hemicellulose that are slow to digest and also need large bioreactors to operate. Therefore, digestibility and high capital cost is also challenging.

In order to increase digestibility, some author recommends specific pretreatment, such as thermal pretreatment, which include thermal hydrolysis, ultrasound or thermochemical treatment.

Ultrasound pretreatment (which is radiation based) or thermochemical pretreatment techniques are most suitable at a temperature of 62 °C or more. Sometime it depends on the species, for example, in case of *Scenedesmus* species, the production of biogas was improved when thermochemical pretreatment was given and as a result biogas production was two times more. Improved anaerobic digestion is also useful in some species, for example, *Oocystis*.

Sometimes combined pretreatment is also helpful in conversion of more biomass to biogas. It requires high temperature treatment (at 130 °C) and then final pretreatment at 170 °C with 8 bar pressure and in 80 min for thermal hydrolysis of algal biomass for better results. This results in better hydrolysis of biomass with high yield of biogas formation as compared to gas obtained at low temperature pretreatment or with ultrasonication technique. Another very good mechanical pretreatment adapted is production of biogas in continuous reactor to improve the overall methane yield.

With combined ultrasonic or microwave pretreatment, there is report of enhanced methane production in continuous bioreactor. Another very successful method is the combined application of enzymatic and thermal pretreatment where protease can be used for digestion of microalgae such as *Chlorella vulgaris* for methane production and the yield was 26% more in continuous reactor. It has been also reported that thermochemical pretreatment combined with alkaline pretreatment can improve the biogas production in the reactors. Cocktail of enzymes also results in improved biogas production. The enzymes which can be mixed are cellulose, glucohydrolase, xylanase, and hemicellulose for intended use. Because of synergistic action of these enzymes, a better digestion of biomass results which could improve biogas production.

Since microalgae alone and directly cannot be used for production of biogas, therefore most of the microalgae which is rich in oil can be first used for extraction of biodiesel and then residual biomass can be used for production of biogas. This process microalgae residue can be mixed with other biowastes for anaerobic digestion which results in improved biogas production up to 74%.

Another technique developed in order to improve the biogas from algae was nutrient starvation during cultivation of microalgae, for example, phosphorus limitation results in improved biogas production so in conclusion, We can say that energy and electricity production can be improved with biomass after adopting suitable pretreatment technique.

Acknowledgements I would like to express my gratitude to my esteemed mentor in Biofuel research Professor Dr. Pradeep Mishra (HOD, Dept. of Chemical Engineering, IT B.H.U.) and Dr. Neha Srivastava for inviting me to write an article on this hot topic, which is need of the hour.

I acknowledge the support of AMRITSAR GROUP OF COLLEGES authorities, Director Finance Madam Ragini (ACET, College), my staff of Agriculture Department for having directly or indirectly supporting me during the writing of this article, especially Dr. Ravindra (GNDU), my wife Dr. Suchi Rai Bhatt (Principal), my son (Master Anshuman Bhatt), and my daughter (Tejaswini Bhatt).

References

Abdeshahian P, Lim JS, Ho WS et al (2016) Potential of biogas production from farm animal waste in Malaysia. Renew Sust Energ Rev 60:714–723

Agrahari R, Tiwari GN (2013) The production of biogas using kitchen waste. Int J Energy Sci 3 (6):48. https://doi.org/10.14355/ijes.2013.0306.05

Amnuaycheewa P, Hengaroonprasan R, Rattanaporn K et al (2016) Enhancing enzymatic hydrolysis and biogas production from rice straw by pretreatment with organic acids. Ind Crop Prod 87:247–254. https://doi.org/10.1016/j.indcrop.2016.04.069

Angelidaki I, Ellegaard L (2003) Codigestion of manure and organic wastes in centralized biogas plants: status and future trends. In: Applied biochemistry and biotechnology – Part A. Enzyme engineering and biotechnology

Appels L, Baeyens J, Degrève J, Dewil R (2008) Principles and potential of the anaerobic digestion of waste-activated sludge. Prog Energy Combust Sci 34(6):755–781

Badshah M, Lam DM, Liu J, Mattiasson B (2012) Use of an automatic methane potential test system for evaluating the biomethane potential of sugarcane bagasse after different treatments. Bioresour Technol 114:262–269. https://doi.org/10.1016/j.biortech.2012.02.022

Battista F, Fino D, Mancini G (2016) Optimization of biogas production from coffee production waste. Bioresour Technol 200:884–890. https://doi.org/10.1016/j.biortech.2015.11.020

Bauer F, Hulteberg C, Persson T, Tamm D (2013) Biogas upgrading – review of commercial technologies, SGC 2013:270. Swedish Gas Technol Centre, SGC

Bharathiraja B, Yogendran D, Ranjith Kumar R et al (2014) Biofuels from sewage sludge – a review. Int J ChemTech Res 6(9):4417–4427

Candia-García C, Delgadillo-Mirquez L, Hernandez M (2018) Biodegradation of rice straw under anaerobic digestion. Environ Technol Innov 10:215–222. https://doi.org/10.1016/j.eti.2018.02.009

Chandra R, Takeuchi H, Hasegawa T, Kumar R (2012) Improving biodegradability and biogas production of wheat straw substrates using sodium hydroxide and hydrothermal pretreatments. Energy 43(1):273–282. https://doi.org/10.1016/j.energy.2012.04.029

Chen GY, Zheng Z, Luo Y et al (2010) Effect of alkaline treatment on anaerobic digestion of rice straw. Huan Jing Ke Xue 31(9):2208–2213

Ganzoury MA, Allam NK (2015) Impact of nanotechnology on biogas production: a mini-review. Renew Sust Energ Rev 50:1392–1404. https://doi.org/10.1016/J.RSER.2015.05.073

Goulding D, Fitzpatrick D, O'Connor R et al (2017) Supplying bio-compressed natural gas to the transport industry in Ireland: is the current regulatory framework facilitating or hindering development? Energy 136:80–89. https://doi.org/10.1016/j.energy.2016.08.037

Granada CE, Hasan C, Marder M et al (2018) Biogas from slaughterhouse wastewater anaerobic digestion is driven by the archaeal family Methanobacteriaceae and bacterial families Porphyromonadaceae and Tissierellaceae. Renew Energy 118:840–846. https://doi.org/10.1016/j.renene.2017.11.077

Gu Y, Zhang Y, Zhou X (2015) Effect of Ca(OH)2pretreatment on extruded rice straw anaerobic digestion. Bioresour Technol 196:116–122. https://doi.org/10.1016/j.biortech.2015.07.004

Gurung A, Karki R, Cho JS et al (2013) Roles of renewable energy technologies in improving the rural energy situation in Nepal: gaps and opportunities. Energy Policy 62:1104–1109. https://doi.org/10.1016/j.enpol.2013.06.097

Hansen KH, Angelidaki I, Ahring BK (1999) Improving thermophilic anaerobic digestion of swine manure. Water Res 33:1805–1810. https://doi.org/10.1016/S0043-1354(98)00410-2

Hassan Dar G, Tandon SM (1987) Biogas production from pretreated wheat straw, lantana residue, apple and peach leaf litter with cattle dung. Biol Wastes 21:75–83. https://doi.org/10.1016/0269-7483(87)90131-5

Hijazi O, Munro S, Zerhusen B, Effenberger M (2016) Review of life cycle assessment for biogas production in Europe. Renew Sust Energ Rev 54:1291–1300

Himmelsbach JN, Raman DR, Anex RP et al (2010) Effect of ammonia soaking pretreatment and enzyme addition on biochemical methane potential of switchgrass. Trans ASABE 53:1921–1928. https://doi.org/10.13031/2013.35791

Horváth IS, Tabatabaei M, Karimi K, Kumar R (2016) Recent updates on biogas production – a review. Biofuel Res J 3:394–402. https://doi.org/10.18331/BRJ2016.3.2.4

Isikgor FH, Becer CR (2015) Lignocellulosic biomass. Polym Chem 6:4497–4559

Kadam R, Panwar NL (2017) Recent advancement in biogas enrichment and its applications. Renew Sust Energ Rev 73:892–903

Kaparaju P, Serrano M, Thomsen AB et al (2009) Bioethanol, biohydrogen and biogas production from wheat straw in a biorefinery concept. Bioresour Technol 100:2562–2568. https://doi.org/10.1016/j.biortech.2008.11.011

Keanoi N, Hussaro K, Teekasap S (2014) The effect of natural water with cow dung and agricultural waste ratio on biogas production from anaerobic co-digestion. Am J Environ Sci 9:529–536. https://doi.org/10.3844/ajessp.2013.529.536

Li Y, Park SY, Zhu J (2011) Solid-state anaerobic digestion for methane production from organic waste. Renew Sust Energ Rev 15:821–826. https://doi.org/10.1016/j.rser.2010.07.042

Li J, Wei L, Duan Q et al (2014) Semi-continuous anaerobic co-digestion of dairy manure with three crop residues for biogas production. Bioresour Technol 156:307–313. https://doi.org/10.1016/j.biortech.2014.01.064

Liew LN, Shi J, Li Y (2011) Enhancing the solid-state anaerobic digestion of fallen leaves through simultaneous alkaline treatment. Bioresour Technol 102:8828–8834. https://doi.org/10.1016/j.biortech.2011.07.005

Liu G, Zhang R, El-Mashad HM, Dong R (2009) Effect of feed to inoculum ratios on biogas yields of food and green wastes. Bioresour Technol 100:5103–5108. https://doi.org/10.1016/j.biortech.2009.03.081

Llore CE, Lo M, López Torres M, Espinosa Lloréns MDC (2008) Effect of alkaline pretreatment on anaerobic digestion of solid wastes. Waste Manag 28:2229. https://doi.org/10.1016/j.wasman.2007.10.006

Mackuľak T, Prousek J, Švorc Ľ, Drtil M (2012) Increase of biogas production from pretreated hay and leaves using wood-rotting fungi. Chem Pap 66:649–653. https://doi.org/10.2478/s11696-012-0171-1

Mittal S, Ahlgren EO, Shukla PR (2018) Barriers to biogas dissemination in India: a review. Energy Policy 112:361–370. https://doi.org/10.1016/j.enpol.2017.10.027

Monlau F, Barakat A, Steyer JP, Carrere H (2012) Comparison of seven types of thermo-chemical pretreatments on the structural features and anaerobic digestion of sunflower stalks. Bioresour Technol 120:241–247. https://doi.org/10.1016/j.biortech.2012.06.040

Mussgnug JH, Klassen V, Schlüter A, Kruse O (2010) Microalgae as substrates for fermentative biogas production in a combined biorefinery concept. J Biotechnol 150:51–56. https://doi.org/10.1016/j.jbiotec.2010.07.030

Mustafa AM, Poulsen TG, Xia Y, Sheng K (2017) Combinations of fungal and milling pretreatments for enhancing rice straw biogas production during solid-state anaerobic digestion. Bioresour Technol 224:174–182. https://doi.org/10.1016/j.biortech.2016.11.028

Nasir IM, Ghazi TIM, Omar R (2012) Production of biogas from solid organic wastes through anaerobic digestion: a review. Appl Microbiol Biotechnol 95:321–329

Nieves DC, Karimi K, Horváth IS (2011) Improvement of biogas production from oil palm empty fruit bunches (OPEFB). Ind Crop Prod 34:1097–1101. https://doi.org/10.1016/j.indcrop.2011.03.022

Otero-González L, Field JA, Sierra-Alvarez R (2014) Fate and long-term inhibitory impact of ZnO nanoparticles during high-rate anaerobic wastewater treatment. J Environ Manag 135:110–117. https://doi.org/10.1016/j.jenvman.2014.01.025

Preeti Rao P, Seenayya G (1994) Improvement of methanogenesis from cow dung and poultry litter waste digesters by addition of iron. World J Microbiol Biotechnol 10:211–214. https://doi.org/10.1007/BF00360890

Recebli Z, Selimli S, Ozkaymak M, Gonc O (2015) Biogas production from animal manure. J Eng Sci Technol 10:722–729. https://doi.org/10.1016/j.proeps.2015.08.144

Ryan F, Caulfield B (2010) Examining the benefits of using bio-CNG in urban bus operations. Transp Res Part D Transp Environ 15:362–365. https://doi.org/10.1016/j.trd.2010.04.002

Sai Ram M, Singh L, Alam SI (1993) Effect of sulfate and nitrate on anaerobic degradation of night soil. Bioresour Technol 45:229–232. https://doi.org/10.1016/0960-8524(93)90116-S

Sambusiti C, Ficara E, Rollini M et al (2012) Sodium hydroxide pretreatment of ensiled sorghum forage and wheat straw to increase methane production. Water Sci Technol 66:2447–2452. https://doi.org/10.2166/wst.2012.480

Sanchis E, Ferrer M, Calvet S et al (2014) Gaseous and particulate emission profiles during controlled rice straw burning. Atmos Environ 98:25–31. https://doi.org/10.1016/j.atmosenv.2014.07.062

Satyanarayan S, Murkute P, Ramakant (2008) Biogas production enhancement by Brassica compestries amendment in cattle dung digesters. Biomass Bioenergy 32:210–215. https://doi.org/10.1016/j.biombioe.2007.09.008

Seadi TA, Rutz D, Prassl H et al (2008) Biogas handbook

Shah MS, Halder PK, Shamsuzzaman ASM et al (2017) Perspectives of biogas conversion into bio-CNG for automobile fuel in Bangladesh. J Renew Energy 2017:1–14. https://doi.org/10.1155/2017/4385295

Singh S, Kumar S, Jain MC, Kumar D (2001) Increased biogas production using microbial stimulants. Bioresour Technol 78:313–316. https://doi.org/10.1016/S0960-8524(00)00143-7

Soam S, Borjesson P, Sharma PK et al (2017) Life cycle assessment of rice straw utilization practices in India. Bioresour Technol 228:89–98. https://doi.org/10.1016/j.biortech.2016.12.082

Song Z, Yang G, Han X et al (2013) Optimization of the alkaline pretreatment of rice straw for enhanced methane yield. Biomed Res Int 2013:1–9. https://doi.org/10.1155/2013/968692

Tasnim F, Iqbal SA, Chowdhury AR (2017) Biogas production from anaerobic co-digestion of cow manure with kitchen waste and Water Hyacinth. Renew Energy 109:434–439. https://doi.org/10.1016/j.renene.2017.03.044

Vervaeren H, Hostyn K, Ghekiere G, Willems B (2010) Biological ensilage additives as pretreatment for maize to increase the biogas production. Renew Energy 35:2089–2093. https://doi.org/10.1016/j.renene.2010.02.010

Weiland P (2010) Biogas production: current state and perspectives. Appl Microbiol Biotechnol 85:849–860

Yan L, Gao Y, Wang Y et al (2012) Diversity of a mesophilic lignocellulolytic microbial consortium which is useful for enhancement of biogas production. Bioresour Technol 111:49–54. https://doi.org/10.1016/j.biortech.2012.01.173

Ye J, Li D, Sun Y et al (2013) Improved biogas production from rice straw by co-digestion with kitchen waste and pig manure. Waste Manag 33:2653–2658. https://doi.org/10.1016/j.wasman.2013.05.014

Yuan X, Wen B, Ma X et al (2014) Enhancing the anaerobic digestion of lignocellulose of municipal solid waste using a microbial pretreatment method. Bioresour Technol 154:1–9. https://doi.org/10.1016/j.biortech.2013.11.090

Zhao J (2013) Enhancement of methane production from solid-state anaerobic digestion of yard trimmings by biological pretreatment. The Ohio State University, Columbus

Zheng Y, Zhao J, Xu F, Li Y (2014) Pretreatment of lignocellulosic biomass for enhanced biogas production. Prog Energy Combust Sci 42:35–53

Ziemiński K, Romanowska I, Kowalska M (2012) Enzymatic pretreatment of lignocellulosic wastes to improve biogas production. Waste Manag 32:1131–1137. https://doi.org/10.1016/j.wasman.2012.01.016

Zwolinski MD, Harris RF, Hickey WJ et al (2013) No Title. Bioresour Technol 5:1–11. https://doi.org/10.1186/1754-6834-6-16

Stoichiometric Analysis of Biogas Production from Industrial Residues

6

Selvaraju Sivamani, B. S. Naveen Prasad,
Zahrah Abdullah Khalfan Al-Sharji,
Khair Abdulhakeem Musallem Al-Rawas,
Azan Salim Dadin Al-Blowshi, Anas Salim Basheer Al-Yafii,
Mahammad Ali Issa Al-Mashani, and
Humoud Suliman Humoud Al-Mamari

Abstract

A stoichiometric analysis of biogas production by anaerobic digestion from cassava wastewater, wheat bran, and sewage sludge is proposed. A wide range of methods are available to study stoichiometry of biochemical reactions. This work reported elemental balances method to solve stoichiometric coefficients in biogas production from cassava wastewater, wheat bran, and sewage sludge. The method could be employed for various substrates for biogas production and for other biochemical reactions.

Keywords

Anaerobic digestion · Biogas · Industrial residues · Stoichiometry

6.1 Introduction

Anaerobic digestion is the process of producing biogas from the organic materials such as animal manure, waste paper, and sewage in the absence of oxygen. In India, more than three million small-scale biogas plants are available. In addition, in the Europe Union, biomass could amount to 1545 million tons per year (Igoni et al. 2008). Incorporation of municipal organic wastes such as food to the anaerobic digestion production helps to raise the amount of produced energy. Also, paper waste can be an additional source of enhanced biogas production. Anaerobic digestion is a multistep biological and chemical process that is beneficial in not only waste management but also energy creation (Weiland 2010).

S. Sivamani (✉) · B. S. N. Prasad (✉) · Z. A. K. Al-Sharji · K. A. M. Al-Rawas ·
A. S. D. Al-Blowshi · A. S. B. Al-Yafii · M. A. I. Al-Mashani ·
H. S. H. Al-Mamari
Chemical Engineering Section, Engineering Department, Salalah College of Technology, Salalah, Sultanate of Oman

N. Srivastava et al. (eds.), *Biofuel Production Technologies: Critical Analysis for Sustainability*, Clean Energy Production Technologies,
https://doi.org/10.1007/978-981-13-8637-4_6

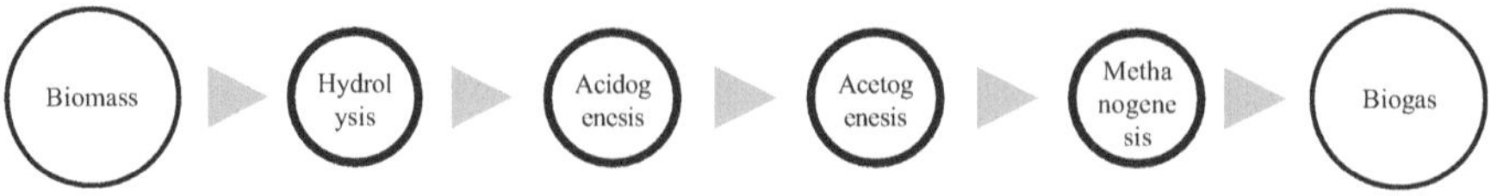

Fig. 6.1 Biogas production process

Table 6.1 Biochemistry of biogas production (Bajpai 2017)

Step	Reaction
Hydrolysis	Carbohydrates + $H_2O \rightarrow$ Glucose Lipids + $H_2O \rightarrow$ Fatty acids Proteins + $H_2O \rightarrow$ Amino acids
Acidogenesis	$C_6H_{12}O_6 + 2\ H_2 \rightarrow 2\ C_2H_5COOH + 2\ H_2O$ $C_6H_{12}O_6 \rightarrow 2\ C_2H_5OH + 2\ CO_2$
Acetogenesis	$C_2H_5COOH + 2\ H_2O \rightarrow CH_3COOH + CO_2 + 3\ H_2$ $C_6H_{12}O_6 + 2\ H_2O \rightarrow 2\ CH_3COOH + 2\ CO_2 + 4\ H_2$ $C_2H_5OH + H_2O \rightarrow CH_3COOH + 2\ H_2$ $C_6H_{12}O_6 \rightarrow 3\ CH_3COOH$
Methanogenesis	$2\ C_2H_5OH + CO_2 \rightarrow 2\ CH_3COOH + CH_4$ $CH_3COOH \rightarrow CH_4 + CO_2$ $CO_2 + 4\ H_2 \rightarrow CH_4 + 2\ H_2O$ $CH_3COOH + H_2SO_4 \rightarrow 2\ H_2O + 2\ CO_2 + H_2S$ $7\ CH_3COOH + 8\ NO_2 \rightarrow 2\ H_2O + 14\ CO_2 + 8\ NH_3$
Overall	$C_aH_bO_cN_dS_e + f\ H_2O \rightarrow p\ CH_4 + q\ CO_2 + r\ NH_3 + s\ H_2S$

6.1.1 Biogas Production Process

The four fundamental steps of anaerobic digestion include hydrolysis, acidogenesis, acetogenesis, and methanogenesis (Fig. 6.1). Throughout the entire process, large organic polymers that make up biomass are broken down into smaller molecules by chemicals and microorganisms. Upon completion of the anaerobic digestion process, the biomass is converted into biogas consisting of carbon dioxide and methane, and digestate (Dioha et al. 2013).

Table 6.1 shows the biochemical reactions taking place in each step during biogas production, and the overall reaction is shown below:

6.1.1.1 Hydrolysis

In general, hydrolysis is a chemical reaction in which the breakdown of water occurs to form H^+ cations and OH^- anions (Sivamani et al. 2018). Hydrolysis is often used to break down larger polymers, often in the presence of an acidic catalyst. In anaerobic digestion, hydrolysis is the essential first step, as biomass is normally comprised of very large organic polymers. Through hydrolysis, these large polymers, namely, carbohydrates, Lipids, and proteins, are broken down into smaller molecules such as simple sugars, fatty acids, and amino acids, respectively (Sivamani et al. 2018).

6.1.1.2 Acidogenesis

Acidogenesis is the next step of anaerobic digestion in which acidogenic microorganisms further break down the monomer products after hydrolysis. These fermentative bacteria produce an acidic environment in the digestive tank while creating volatile fatty acids, carbonic acids, alcohols, as well as trace amounts of other by-products. While acidogenic bacteria further break down the organic matter, it is still too large and unusable for the ultimate goal of methane production, so the biomass must next undergo the process of acetogenesis (Parawira et al. 2004).

6.1.1.3 Acetogenesis

In general, acetogenesis is the creation of acetic acid or acetate, from carbon and energy sources by acetogens. These microorganisms catabolize many of the products created in acidogenesis into acetic acid, CO_2, and H_2. Acetogens break down the volatile fatty acids and alcohols to a point at which methanogens can utilize much of the remaining materials to create methane as a biofuel (Kalia 2007).

6.1.1.4 Methanogenesis

Methanogenesis constitutes the final stage of anaerobic digestion in which methanogens create methane from the final products of acetogenesis as well as from some of the intermediate products from hydrolysis and acidogenesis. While CO_2 can be converted into methane and water through the reaction, the main mechanism to create methane in methanogenesis is the path involving acetic acid. This path creates methane and CO_2, the two main products of anaerobic digestion (Ziemiński and Frąc 2012; An et al. 2008).

6.1.2 Factors Affecting Biogas Yield

The factors affecting the biogas yield are carbon/nitrogen ratio, temperature, pH, dilution and consistency of feed, loading rate, hydraulic retention time, toxicity, agitation, and additives (Mahanta et al. 2005).

6.1.2.1 Carbon/Nitrogen (C/N) Ratio

The relationship between the amount of carbon and nitrogen present in organic materials is expressed by the carbon/nitrogen (C/N) ratio. A suitable C/N ratio plays an important role for the proper proliferation of the bacteria for the degradation process (Augenstein et al. 1976). Depending upon the relative richness in carbon and nitrogen content, feed material can be classified as nitrogen-rich or carbon-rich. It is generally found that during digestion, microorganisms utilize carbon 25–30 times faster than nitrogen, that is, carbon content in feedstock should be 25–30 times of the nitrogen content (Report No. ETSU B 1118 1986; Barnett et al. 1978; Fry and Merrill 1973). To meet this requirement, constituents of feedstock are chosen in such a way to ensure a C/N ratio of 25:1 to 30:1 and concentration of dry matter as

7–10%. Even in situations where C/N ratio is close to 30:1, the biomass can undergo efficient anaerobic digestion only if waste materials are also biodegradable at the same time (Singh 1974; SPOBD 1979).

6.1.2.2 Temperature

Most digesters installed in the field lack mechanisms for temperature control and removal of dissolved oxygen. Hence, efficiency of these digesters is reported to be low, particularly during the winter months. There are different temperature ranges during which mesophilic and thermophilic bacteria are most active causing maximum gas yield. Generally, mesophilic bacteria are most active in the temperature range 35–40 °C and thermophilic bacteria in the range 50–60 °C. Choice between the mesophilic and thermophilic digestions is governed by the natural climatic conditions in which the plant is located. Though, it is possible to create conditions for thermophilic digestion by external heat, but such a method is generally uneconomical. Length of digestion period is linked with the digester temperature.

The methanogens are inactive in extreme high and low temperatures, while the optimum temperature is 35 °C. When the ambient temperature decreases to 10 °C, gas production virtually stops. Satisfactory gas production takes place in the mesophilic range (30–40 °C). Proper insulation of digester helps to increase gas production during the cold weather. When the digester operates at a temperature of 15 °C, it takes nearly a year for the digestion cycle to complete. However, if the temperature is approximately 35 °C, the cycle can be easily completed in less than a month. When the digester temperature is maintained at 25 °C, it takes approximately 50 days for digestion of cattle waste. But, if the temperature ranges between 32 and 38 °C, digestion is complete within 28 days. Mahanta et al. carried out experiments to analyze the effect of temperature variation on anaerobic digestion of cattle wastes (NAS 1977). Smith et al. suggested that at low temperature, biogas plants with some design modifications could also function quite effectively as in a warm climate (Mahanta et al. 2004a).

6.1.2.3 pH

During anaerobic digestion, microorganisms require a natural or mildly alkaline environment for efficient gas production (Smith et al. 1982). An optimum biogas production is achieved when the pH of feed mixture in the digester is between 6.25 and 7.50 (Mahanta et al. 2004b; Wise 1987). The pH in a biogas digester is also a function of retention time. In the initial period of digestion, as large amounts of organic acids are produced by acidogenic bacteria, the pH inside the digester can decrease below 5. This inhibits or even stops the process. Methanogenic bacteria are very sensitive to pH and do not thrive below 6.5. Later, as the digestion process continues, concentration of NH_3 increases due to the digestion of N_2, which can increase the pH to above 8. When the CH_4 production level is stabilized, the pH range remains between 7.2 and 8.2. According to studies in China, during the period when ambient temperature varies between 22 and 26 °C, it takes approximately 6 days for pH to acquire a stable value (SPOBD 1979). Similarly, during the period when ambient temperature ranges between 18 and 20 °C, it takes approximately 14–18 days for pH to attain a stable value.

6.1.2.4 Dilution and Consistency of Feed

All waste materials fed to a biogas plant consist of solid substance, volatile organic matter and nonvolatile matter, and water. During anaerobic digestion process, volatile solids undergo digestion and nonvolatile solids remain unaffected. According to a finding by The Energy and Resources Institute (TERI), fresh cattle waste consists of approximately 20% total solids (TS) and 80% water (TERI 1987). TS, in turn, consists of 70% volatile solids and 30% fixed solids. For optimum gas yield through anaerobic digestion, normally, 8–10% TS in feed is required. This is achieved by making slurry of fresh cattle dung in water in the ratio of 1:1. However, if the dung is in dry form, the quantity of water has to be increased accordingly to arrive at the desired consistency of the feed (i.e., ratio could vary from 1:1.25 to even 1:2). If the dung is too diluted, the solid particles will settle down into the digester and if it is too thick, the particles impede the flow of the gas formed at the lower part of the digester. In both cases, gas production will be less than optimum. It is also necessary to remove inert materials such as stones from the feed before sending the slurry into the digester. Otherwise, the effective volume of digester will decrease.

6.1.2.5 Loading Rate

Loading rate is defined as the amount of raw materials fed per day per unit volume of digester. It is an important parameter that affects gas yield. If the plant is overfed, acids will accumulate and methane production will be inhibited since bacteria cannot survive in acidic conditions. Similarly, if the plant is underfed, the gas production will also be low because of alkaline solution, which is also not a favorable condition for anaerobic bacteria. The effect of daily and alternate day loadings on biogas yield was also studied (Gore 1981). It was found that a 50 kg charge on daily basis and 100 kg charge on alternate day basis produced 2.9043 and 2.9285 m^3 of gas, respectively. Also, for a particular size of plant, there is an optimum feed of charge rate that will produce maximum gas and further quantity of charge will not proportionately produce more gas. A daily loading rate of 16 kg of volatile solids per m^3 of digester produces 0.04–0.074 m^3 of gas per kg of raw dung fed (Moharao 1975). The recommended loading rates for plants working on night soil range from 1.04 to 2.23 kg of volatile solids per m^3 of digester (Moharao 1974). Higher loading rates are recommended only in cases where mean ambient temperature is high.

6.1.2.6 Hydraulic Retention Time

Hydraulic retention time (HRT) is the average period that a given quantity of input material remains in the digester to be acted upon by the methanogens. In a cattle dung plant, the retention time is calculated by dividing total volume of the digester and volume of feed per day. The rate of gas generation is initially high and then, gradually, declines as the digestion approaches completion. Thus, the time required for 70–80% digestion is considerably less than that needed to achieve complete digestion. HRT is chosen to achieve at least 70–80% digestion. Langrage (Langrage

1979) suggested that HRT depends upon the interior temperature of the digester. Higher the temperature of the digester, lower is the retention time. HRT varies between 20 and 120 days, depending upon the design and operating temperature of the digester. For digesters operating in countries of tropical region, HRT is usually taken as 40–60 days. In countries of colder region, digesters are designed for HRT of about 100 days (UN Guidebook on Biogas Development 1980).

6.1.2.7 Toxicity

Mineral ions, heavy metals, and detergents are toxic materials that inhibit the normal growth of pathogens in the digester. Small quantity of minerals (e.g., sodium, potassium, calcium, magnesium, ammonium, and sulfur) also stimulates the growth of bacteria, while high concentration leads to toxic effects. For example, presence of ammonium from 50 to 200 mg/L stimulates the growth of anaerobic microbes, whereas its concentration above 1500 mg/L produces toxic compounds. Similarly, heavy metals, such as copper, nickel, chromium, zinc, lead etc., in small quantities, are essential for the growth of bacteria, but their higher concentration has toxic effects (Moharao 1975). Detergents, including soap, antibiotics, organic solvents, etc. also inhibit the activity of methane-producing bacteria, and hence, addition of these substances in the digester should be avoided.

6.1.2.8 Agitation

Agitation or mixing of digester contents significantly helps to ensure intimate contact between microorganisms, which leads to improved fermentation efficiency. Coppinger (Coppinger 1979) suggested that effect of varying degrees of mixing of digester contents improves biogas production. The major problem associated with the different designs of biogas plant is that a thick layer of scum formation appears at the top of the digester which blocks the gas from coming out of the upper free portion of the digester. Thus, no gas is available at the utility point. The effects of recirculation of gas to break the scum formation were investigated by Mahanta et al. They found that recirculation of gas improves the biogas yield. The recirculation of gas increases the biogas production by three times. The gas production with circulation is much more than that without recirculation at the same pH.

6.1.2.9 Additives

The additives seem to play an important role in biogas yield. Addition of 5% commercial charcoal to cattle dung slurry on dry-weight basis raised the yield by 17 and 35% in batch and semicontinuous processes, respectively. Madamwar and Mithal (Madamwar and Mithal 1986) performed two sets of experiments: one at controlled temperature of 38 °C and the other at ambient temperature of 15 °C to find the impact of adding pectin to cattle dung slurry as feed on biogas yield. Pectin not only enhances gas yield but also imparts process stability during the periods of fluctuating temperature. The impact of adding inert materials, such as vermiculite, charcoal, and lignite bovine excreta, as feed on biogas yield has also been reported (Geeta et al. 1986). These additives increased biogas yield by 15–30%. Pebbles, glass marbles, and plastic mesh when suspended in digester slurry reportedly led to an increase in the gas yield by 10–20%.

Prasad (Prasad 1985) studied the effects of adding bagasse, Gulmohar leaves, wheat straw, groundnut shells, and leguminous plant leaves as additives to cattle dung on the biogas yield, gas composition, and extent of biodegradation. These additives were separately mixed with cattle dung in the ratio of 1 part (oven dry) to 10 parts of fresh dung containing 19.2% of TS on weight basis. Anaerobic fermentation was carried out under batch process in bottles in laboratory at ambient temperature between 30 and 32 °C for 9 weeks. The volume of biogas generated in 24 h was measured every day and gas composition analyzed periodically. It has been concluded that addition of additives is advantageous for obtaining a high gas yield.

6.2 Materials and Methods

6.2.1 Materials

Cassava was purchased from local supermarket in Salalah, Sultanate of Oman. Wheat bran was kindly provided by Salalah Mills Company SAOG, Raysut, Sultanate of Oman. Sewage sludge was kindly supplied by Salalah Sanitary Drainage Services Company SAOC, Raysut, Sultanate of Oman. Samples were used without further treatment.

6.2.2 Methods

6.2.2.1 Preparation of Cassava Wastewater

Fresh cassava was peeled and 3 kg of peeled cassava root was weighed. The roots were washed and cut into pieces. The cut roots were crushed with water in blender, and the final mixture was filtered. The filtrate was used as cassava wastewater for further analysis without any further dilution (Liu et al.).

6.2.2.2 Analytical Procedures

The total carbohydrate, lipid, and protein contents in the samples were estimated by anthrone (Hedge et al. 1962), Soxhlet extraction (Saim et al. 1997), and micro-Kjeldahl (Pellet and Young 1980) methods.

6.2.2.3 Stoichiometric Analysis

The procedure for stoichiometric analysis of biogas production is as follows:

1. The contents of carbohydrates, lipids, and proteins were estimated in cassava wastewater, wheat bran, and sewage sludge.
2. The percentages of carbon, hydrogen, oxygen, nitrogen, and sulfur present in carbohydrates, lipids, and proteins are known from the literature.

3. The concentrations of carbon, hydrogen, oxygen, nitrogen, and sulfur present in cassava wastewater, wheat bran, and sewage sludge were calculated.
4. The gram atoms of carbon, hydrogen, oxygen, nitrogen, and sulfur present in cassava wastewater, wheat bran, and sewage sludge were calculated.
5. Consider the overall reaction for biogas production,

$$C_aH_bO_cN_dS_e + f\,H_2O \rightarrow p\,CH_4 + q\,CO_2 + r\,NH_3 + s\,H_2S$$

Writing balances for each element, we get

$$\text{Carbon} : a = p + q \tag{6.E1}$$

$$\text{Hydrogen} : b + 2f = 4p + 3r + 2s \tag{6.E2}$$

$$\text{Oxygen} : c + f = 2q \tag{6.E3}$$

$$\text{Nitrogen} : d = r \tag{6.E4}$$

$$\text{Sulfur} : e = s \tag{6.E5}$$

Knowing the values of a, b, c, d, and e from step (4) above, the values of f, p, q, r, and s can be calculated solving five simultaneous Eqs. (6.E1, 6.E2, 6.E3, 6.E4 and 6.E5) (Tchobanoglous et al. 1991).

6.3 Results and Discussion (Tables 6.2, 6.3 and 6.4)

mg atom of carbon in cassava wastewater a_1	= 1139.37/12	= 94.95
mg atom of hydrogen in cassava wastewater b_1	= 88.25/1	= 88.25
mg atom of oxygen in cassava wastewater c_1	= 1380.83/16	= 86.30
mg atom of nitrogen in cassava wastewater d_1	= 16.29/14	= 1.16
mg atom of sulfur in cassava wastewater e_1	= 1.09/32	= 0.03
mg atom of carbon in wheat bran a_2	= 115.51/12	= 9.63
mg atom of hydrogen in wheat bran b_2	= 9.94/1	= 9.94
mg atom of oxygen in wheat bran c_2	= 94.09/16	= 5.88
mg atom of nitrogen in wheat bran d_2	= 8.66/14	= 0.62
mg atom of sulfur in wheat bran e_2	= 0.58/32	= 0.02
mg atom of carbon in sewage sludge a_3	= 141.44/12	= 11.79
mg atom of hydrogen in sewage sludge b_3	= 12.34/1	= 12.34
mg atom of oxygen in sewage sludge c_3	= 101.49/16	= 6.34
mg atom of nitrogen in sewage sludge d_3	= 12.13/14	= 0.87
mg atom of sulfur in sewage sludge e_3	= 0.81/32	= 0.03

For cassava wastewater, substituting the mg atoms in Eqs. (6.E1, 6.E2, 6.E3, 6.E4 and 6.E5),

Table 6.2 Biochemical composition of carbohydrates, lipids, and proteins in cassava wastewater, wheat bran, and sewage sludge

Constituents	Carbohydrates (ppm)	Lipids (ppm)	Proteins (ppm)
Cassava wastewater	2507.36	19.74	98.72
Wheat bran	146.28	30.0	52.50
Sewage sludge	148.00	46.7	73.50

Table 6.3 Percentage of carbon, hydrogen, oxygen, nitrogen, and sulfur in carbohydrates, lipids, and proteins

Elements	Carbohydrates	Lipids	Proteins
Carbon	42.7	84.6	52.7
Hydrogen	3.2	5.1	7.1
Oxygen	54.1	10.3	22.6
Nitrogen	–	–	16.5
Sulfur	–	–	1.1

$$94.95 = p + q \tag{6.E6}$$

$$88.25 + 2f = 4p + 3r + 2s \tag{6.E7}$$

$$86.30 + f = 2q \tag{6.E8}$$

$$1.16 = r \tag{6.E9}$$

$$0.03 = s \tag{6.E10}$$

Solving above equations, the values of f, p, q, r, and s were found to be

$$f = 30.62$$

$$p = 36.49$$

$$q = 58.46$$

$$r = 1.16$$

$$s = 0.03$$

Similarly, for wheat bran,

$$9.63 = p + q \tag{6.E11}$$

$$9.94 + 2f = 4p + 3r + 2s \tag{6.E12}$$

Table 6.4 Concentration of carbon, hydrogen, oxygen, nitrogen, and sulfur in cassava wastewater, wheat bran, and sewage sludge

	Cassava wastewater			Wheat bran			Sewage sludge		
Elements	Carbohydrates	Lipids	Proteins	Carbohydrates	Lipids	Proteins	Carbohydrates	Lipids	Proteins
Carbon (ppm)	1070.64	16.70	52.03	62.46	25.38	27.67	63.20	39.51	38.74
Hydrogen (ppm)	80.24	1.01	7.01	4.68	1.53	3.73	4.74	2.38	5.22
Oxygen (ppm)	1356.48	2.03	22.31	79.14	3.09	11.87	80.07	4.81	16.61
Nitrogen (ppm)	–	–	16.29	–	–	8.66	–	–	12.13
Sulfur (ppm)	–	–	1.09	–	–	0.58	–	–	0.81

$$5.88 + f = 2q \tag{6.E13}$$

$$0.62 = r \tag{6.E14}$$

$$0.02 = s \tag{6.E15}$$

Solving above equations, the values of f, p, q, r, and s were found to be

$$f = 4.68$$

$$p = 4.35$$

$$q = 5.28$$

$$r = 0.62$$

$$s = 0.02$$

Similarly, for sewage sludge

$$11.79 = p + q \tag{6.E16}$$

$$12.34 + 2f = 4p + 3r + 2s \tag{6.E17}$$

$$6.34 + f = 2q \tag{6.E18}$$

$$0.87 = r \tag{6.E19}$$

$$0.03 = s \tag{6.E20}$$

Solving above equations, the values of f, p, q, r, and s were found to be

$$f = 6.20$$

$$p = 5.52$$

$$q = 6.27$$

$$r = 0.87$$

$$s = 0.03$$

References

An Y, Yang F, Chua HC, Wong FS, Wu B (2008) The integration of methanogenesis with shortcut nitrification and denitrification in a combined UASB with MBR. Bioresour Technol 99 (9):3714–3720

Augenstein DC, Wise DL, Wentworth RL, Cooney CL (1976) Fuel gas recovery from controlled landfill of municipal wastes. Res Recover Conserv 2:103

Bajpai P (2017) Basics of anaerobic digestion process. In: Anaerobic technology in pulp and paper industry. Springer, Singapore, pp 7–12

Barnett A, Pyle L, Sibramaniam SK (1978) Biogas technology in the third world: a multidisciplinary review. International Development Research Center (IDRC), Ottawa, p 51

Coppinger ER (1979) The operation of a 50,000 gallon anaerobic digester at the Monroe state dairy farm. Ecotope Group, 2332, East Madison/Seattle

Dioha IJ, Ikeme CH, Nafi'u T, Soba NI, Yusuf MBS (2013) Effect of carbon to nitrogen ratio on biogas production. Int Res J Nat Sci 1(3):1–10

Fry LJ, Merrill R (1973) Methane digesters for fuel gas and fertilizer. Newsletter No. 3. New Alchemy Institute, Santa Cruz

Geeta GS, Raghavendra S, Reddy TKR (1986) Increase of biogas production from bovine excreta by addition of various inert materials. Agric Wastes 17(2):153–156

Gore JA (1981) More gas from gobar gas plants. Indian Farming

Hedge JE, Hofreiter BT, Whistler RL (1962) Carbohydrate chemistry. Academic Press, New York, p 17

Igoni AH, Ayotamuno MJ, Eze CL, Ogaji SOT, Probert SD (2008) Designs of anaerobic digesters for producing biogas from municipal solid-waste. Appl Energy 85(6):430–438

Kalia VC (2007) Microbial treatment of domestic and industrial wastes for bioenergy production. Appl Microbiol (e-Book). National Science Digital Library NISCAIR, New Delhi, India. http://nsdl.niscair.res.in/bitstream/123456789/650/1/DomesticWaste.pdf

Langrage B (1979) Biomethane 2: principles – technique utilization, *EDISUD*, La Calade, 13100 Aix-en-Provence, France

Liu, K., Wei, B., Su, Z., Yan, D., & Qin, X. Simulated Test Studying on CN-Containing Cassava Starch Industry Wastewater Treated by Anaerobic Process

Madamwar DB, Mithal BM (1986) Effect of pectin on anaerobic digestion of cattle dung. Biotechnol Bioeng 28(4):624–626

Mahanta P, Dewan A, Saha UK, Kalita P (2004a) Effect of temperature and agitation on the performance of biogas digesters. In: Proceedings of 2nd BSME-ASME international conference on thermal engineering, vol II, Dhaka, pp 871–879

Mahanta P, Dewan A, Saha UK, Kalita P (2004b) Influence of temperature and total solid concentration on the gas production rate of biogas digester. J Energy South Afr 15(4):112–117

Mahanta P, Saha UK, Dewan A, Kalita P, Buragohain B (2005) Biogas digester: a discussion on factors affecting biogas production and field investigation of a novel duplex digester. J Sol Energy Soc India 15(2):1–12

Moharao GJ (1974) Scientific aspects of cow dung digestion. Khadi Gramodyog 20(7):340–347

Moharao GJ (1975) Aspects of night soil digestion, sewage farming and fish culture. A working paper. All India Institute of Hygiene and Public Health

NAS (1977) Methane generation from human, animal and agricultural wastes, report of an AdHoc panel of the Advisory Committee on Technology Innovation. National Academy of Sciences, Washington, DC

Parawira W, Murto M, Read JS, Mattiasson B (2004) Volatile fatty acid production during anaerobic mesophilic digestion of solid potato waste. J Chem Technol Biotechnol 79 (7):673–677

Pellet PL, Young VR (1980) Nutritional evaluation of protein foods. In: Food and nutrition, vol 4. United Nations University

Prasad CR (1985) Utilisation of organic wastes in biogas plant. Khadi Gramodyog:514–518

Report No. ETSU B 1118 (1986) Research into the development of prototype units for the production of biogas methane from farm wastes and energy crops. Department of Microbiology, University College, Cardiff

Saim NA, Dean JR, Abdullah MP, Zakaria Z (1997) Extraction of polycyclic aromatic hydrocarbons from contaminated soil using Soxhlet extraction, pressurised and atmospheric microwave-assisted extraction, supercritical fluid extraction and accelerated solvent extraction. J Chromatogr A 791(1–2):361–366

Singh RB (1974) Biogas plant: generation of methane from organic wastes. Gobar gas research station, Ajitmal/Etawah, p 33

Sivamani S, Chandrasekaran AP, Balajii M, Shanmugaprakash M, Hosseini-Bandegharaei A, Baskar R (2018) Evaluation of the potential of cassava-based residues for biofuels production. Rev Environ Sci Biotechnol:1–18

Smith PH, Frank JR, Smith WH, Strub A, Charter P, Schleser G (1982) Biomass feedstocks for methane production. In: Proceedings of 2nd E.C. conference. Applied Science Publishers, New York, pp 122–126

SPOBD (1979) Biogas technology and utilization. Chengdu seminar. Sichuan Provincial Office of Biogas Development, Sichuan, People's Republic of China

Tchobanoglous G, Burton FL, Stensel HD (1991) Wastewater engineering. Management 7:1–4

TERI (1987) Fixed dome biogas plants: a design, construction and operation manual. TERI, New Delhi, p 3

UN Guidebook on Biogas Development (1980) United Nations, New York, p 99

Weiland P (2010) Biogas production: current state and perspectives. Appl Microbiol Biotechnol 85 (4):849–860

Wise DL (1987) Global bioconversions, vol IV. CRC Press, Boca Raton, pp 178–189

Ziemiński K, Frąc M (2012) Methane fermentation process as anaerobic digestion of biomass: Transformations, stages and microorganisms. Afr J Biotechnol 11(18):4127–4139

Bioethanol Production: Generation-Based Comparative Status Measurements

7

Bikash Kumar, Nisha Bhardwaj, Komal Agrawal, and Pradeep Verma

Abstract

Bioethanol is a major renewable biofuel obtained from different waste biomass. It can potentially substitute the depleting and pollution-causing fossil fuels. It can endow with energy security along with environmental protection over fossil fuels. Biofuels can be classified into four different generations (G), i.e., first generation (1G), second generation (2G), third generation (3G), and fourth generation (4G) based on the groups of feedstocks used. Bioethanol can be produced from all groups of feedstocks; therefore, ethanol obtained from respective group can be named after that generation, i.e., 1G-, 2G-, 3G-, and 4G-based bioethanol. Different microorganisms which can efficiently convert waste biomass into bioethanol are studied, and several biotechnological techniques have been applied for enhancing the production. Similarly, different pretreatment technologies, fermentation processes, and experimental design have been implemented for maximally utilizing the waste and converting it to bioethanol. There are several factors which affect various steps of bioethanol production which affect the final ethanol yield. Therefore, this chapter gives an insight onto current status measurements of 1G, 2G, 3G, and 4G bioethanol production with a focus on using different feedstock and associated technologies, role of microorganisms, factors affecting overall bioethanol production, and current global scenario along with limitations and future prospects.

Keywords

Bioethanol · Biomass · Status measurements

B. Kumar · N. Bhardwaj · K. Agrawal · P. Verma (✉)
Bioprocess and Bioenergy Laboratory, Department of Microbiology, Central University of Rajasthan, Ajmer, Rajasthan, India
e-mail: pradeepverma@curaj.ac.in

N. Srivastava et al. (eds.), *Biofuel Production Technologies: Critical Analysis for Sustainability*, Clean Energy Production Technologies,
https://doi.org/10.1007/978-981-13-8637-4_7

List of Abbreviation

1G	first generation
2G	second generation
3G	third generation
4G	fourth generation
SPR	sweet potato residues
VHG	very high gravity fermentation
SHF	separate hydrolysis and fermentation
SHCF	separate hydrolysis and co-fermentation
SSF	simultaneous saccharification and fermentation

7.1 Introduction

Presently environmental and energy security concern are synergistically associated two important factors controlling the sustainable growth of the human being all over the world. With rapid industrialization, the dependence on fossil fuels has increased. These fossil fuels helped in meeting the global energy need, but at the same time, it has been contributing to pollution and global warming through greenhouse gas (GHG) emissions (Vanhala et al. 2016). In addition to this, the fossil fuels (nonrenewable) reservoirs are also depleting, and at the current rate of consumption, it's expected to end by the next 40–50 years (Chen et al. 2016). With these concerns in mind, different stakeholders working in the area of energy generation and environmental pollution have suggested different approaches such as the use of renewable resources, i.e., solar, wind, tidal, and biomass-based energy generation approach. Among these methods, solar, wind, and tidal are being exploited; however, huge cost and land involved for these techniques and also changing environmental patterns have limited their application. Therefore, biomass-based bioenergy especially bioethanol is a cost-effective and eco-friendly alternative to fossil fuel (Balat 2009; Demirbas 2009). In an attempt to minimize the dependence on fossil fuel, several developed countries, e.g., the USA, China, Brazil, and Canada, and several EU nations have made substantial progress in bioethanol production. Developing nations such as India have also made a substantial stride in the area of bioethanol generation. Although gasoline gives 44% better energy yield as compared to ethanol, it has certain disadvantageous properties as mentioned in Table 7.1.

There are different feedstocks used for bioethanol production, and its composition determines the overall yield and process cost involved. The other aspects associated with the feedstock are its

availability, transportation, and their efficiency as alternatives to fossil fuels. Based on the above parameters, various literature surveys suggest that a wide range of biomass have been used by different research groups. Biofuels can be divided into different generations, i.e., first, second, third, and fourth generations on

Table 7.1 Distinguishing features of ethanol and gasoline with advantage of ethanol over gasoline

Properties	Ethanol	Gasoline	Advantage of ethanol over gasoline	References
Octane number	High	Low	High octane level thus lowers engine knock	Nigam and Singh (2011)
Evaporation enthalpy	High	Low	Improved performance of ethanol-blended gasoline	Al-Hasan (2003); Hara and Tanoue 2006; Naik et al. (2010)
Laminar flame speed	High	Low	Improved performance of ethanol-blended gasoline	Bayraktar (2005); Hara and Tanoue 2006; Naik et al. (2010)
Heat of vaporization (HV)	High	Low	High (HV) lead to better volumetric efficiency; this improves power output	Lynd (1996)
Oxygenation	34.7 % oxygen	No oxygen	High oxygenation minimizes nitrogen oxides and particulates matter emission	Kar and Deveci (2006)
Sulfur content	Negligible sulfur	High sulfur content	Low sulfur inhibits emission of carcinogenic sulfur dioxide and minimizes event of acid rain	Pickett et al. (2008)
Octane enhancer	Bioethanol can be an alternative to chemical-based octane enhancer	Use of octane enhancer methyl tertiary butyl ether (MTBE)	Bioethanol blending with gasoline acts as octane enhancer and thus limits the use of MTBE (US energy policy 2001)	Yao et al. (2009)

the basis of the type of biomass used (Fig. 7.1). Bioethanol is a major class of biofuel which can be obtained from all four generations of feedstocks. Thus, the bioethanol obtained from food crops (grains and sugarcane) is called as first-generation (1G) bioethanol. The lignocellulosic biomass (forest and agricultural residues)-based ethanol is called second-generation (2G) bioethanol, and algal-based ethanol is considered as third-generation (3G) bioethanol (Thatoi et al. 2016). One of the recently developed approaches is the fourth generation (4G) which is basically based on the capture of free CO_2 from the environment and its storage in biomass materials using a process such as oxy-fuel combustion. Different feedstocks vary in their compositional analysis which greatly influences the type of process adapted for ethanol generation from the feedstocks. Composition of different renewable feedstocks for different generations of bioethanol is explained through Fig. 7.1. Researchers are making every possible attempt for economical production of

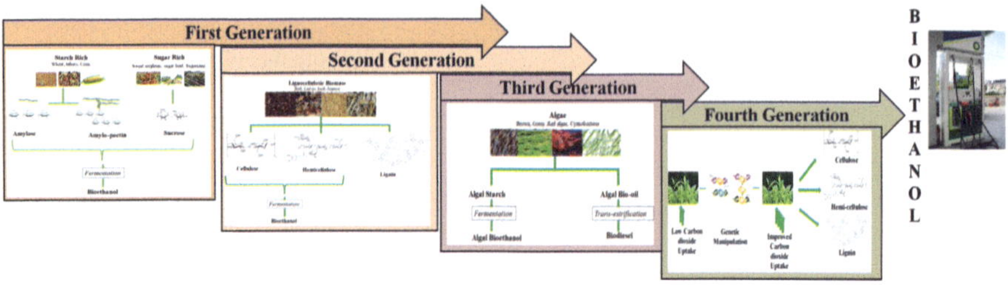

Fig. 7.1 Schematic representation of the different generations of renewable feedstocks, their compositions, and approach used for bioethanol production

bioethanol at pilot scale or large scale. Bioethanol production utilizing 1G and 2G feedstocks has been developed. However, 1G feedstocks compete with food crops and thus can lead to food crisis. The lignocellulosic biomass recalcitrance acts as a hindrance for making the overall lignocellulosic-based biorefinery an economical option. 3G and 4G bioethanols are still in a very immature stage limited to lab scale. The chapter will provide a comprehensive overlook on generation-based status measurement with special emphasis on different feedstocks and associated technologies (pretreatment, hydrolysis, fermentation, and distillation) which are playing a significant role on the overall development of bioethanol production. The global scenario of four generations of ethanol is also discussed in short. Thus, this chapter can provide a thoughtful insight toward future commercial bioethanol production strategies on large scale.

7.2 Renewable Feedstocks for Different Generations of Bioethanol

7.2.1 First-Generation (1G) Feedstocks

Food crops (grains, sugarcane, tubers, and roots) are used as feedstock for the production of 1G bioethanol. 1G feedstock-based bioethanol production technology is commercially utilized to minimize the global environmental pollution and provide energy security. Different food crops available for generation of bioethanol are discussed below.

7.2.1.1 Starch-Based Biomass for Bioethanol Production

Feedstock rich in starch is readily available around the world, stored for a long period of time and easily converted to bioethanol with high yield. The starch-based major feedstocks are cereals (60–80% starch), green/immature fruits (70% starch), legumes (25–50% starch), and tubers and roots (60–90% starch) (Santana and Meireles 2014). Corn, tuber (cassava, potato, sweet potato), wheat, yam, sorghum grains, and aroids are used extensively as sources for bioethanol (Mohapatra et al. 2019) Different starch and sugar crops with their major carbohydrate composition and final ethanol yield are also tabulated in Table 7.2.

Table 7.2 1G and 2G biomass and carbohydrate composition along with their ethanol production potential

Generation	Biomass		Carbohydrate composition	Ethanol yield		Ethanol production potential (million liters)	References
	Type	Name of biomass		Theoretical (g/L)	Practical (g/L)		
1G	Starch based	Sweet potato	10–20%	38-45	89	270.23	Lareo et al. (2013)
		Cassava	20–25%	42.4	106	330.76	Rajandran (2013)
		Potato	15–20%	34.9	42.30	290.65	Lareo et al. (2013)
	Sugar based	Sugar beet	25–30%	39.40	45.72	251.30	Salazar-Ordóñez et al. (2013)
		Sugarcane	12–16%	90	89.73	402.30	Liang et al. (2008); Cardona et al. (2010)
2G	Agricultural residues	Bagasse	20–30%	36.5	41.06	290	Maryana et al. (2014)
		Corn stover	30–45%	112.7[a]	64.4%	302.83	(Humbird and Aden (2009); Liew et al. (2014))
	Grasses	Switch grass	30–45%	46.5	54.06	32,915[b]	Keshwani and Cheng (2009)
		Elephant grass	22–24%	23.4	36.4	23,700[b]	Aiyejagbara et al. (2016)
		Moroccan grass	27–38%	17.62	23.11	6762[b]	Semhaoui et al. (2017)

[a]gal/ton
[b]L/ha/year

7.2.1.1.1 Corn for Bioethanol Production

Commercially, corn is extensively used as feedstock for bioethanol production, and its application has increased rapidly. The USA is the leading corn producer with annual production of 116.32 billion gallons and utilizes 28% of it, i.e., 42.4 billion gallons for ethanol generation (AFDC 2016). In year 2014, the USA alone generated around 14.3 billion gallons of bioethanol from corn (RFA 2015). The corn yield is also very high in Asia followed by Europe and South America (Kim and Dale 2004). The factors affecting the corn-based ethanol yield are corn variety, quality of corn used, kernel composition, presence or absence of mycotoxins, and hardness of endosperms. The ethanol yield ranges from 3 to 23% with more yield from free sugar-rich kernels (Singh 2012).

7.2.1.1.2 Grains for Bioethanol Production

The climatic condition of an area also limits the application of different crops for bioethanol production. The climatic condition of Canada limits corn growth and favors wheat; therefore, it is most readily used crop for grain-based bioethanol production. The other crops used for bioethanol production were triticale, barley, and oat (Saunders et al. 2011). McLeod et al. (2010) tested the suitability of grains (i.e., 31 lines and cultivars of wheat, triticale, barley, and oats) as feedstock for production of ethanol in Western Canada. Starch, pentoses, and β-glucan yield from different cultivars were determined to estimate ethanol yields. It was observed that the ethanol yield was maximum for wheat followed by triticale, barley, and oat. Muktham et al. (2016) suggested that wheat can be used to replace barley for bioethanol production. Belboom et al. (2015) also suggested that wheat-based ethanol can generate 42.5–61.2% less GHG emissions as compared when the same amount of gasoline is used.

7.2.1.1.3 Tubers and Roots for Bioethanol Production

Different tubers and roots, i.e., sweet potatoes, potatoes, Jerusalem artichoke, and cassava, have high concentrations of stored starch. This high starch storage potential makes it as suitable raw materials for 1G bioethanol (Hoover 2001). There are several advantages using tubers and roots for ethanol production such as:

(i) Economic harvesting.
(ii) On-farm processing.
(iii) Availability of cost-effective ethanol conversion techniques.
(iv) Can be grown in different soil types, even in low fertile soil thus leaving fertile one for other crops (Thatoi et al. 2016).
(v) Can be grown in a variety of climates (tropical, subtropical, semi-arid conditions).
(vi) Annual nature of most tubers and root crops makes it a suitable feedstock for bioethanol (Ray and Swain 2011).

The annual production of cassava ranks sixth after rice, wheat, corn, potatoes, and barley in the developing countries. Its ability to grow in low fertile soil, in all types of climatic condition, all year around, and high starch yield per hectare (~36.3 tons) make it suitable for bioethanol feedstock (Behera and Ray 2015). Potato is a seasonal crop grown primarily in a temperate climate with abundant availability in northern hemisphere region. Potato is ranked as fourth important crop around the world. China is the leading producer of potatoes (20%); Russia and India follow the list with 12% and 8%, respectively (Thatoi et al. 2016). Sweet potato is an annual crop grown in tropical and subtropical regions. China is the leading producer of sweet potatoes (80%) followed by Uganda and Nigeria (3%) (Duvernay et al. 2013). Sweet potato is ranked seventh as favorite food crop grown around the world. Different varieties of sweet potatoes are grown such as white, orange, and purple. The white variety of sweet potato containing high starch is less sweet, so it is less preferred as food source, thus making it an appropriate substrate for production of bioethanol (Thatoi et al. 2016). Sweet potatoes are even manipulated for its high starch content and low liquefaction temperature, thus making it more adaptive for ethanol production. Sweet potatoes K 9807.1 has 4500–6500 L ethanol yield per hectare which is 1.6 times higher than corn (Lareo et al. 2013).

7.2.1.2 Sugar-Containing Feedstock for Bioethanol Production

The sugar-based bioethanol is produced by sugar-based feedstocks, i.e., sugarcane, beet sugar, sweet sorghum, fruits, and energy crops. Brazil, Germany, France, and India are the leading producers of sugar-rich feedstock. The sugarcane (C4 grass) has high solar radiation to biomass conversion capability with per hectare yield of 62–74 tons (Sindhu et al. 2016). Brazil is the leading producer of sugarcane-based bioethanol (79% from sugarcane) and replaced almost 42 % of its gasoline need with sugarcane-based bioethanol. In 2015–2016, Brazilian produced 8 billion gallons bioethanol and used mostly to meet its domestic needs by blending with gasoline from 18 to 27.5% (SugarCane.org 2019). Another C4 plant i.e. sweet sorghum, has high yield (54–111 tons/ha), high carbon assimilation efficiency (50 $g/m^2/day$), accumulation of extractable sugar in high concentration in their stalks, have high water use efficieny, can be grown in tropical as well as temperate climate and can be grown from seed (added advantage over sugarcane) (Laopaiboon et al. 2007). Ekefre et al. (2017) evaluated three different cultivars of sweet sorghum as ethanol feedstock. Different parameters such as diameter, density, and height of stalk along with overall biomass and juice yield were compared for enhanced ethanol yield. The cultivar Theis resulted in maximum ethanol yield of 7619 L/ha followed by M81 E (6106 L/ha) and Dale (5077 L/ha) (Ekefre et al. 2017). Sugar beet is used for bioethanol production extensively in France and introduced in India and in trail stages in tropical countries (Marx et al. 2012). Sugar beet has a high yield of 54–111 tons/ha with low requirement of rainfall (Vohra et al. 2014). Several researches are ongoing for application of sugar beet and sugar beet processing waste for ethanol generation and found that sugar beet can be an economical option (Dziugan et al. 2013).

The fruits which are usually discarded due to its physical appearance and low quality can be used as feedstocks for production of ethanol as these fruits are rich in soluble sugar and do not require any complex pretreatment before fermentation by yeast (Chandrasekaran and Bahkali 2013; Chniti et al. 2014). Fruit juice or fruit processing industries or waste from pineapple and grape pomace can be a feasible option for production of bioethanol (Ban-Koffi and Han 1990; Rodriguez et al. 2010; Upadhyay et al. 2010).

7.2.1.3 Limitations of 1G Feedstocks

The 1G bioethanol generated from food crops is the most developed technology providing relief in GHG emission; however, it has its limitation which arises due to concern associated with food security and availability of cultivable land and can also lead to breakage of food chain due to non-availability and soaring prices of food crops. This claim may be supported by the work done by Shikida et al. (2014) where they concluded an interesting point that the food crops used by the USA, Brazil, China, India, and the Netherlands for bioethanol generation can feed up to 200 million people starving in their own nation. Similarly, Rulli et al. (2016) evaluated that the amount of food crops used for production of 1 Terajoule of 1G bioethanol can be used to feed 110 people. The growth of food crops for bioethanol production can compete with 3% global water requirement for food production (Agência Nacional doPetróleo 2015). Therefore, based on the above observation, there is an urgent need to revisit the potential feedstocks for future ethanol generation.

7.2.2 Second-Generation (2G) Feedstock for Bioethanol Production

The food versus fuel, food-fuel-land-water nexus, and environmental impacts of large-scale 1G bioethanol have led to a search for an alternative. 2G feedstock-based bioethanol from nonfood lignocellulosic feedstocks are considered as a feasible option because lignocellulosic biomass is abundant all over the world which can be used without competing with land, water, and food requirement of animal kind. Annually, global production of plant biomass is around 200×10^9 ton per year, of which around 8×10^9–20×10^9 can be used for generation of biofuel such as bioethanol, biogas, and bioelectricity (Kuhad and Singh 1993; Saini et al. 2015). It can even mitigate the problem of pollution which is usually generated due to burning of these lignocellulosic wastes. Overall lignocellulosic feedstock-based biorefineries are the need of the hour for both rural and urban areas as it can provide energy security, mitigate environmental concern, promote agriculture, provide employment opportunities, and save foreign exchange (earlier used to procure petroleum) and have large-scale socioeconomic impact (Zafar 2018). Different lignocellulosic biomass used for 2G bioethanol production can be divided into several groups such as woody biomass, i.e., forest residues, or non-woody biomass, i.e., agricultural residues, energy crops, aquatic plants, and municipal waste. These various biomasses are discussed below.

7.2.2.1 Woody Biomass for Bioethanol Production

Forest biomass can be collectively considered as woody biomass which mainly includes hardwood and softwoods along with forest residues such as dead leaves, dead branches, sawdust, woodchips, and pruning and bark thinning residues. The USA has forest cover of around 310 million hectares and generates around 370 million tons of woody biomass per year (Robert et al. 2005). Softwood trees are the evergreen plants of cedar, cypress, hemlock, pine, reed, and cedar, and these are fast-growing trees and possess low densities. Hardwood trees include acacia, alder, aspen, beech, cottonwood, eucalyptus, poplar, oak, and willow and are mainly found in the northern hemisphere. The hardwoods and softwoods differ in physical and chemical composition (Romani et al. 2011). The woody biomass consists of cellulose microfibrils reinforced together with hemicellulose and lignin (Alvira et al. 2010). These complex structures are usually formed by tracheids and vessels (plant parts responsible for translocation and transportation of water in plants) in the middle covered by layers of complex microfibrils around them. These structural complexities provide toughness to wood and require different pretreatment methods before its conversion to bioethanol (Zhu et al. 2010).

7.2.2.2 Non-woody Biomass for Bioethanol Production

Non-woody biomass usually comprises of agricultural residues, aquatic plants, native energy crops, and municipal waste. Apart from huge availability, non-woody biomass is relatively easy, cheap, and less energy intensive in conversion to bioethanol as compared to woody biomass. The detailed account of non-woody biomass is given below.

7.2.2.2.1 Agricultural Residues for Bioethanol Production

After harvesting of the agricultural crops such as corn, wheat, rice, sugarcane, cassava, and palm oil, there are huge amount of residues generated such as straws, stover, bagasse, and empty fruit brunches which can contribute to the world biomass and potential feedstock for bioethanol production.

7.2.2.2.1.1 Cereal Straws for Bioethanol Production

The dry stalks which remain after harvesting of cereal grains such as rice, wheat, barley, oat, corn, and sorghum are called as cereal straws. It consists of 33–47% cellulose along with 20–30% hemicelluloses and can be considered as an important substrate for bioethanol generation. The harvesting of wheat from 1 hectare of field usually generates 1–3 tons of straw and the annual world rice straw production is around 731 million ton/year, and as wheat and rice are stapled food for several countries, there is huge availability of wheat/rice straw for bioethanol production (Saini et al. 2015). Annual cereal straw production collectively from Europe, the USA, India, and China is 1580.2 million ton/year (Tye et al. 2016).

7.2.2.2.1.2 Corn Stover for Bioethanol Production

The residues generated after harvesting of corn from corn plants are collectively called as corn stover. The corn stover mainly constitutes cobs, husks, leaves, and stalks and produced at a rate of around 4 ton/ acre of harvested land (Kim and Dale 2004). Corn stover is mainly produced in North and South American nations. However, it plays a crucial role in restocking organic matter in soil, but with sustainable management and safeguards, a suitable amount of corn stover can lead to production of 80 million gallons of bioethanol (Liew et al. 2014).

7.2.2.2.1.3 Bagasse for Bioethanol Production

Bagasses are obtained after processing of the starchy tuber crop "cassava and sorghum" and sugar-rich crop "sugarcane" cultivated mainly in Asian, Latin American, and African countries. India is home to almost 1100 cassava processing units which generate a huge amount of bagasse after processing of approximate 8.74 million tons of cassava (Mohapatra et al. 2019). These bagasses are also rich in carbohydrate polymer, i.e., 30–35% (Ray and Swain 2011). It is estimated that 1 ton of cassava bagasse can lead to the generation of 114 L of bioethanol (Sangodoyin and Amori 2013). The fifth most important cereal crop grown in the world is sorghum. A by-product obtained after sorghum processing, i.e., sorghum bagasse, consists of high amount of carbohydrate polymer (35.5% hemicellulose, 44.4 % cellulose) and very low lignin (3.5%) and thus can be converted to ethanol very easily due to low lignin recalcitrance (Dogaris et al. 2009).

The leftover residue after extraction of juice from sugarcane is known as sugarcane bagasse; it is a cheap residue and has an estimated production of 317–380 million ton/year (Sánchez 2009). It has high carbohydrate content of ~70–80% (cellulose and hemicelluloses) and ~20–30% lignin with low 1.9% ash content (Peng et al. 2009). This bagasse may be immediately available at their processing site where crops are processed. The integrated biorefinery where processing of crop followed by its conversion to bioethanol is a cost-efficient method (Furlan et al. 2013).

7.2.2.2.1.4 Sweet Potato Residues (SPRs) for Bioethanol Production

As discussed in starch-based biomass, sweet potato accounts for 10% residues after extraction of starch and can be used as bioethanol generation substrate. China is the leading sweet potato producer with an annual production of ~71 million tons. The processing units for sweet potato in China produce around 2 million tons of sweet potato residues which because of their high viscosity remain unutilized and can act as environmental pollutants. By enzymatic breakdown, these residues can be substantially used for sugar and ethanol generation (Izmirlioglu and Demirci 2012).

7.2.2.2.1.5 Oil-Palm Biomass for Bioethanol Production

Once the oil is extracted from the palm fruit bunches, a large amount of empty fruit bunches, fronds, and trunks are left out. On per hectare basis, 40 tons of empty fruit brunch, 10.5 tons of fronds, and 2.8 tons of trunks are generated per year; these residues can act as substrates for bioethanol production. The high cellulosic content of fronds (31.0–32.0%) and trunks (39.9–41.0%) makes it more suitable for cellulosic ethanol generation. However, even the cellulose content in empty fruit bunches are less, i.e., 7.7–14.7%, but applying enzymes such as amylase and cellulase can result in 96.3% glucose yield with ethanol yield of approximately 93.5% (Eom et al. 2015)

7.2.2.2.1.6 Bast Fiber for Bioethanol Production

Bast crops are naturally occurring non-woody plant and can be potentially considered as energy crop due to several advantages such as it's capability of growing on waste or even brackish water, high CO_2 absorption capacity (1.9 tons of CO_2 for 1 ton cellulose produced), high cellulose content and high productivity even in less fertile condition. The outer layer of different bast crops such as sun hemp, jute, ramie, flax, and industrial hemp constitutes around one third of their weight and can be obtained as fibrous bundles that can be used as suitable feedstocks for production of bioethanol. Among different bast crops, industrial hemps are considered as the most sustainable substrates for production of bioethanol (Cherney and Small 2016). Annual production of hemp is 1×10^5 tons, and it has very high 70–90% cellulosic content by weight. Similarly, the cellulosic content for different bast crops is also very high for flax, ramie, and jute, i.e., 60–80%, 68–76%, and 51–84%, respectively (Paridah et al. 2011).

7.2.2.2.2 Energy Crops for Bioethanol Production

7.2.2.2.2.7 Native Grasses for Bioethanol Production

In order to maintain low-cost ethanol production, there is a need of an uninterrupted and consistent supply of raw materials. The native grasses are one alternative which has a short growth period, minimal cultivable land, fertilizers/pesticides, and water requirement. These grasses also have huge carbon storage capacity due to its C4 carbon fixation ability. These plants are perennial in nature, grown mostly in warm and temperate regions, and generate large biomass by huge carbon capture around the year (Lewandowski et al. 2003). Thus, these native grasses have all inherent properties to be considered as energy crops. C4 grass such as coastal bermuda grass, napier grass, saw grass, and switch grass are potential substrates for bioethanol production.

Miscanthus giganteus (saw grass) has carbohydrate content of 40–60% cellulose and 20–40% hemicelluloses, making it capable of generating five to eightfold more ethanol as compared to corn (Brosse et al. 2012). Similarly, *Panicum virgatum* (switch grass) has lignocellulosic content of 37–40% cellulose, 25–29% hemicelluloses, and 18–25% lignin. *Pennisetum purpureum* (napier grass) has carbohydrate content, i.e., 40–50% cellulose and 20–40% hemicelluloses. It produces huge biomass under limited nitrogen supply (presence of diazotrophic nitrogen-fixing bacteria) and highly efficient CO_2 fixation and has fast-growing capability. *Cynodon dactylon* (bermuda grass) has high carbohydrate content, i.e., 40–55% of cellulose and hemicellulose accompanied with a high yield of 14.1–24.2 ton/ha (Takara and Khanal 2015). These C4 plants have high carbohydrate content and fast-growing capability, making it a suitable substrate for ethanol generation.

Some of the C3 grasses such as *Medicago sativa* (alfalfa), *Phalaris arundinacea* (reed canary grass), *Arundo donax* (giant reed), and *Dactylis glomerata* (cocksfoot grass) are also reported to have high hemicellulosic content and thus used as bioethanol-generating substrate. Njoku et al. (2013) utilized hemicellulose fraction of cocksfoot grass for generation of ethanol with yield 89–158 mL/kg of dry biomass.

7.2.2.2.3 Municipal Solid Wastes

Due to rapid industrialization, there is huge increase in solid wastes (1.3×10^9 ton in 1990 to 2.3×10^9 ton in 2000) generated from different residential and nonresidential establishments. These are usually recyclable biomass generated from food wastes and paper mill sludge (Hadar 2013). However, the application of these wastes is limited because of differences in composition and microbial contamination and limited potential in small regions. These are usually used for bio-oil production using pyrolysis and solid waste management's approaches for generation of value-added organic products.

7.2.3 Third-Generation Feedstock for Bioethanol Production

The algae are simple chlorophyll-containing photosynthetic organisms. These are either phototrophic, i.e., utilizing atmospheric CO_2 to nutrients such as carbohydrate, or heterotrophic, i.e., utilizing organic carbon sources (Wen and Chen 2003). Algal biomass is considered as an alternative to 1G and 2G feedstock due to high productivity, easy cultivation techniques, can use waste water for cultivation, and convenient harvesting. The algal biomass serves three major purposes, i.e., bioethanol production (algal polysaccharides), biodiesel production (algal bio-oils), and simultaneous waste water treatment. Microalgae and macroalgae are two major groups of algae and have huge potential for bioethanol production.

7.2.3.1 Microalgae as Feedstock for Bioethanol Production

Microalgae consist of unicellular prokaryotic or eukaryotic photosynthetic microorganisms. They have simple colony structures and are capable of surviving under stressed condition (Mata et al. 2010). The total global production of dry algal biomass is 10,000 ton/year, of which around 7000 ton/year is produced in open systems (Gris et al. 2013; Lee and Lee 2016). Different species of microalgae used for production of bioethanol are *Chlorella*, *Spirulina*, and *Dunaliella*. The major polysaccharide yields are arabinose, galactose, glucose, rhamnose, and xylose with bioethanol yield potential of 0.234 g/g dry algal biomass having high 11.7 g/L titer (Ho et al. 2013). Microalgae can also be used for generation of bio-butanol, acetone, biogas (Marin et al. 2018), eicosapentaenoic acid (Cheng-Wu et al. 2001), omega-3 oil, livestock feed (Besada et al. 2009), pharmaceuticals, and cosmetics (Spolaore et al. 2006). These sub-products are of high value and thus can facilitate in minimizing the cost of the overall bioethanol production process (Demirbas 2011). The chemical composition of the microalgae is affected extensively by the cultivation type and cultivation condition (Burton et al. 2009).

7.2.3.2 Macroalgae as Feedstock for Bioethanol Production

Seaweed or macroalgae have in-habituated to marine form during the course of evolution. They are broadly classified as Rhodophyceae (red algae), Phaeophyceae (brown algae), and Chlorophyceae (green algae) based on the type of pigments they produce (Jung et al. 2013). Major species of red algae are *Eucheuma* sp. *Eucheuma denticulatum*, *Gracilaria verrucosa*, and *Kappaphycus alvarezii*, with very high annual production of 8.98 million metric ton/year (Lee and Lee 2016). The major carbohydrate polymers obtained from red algae are carrageenan, apart from agar and cellulose which can be used as bioethanol feedstock. Several reports suggest bioethanol yield of 45–236 mg/g of dry biomass with 0.5–4.72 g/L high titer (Meinita et al. 2013;Wu et al. 2014). The major species of brown algae are *Laminaria japonica*, *Sargassum fusiforme*, and *Undaria pinnatifida* with estimated production of nearly 0.68 million metric ton per year (Lee and Lee 2016). The major carbohydrate polymers for brown algae are cellulose, fucoidan, mannitol, alginate, and laminarin. Different biotransformation procedures are developed for utilization of alginate and mannitol for generation of ethanol. The bioethanol yield obtained from different brown algae varies from 152 mg up to 362 mg per gram of dry biomass with 0.196–37.8 g/L titer (Wargacki et al. 2012; Kim et al. 2013; Enquist-Newman et al. 2014). The major green algae strains used for bioethanol production are *Caulerpa* sp., *Codium fragile*, *Enteromorpha clathrata*, *Monostroma nitidum*, and *Ulva fasciata* with starch and cellulose as major polysaccharide content. Among different macroalgal groups, the annual production of green algae is minimum, i.e., 21.5 thousand metric ton per year (Lee and Lee 2016). The ethanol yield for the green algae is also very low, i.e., 0.09 with comparable titer of 9.31 for *Ulva fasciata* (Trivedi et al. 2013). Due to high photosynthetic efficiency, macroalgae generate huge carbohydrate polymers with no or less lignin and hemicelluloses. Therefore, the carbohydrate polymers are readily available can be subjected to hydrolysis without any prior pretreatment (John et al. 2011). Also, the short growth time, that can be one added advantage for making the large biomass available in short time.

7.2.4 Fourth-Generation Feedstock for Bioethanol Production

Fourth-generation feedstocks are bio-oils, genetically modified microbes, and plants with high carbon capture and sequestration efficiency. These feedstocks are engineered to capture more CO_2 during the growth of the feedstock. The processing of this feedstock also involves processes which can enhance carbon capture which can be stored in geological formations (e.g., exhausted oil fields) or as mineral storage in the form of carbonates. Therefore, fourth-generation feedstock-based bioethanol generation technology is often termed as "bioenergy with carbon storage" or carbon negative techniques (CBU 2007; Rubens 2008).

Fourth-generation feedstocks are used in different approaches such as (i) petroleum-like hydroprocessing and advanced biochemistry approach of conversion of bio-oils to bioethanol, (ii) innovative processes of Joule's "solar-to-fuel" method, (iii) genetic modification of feedstock with the ability to increase carbon capture capacity, and (iv) synthetic biology approach for genomically synthesized microbes (algae and cyanobacteria). Fourth generation defies any other category of biofuel (Kagan 2015; Aro 2016). The schematic representation of different steps involved in fourth-generation feedstock-based bioethanol generation is shown in Fig. 7.2

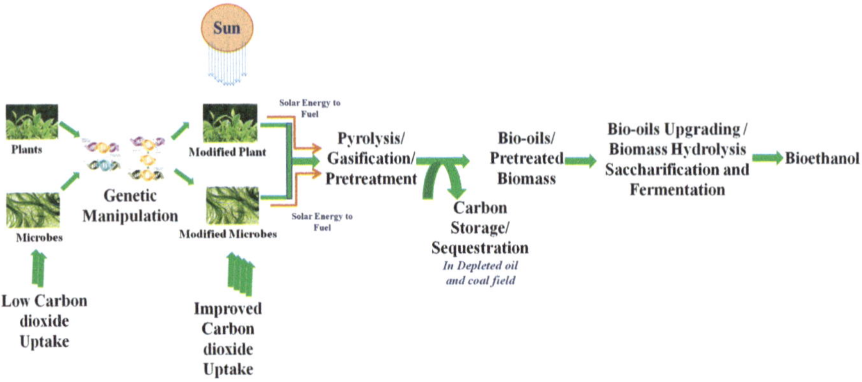

Fig. 7.2 Schematic representation of steps involved in fourth-generation feedstock-based bioethanol generation

However, several limitations such as intensive technology associated with the genetic modification of plants and microbes, the technological limits and ethical concern associated with the synthetically synthesized microbes, and carbon sequestration and capture techniques are in its preliminary stages and therefore minimize the application of this feedstock (Rubens 2008; Aro 2016).

7.3 Process Associated with Bioethanol Production from Different Generations (1G, 2G, 3G, and 4G) Feedstocks

7.3.1 Pretreatment of Different Feedstocks for Bioethanol Production

During biomass to bioethanol generation, one of the rate-limiting processes is biomass pretreatment as it is costly and involves several complexities. The pretreatment alters the macroscopic, microscopic, and submicroscopic structure of lignocellulosic and algal biomass. In the case of lignocellulosic biomass, pretreatment primarily removes lignin and hemicelluloses. However, in the case of 1G and 3G feedstocks, lignin is absent and therefore removal of lignin is not necessary. Complex polymers such as microfibrillar and matrix polysaccharides along with proteoglycans provide recalcitrance to algal biomass (Mishra et al. 2017; Kumari and Singh 2018a). Thus apart from lignin removal, the pretreatment also helps in depolymerizing the starch, cellulose, and hemicellulosic fractions, enhancement in surface area, improving biomass porosity, and decrease in crystallinity of cellulose (Sørensen et al. 2008). However, under harsh pretreatment conditions, formation of fermentation inhibitory compounds such as acetic acid, formic acid, and some furanic compounds can take place (Hargreaves et al. 2013). Therefore, the choice of an optimum pretreatment method is required for enhanced and cost-efficient production of bioethanol by improving the accessibility of

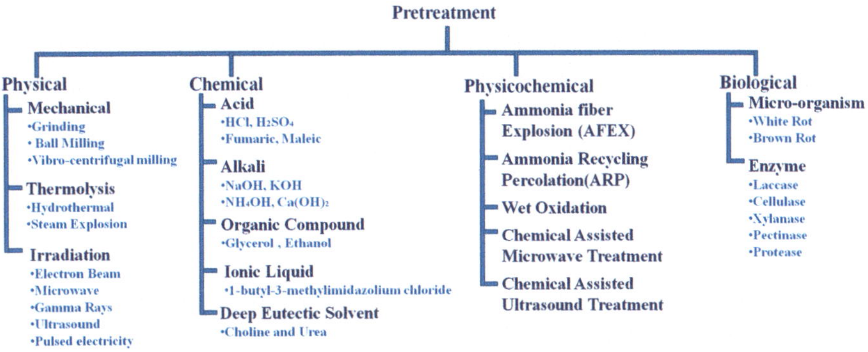

Fig. 7.3 Classification of pretreatment methods used for bioethanol production from different feedstocks

enzymes for carbohydrate hydrolysis and enhancing the rate of fermentation. Therefore, selection of an optimal pretreatment strategy may have the following characteristics: (i) high sugar and carbohydrate content after pretreatment, (ii) improved enzymatic accessibility, (iii) minimum inhibitory generation and its neutralization, (iv) assessing pretreated samples for value-added product generation, and (v) selecting the source of bioethanol (liquid hydrolysate or water-insoluble solids) (Agbor et al. 2011; Lee et al. 2013). Several pretreatment methods are investigated for pretreatment of feedstocks that can be broadly categorized into four groups, viz., physical, chemical, physicochemical, and biological (Fig. 7.3). Physical pretreatment involves breakdown of biomass cellular structure using mechanical, thermolysis, and irradiation-based system. Different mechanical methods are grinding such as ball, hammer, and vibro-centrifugal milling, etc., thermolysis such as hydrothermal and steam explosion, and irradiation-based methods like electron beam, microwave, gamma rays, ultrasound, pulsed electricity, etc., which are extensively reported for efficient lignocellulosic biomass pretreatment with enhanced production of bioethanol (Kumari and Singh 2018b). The chemical pretreatment involves the application of acid (H_2SO_4, HCl, H_3PO_4, and HNO_3), alkali (NaOH, KOH, NH_4OH and $Ca(OH)_2$), organic acid (fumaric and maleic acids), organic compounds (glycerol, ethanol, ethyl glycol), ionic liquids, and deep eutectic solvents (choline, urea), etc. Most widely used physicochemical methods are chemical-assisted steam explosion, microwave and ultrasound treatment, viz., ammonia fiber explosion, wet oxidation, peroxide-assisted microwave treatment (Verma et al. 2011), CO_2 explosion and ammonia recycling percolation (Rabemanolontsoa and Saka 2016), etc. Biological pretreatment of different feedstocks can be performed using different microorganisms such white and brown rot fungi and different enzymes of microbial origin. Among different microorganisms, selective lignin-degrading microorganisms such as white rot fungi are used for improved delignification (Sarkar et al. 2012). Other biological method is use of different enzymes such as cellulase, amylase, and amyloglucosidase for hydrolysis of polysaccharides and proteases for the hydrolysis of glycoproteins present in cell wall of lignocellulosic and algal biomass (Pirwitz et al. 2016). Each method has several advantages and disadvantages; thus while choosing an ethanol generation process from a particular biomass, these measures must be evaluated for maximum sugar/ethanol yield and minimum cost. Different pretreatment methods used for generation of ethanol from 1G, 2G, and 3G are tabulated in Table 7.3 (1G and 2G feedstock) and Table 7.4 (3G feedstock) and are explained as mentioned below

7.3.2 Hydrolysis of Carbohydrate Polymer to Fermentable Sugar for Bioethanol Production

The hydrolysis is one of the primary requirements of converting the polymeric carbohydrates (starch, cellulose, and hemicelluloses) into fermentable sugar which can be converted to ethanol by fermentation. However, ethanologenic organism such as yeast cannot utilize these carbohydrate polymers in their native state (Liu et al. 2016). Therefore, there is a need to find a method to hydrolyze the polymeric structure to simpler form. Different acid- or enzyme-based methods are suggested for the breakdown of starch, hemicelluloses, and cellulose.

7.3.2.1 Hydrolysis of Starch to Fermentable Sugar for Bioethanol Production

Starch is the major component in the food crops and in some algal biomass used as feedstock for bioethanol generation. Chemical (acid)-based hydrolysis and physical and enzymatic methods are usually used for breaking down the starch into soluble sugars. During starch hydrolysis, the major constituents of starch, i.e., amylose and amylopectin, are broken down by biological or chemical agent to generate soluble sugars such as maltose, glucose, and maltotriose. The acid hydrolysis is an efficient method for conversion of complex polymers to fermentable sugars. Due to severity of the acid-assisted hydrolysis, a need for acid recovery and generation of undesirable inhibitory products (such as 5-hydroxymethylfurfurals (5-HMF), levulinic and formic acid) minimizes the utilization of acid-based hydrolysis (Yang et al. 2011). The inhibitory by-products are toxic for growth of microbial cells (yeast), negatively affecting the overall ethanol generation (Loow et al. 2016). Physical method such as high pressure and temperature-based extrusion method were also tried for the hydrolysis of starch. However, results show that due to catalytic inhibition effect on α-amylase under high pressure, rate of starch to sugar conversion is slow resulting in low fermentable sugar yield (Buckow et al. 2007).

The enzyme-based hydrolysis is the most sorted method for conversion of starch to fermentable sugars due to its biological origin, simple, highly specific and eco-friendly nature, and high efficiency (Duvernay et al. 2013). The thermostable α-amylase and glucoamylase regulate two major steps of starch to fermentable sugar conversion, i.e., liquefaction and saccharification, respectively (Lamsal et al. 2011; Zabed et al. 2016a). Liquefaction is performed at higher temperature (85–165 °C) where "α-amylase," an endoenzyme, cleaves the α-1 $\rightarrow$ 4 glycosidic linkages of starch converting it to shorter chains such as dextrins, maltose, and maltotriose (Pandey et al. 2000). The thermophilic *Bacillus licheniformis* and recombinant *Escherichia coli* are usually used for production of heat-stable α-amylase (Sanchez and Cardona 2008; Rakin et al. 2009). The saccharification of this liquefied starch to glucose using enzyme glucoamylase is the second step after liquefaction and usually carried out at relatively lower temperature (30–50 °C) (Plumier et al. 2015). The major source of the glucoamylase is *Aspergillus niger* or *Rhizopus* species (Rakin et al. 2009)

7.3.2.2 Hydrolysis of Holocellulose to Fermentable Sugar for Bioethanol Production

The pretreatment of biomass usually results in water-insoluble solid and liquid fraction. Based on the compositional analysis of the water-soluble and -insoluble fractions, these fractions are subjected to different hydrolysis methods to depolymerize celluloses and hemicelluloses completely to fermentable sugar (Girio et al. 2010). The acid-assisted hydrolysis is performed using dilute and concentrated acid. The major acids used for this method are sulfuric acid certain inorganic acids such as HCl, HNO_3 or H_3PO_4, and CF_3COOH (Girio et al. 2010). The dilute acid hydrolysis is performed at high temperature for short duration, whereas the concentrated acid hydrolysis performed at low temperature for long reaction time (Chandel et al. 2007). The hydrolysis of hemicelluloses using dilute acid can be usually performed at low temperature as depolymerization of hemicelluloses is maximum at low temperature, whereas the conversion of cellulose to glucose is usually performed at high temperature (230–240 °C). For dilute acid treatment, the acid concentration usually ranges from 0.5% to 1.5% (Balat 2011). The acid concentration for concentrated acid hydrolysis usually ranges between 41% and 100% (Fengel and Wegener 2011) with longer residence time and moderate process temperature as compared to dilute acid hydrolysis (Zhang et al. 2007). The limitation associated with acid-based hydrolysis is a requirement of specialized non-corrosive equipments, requirement for recovery or neutralization of acid, and inhibitory product generation which is toxic for subsequent fermentation stage. Several advantages are associated with the enzymatic hydrolysis such as low temperature requirement (45–50 °C), no corrosion of equipment, and no or less sugar degradation and inhibitory products (Duff and Murray 1996). Therefore, the cellulose and hemicelluloses can be hydrolyzed using cellulase and hemicellulase enzyme, respectively. Subjecting the *C. vulgaris* and *Scenedesmus* sp. to commercial enzymatic cocktails (Viscozyme, Celluclast, and Pectinex) resulted in 84% total sugar (Mahdy et al. 2014).

7.3.3 Ethanol Fermentation Using Different Feedstocks

Alcoholic fermentation can be defined as the biochemical conversion of fermentable sugars to alcohol in the presence of microorganisms (Balat and Balat 2009). A model reaction for representing the ethanol generation is conversion of one molecule of glucose to form two molecules of alcohol and two molecules of carbon dioxide. Theoretical ethanol yield from conversion of 1 kg of glucose is around 0.51 kg, and the remaining 0.49 kg is CO_2. However, the actual yield is comparatively low as microbes require glucose to meet their nutrient need (Demirbas 2005). *Saccharomyces cerevisiae* is the most widely used organism apart from some bacteria, i.e., *Zymomonas mobilis* and recombinant *E. coli*. The properties of ethanologenic microbes are discussed later under different headings.

7.3.3.1 Solid-State Fermentation for Bioethanol Production

During solid-state fermentation, the substrate is fermented in its natural state and does not require additional water or extraction of juice from crops, and thus it is economically and technologically easy (Zabed et al. 2017). The solid-state fermentation has several other advantages such as low sterilization cost, easier aeration (large surface area), and lower contamination risks, which make the process an ideal choice for bioethanol generation (Yu et al. 2008). The limitations associated with the solid-state fermentations are poor heat transfers, design of reactors and its operation on large scale and difficulty in agitation of substrate (Li et al. 2013). The major advantage of this process, i.e., low water requirement, may sometimes act as a disadvantage which limits the microbial growth under low humidity. In order to overcome these limitations, several technological advancements such as design of special rotary drum bioreactors can help in proper agitation and mass heat transfer and result in enhanced production of bioethanol under short fermentation time (Han et al. 2010; Wang et al. 2010).

7.3.3.2 Very High Gravity Fermentation (VHG) for Bioethanol Production

There is a need of fermentation process with high initial substrate concentration to enhance the concentration of ethanol in fermentation broth that can result in decreases in overall energy consumption during distillation process (Alkasrawi et al. 2002). Therefore, normal fermentation has fixed initial solid concentration of 20–25% for starch and 10–15% for lignocellulosic biomass and thus have low ethanol yield. Advancement of gravity fermentation that is “very high gravity” (VHG) fermentation is a promising technology that can overcome several limitations of the normal fermentation process. It can lead to several benefits in the form of savings associated with water and energy requirements through high initial solid concentration of approximately 270 g/L or more solids (Bayrock and Ingledew 2001; Puligundla et al. 2011). Several studies demonstrated that VHG can be exploited for enhanced and prolonged growth of *Saccharomyces cerevisiae* under fermentation-induced oxidative stress or low oxygen level (Burphan et al. 2018). Other advantages associated with the VHG are high fermentation efficiency, better utilization of fermenter space, low water requirement, low energy/power consumption, low distillation cost, reduced contamination risk, and high ethanol yield (Lim et al. 2013). Apart from benefits, there are certain limitations associated with the high solid loading; it causes high sugar concentration in fermentation broth which can cause high osmotic pressure on cell leading to the cell lysis and loss of cell viability, negatively affecting the ethanol yield. This effect of osmotic stress can be reduced to supplementing different nitrogenous and fat containing components such as ammonium ions (Laopaiboon et al. 2009), oil seed meal (Sankh et al. 2011), soya flour (Xiao et al. 2010), urea (Pradeep and Reddy 2010), and yeast extract (Chang et al. 2011), etc. which release free amino nitrogen or fatty acid-enhancing cell growth and viability (Kawa-Rygielska and Pietrzak 2014).

7.3.3.3 Integration of Saccharification and Fermentation for Enhanced Bioethanol Production

The ethanol generation from different biomasses involves several steps. In order to minimize the capital and operational cost, there is a need to minimize the steps involved in biomass to production of bioethanol. The key process, i.e., hydrolysis and fermentation, is usually performed separately. 1G and 2G biomass are rich in cellulose, so separate hydrolysis of cellulose followed by fermentation of hydrolyzed sample is usually performed. In commercial bioethanol production from lignocellulosic biomass, different strategies were followed for hydrolysis and fermentation of cellulosic and hemicellulosic component.

7.3.3.3.1 Separate Hydrolysis and Fermentation Process (SHF) for Bioethanol Production

It is a three-step process where joint liquid obtained from hemicelluloses and cellulose hydrolysis reactors is first passed to hexose fermentation reactors; the broth from this reactor is distilled for ethanol generation. The leftover broth containing xylose and other pentose passed through pentose's reactor and converted to ethanol using pentose-utilizing ethanologenic microbes followed by distillation (Hamelinck et al. 2005).

7.3.3.3.2 Separate Hydrolysis and Co-fermentation (SHCF) for Bioethanol Production

Further development reduced saccharification and fermentation to two-step process where hemicellulose and cellulose hydrolyzed separately and fermentation of both hexoses and pentoses was carried out in the same reactors called as separate hydrolysis and co-fermentation(SHCF) process (Girio et al. 2010). In SHF and SHCF process, the hydrolysis of cellulose and hemicelluloses can lead to accumulation of sugars that can inhibit the enzymatic action of hydrolyzing enzyme, thus causing incomplete breakdown of carbohydrate polymer (Mojović et al. 2006). When the complete hydrolysate is transferred to fermenter, the high sugar content in broth may cause osmotic stress to yeast, thus causing decrease in final bioethanol yield (Nikolić et al. 2010)

7.3.3.3.3 Simultaneous Saccharification and Fermentation (SSF) for Bioethanol Production

The osmotic stress and end product's inhibitory effect can be countered by a novel method of simultaneous saccharification and fermentation. Currently, SSF is the most sought-after technique at both laboratory and industrial scale where both cellulose-hemicelluloses hydrolysis and fermentation were performed in single reactor. SSF minimized microbial contamination as well (Srichuwong et al. 2009; Kaur et al. 2018). As two processes are combined, therefore, there is a low need of enzyme and low overall cost. The main limitation of the process is the difference in temperature requirement for saccharification (50 °C) and fermentation (usually 32–37 °C) (Danquah et al. 2011; Sarkar et al. 2012). However, several recombinant strains capable of producing hydrolyzing enzymes and mediating fermentation are developed for mediating the SSF and overcoming the limitations of temperature variance (McBride et al. 2018).

7.3.4 Distillation Process for Recovery of Bioethanol

The fermentation broth is subjected to distillation where water or any other impurities are removed for generation of anhydrous high-quality ethanol that can be used as transportation fuel. The principle of removing impurities through distillation uses the difference in boiling point of different constituents of the mixture. The boiling point (BP) of ethanol is 78.2 °C, which get vaporized and get separated from other component (water BP: 100 °C). The different distillation processes used for obtaining anhydrous ethanol are adsorption distillation, azeotropic distillation, chemical dehydration, diffusion distillation, extractive distillation, membrane distillation, and vacuum distillation.

The variation in molecular size of ethanol and water is used during adsorption distillation. The mixture is passed through the molecular sieve that allow ethanol to pass while entrap the water. Azeotropic distillation utilizes the application of binary azeotropic mixtures which disobeys Raoult's law (Kumar et al. 2010). An additional chemical called entrainer is used to amend the relative volatilities of azeotropes and help in the recovery of azeotropes that can be reused. In dehydration distillation process, a hygroscopic chemical substance (calcium oxide, quicklime) is introduced into the ethanol-water mixture causing rapid dehydration of the mixture (in liquid or vapor phase) by forming soluble calcium hydroxide leaving ethanol on the surface. The method of diffusion distillation involves ethanol recovery by diffusing the mixture through the voids of inert gas followed by condensation. Extractive distillation is based on variation in volatility of different components of the mixture, and thus an additional solvent mixture is added. The highly volatile components (lighter component) are present at the top of the column, while other components with low volatility (heavier component) along with additional solvent are at the bottom. The process involves two distillation runs where during the first run, light component is extracted and the left out heavy component is recovered during the second distillation run (Ravagnani et al. 2010; Gryta 2012). In membrane distillation, the principle of permeability is used where semi-permeable membrane is used for mass transfer through a semi-permeable membrane. The volatile component is crossed through the membrane as the surface tension associated with the membrane blocks the feed. Usually, membranes are made up of polymers; however in some cases, metals and ceramic material are also used (Wang and Chung 2015). Efficiency of the distillation also affects the final yield and overall cost of production of bioethanol (Aditiya et al. 2016).

7.3.5 Microorganisms Used in Ethanol Generation

During conversion of the biomass to bioethanol, various microorganisms are used during different steps such as pretreatment, hydrolysis, and fermentation. The selection of these microorganisms usually depends upon their suitability of the particular process, cost involved for maintaining them, and physical condition required for the growth of microbes. Usually in processes such as pretreatment, detoxification, and hydrolysis, biological products such as enzyme are often used. But for the ethanol generation, ethanologenic microbes are always used. As the ethanol (alcohol) generation is one of the oldest processes developed by human and thus for the ethanol generation on large scale, ethanologenic microbes must have different characteristic properties such as (i) requirement of inexpensive media, (ii) high growth rate, (iii) tolerance to stress condition such as high ethanol (>40.0 g/L)/sugar (above 20%) concentration, (iv) very high ethanol yield (>90.0%), (v) high productivity of ethanol (>1.0 g/L/h), and (vi) capability to minimize growth of contaminants (Dien et al. 2003; Zabed et al. 2017). Yeast, namely, *Saccharomyces cerevisiae*, is the most commonly used microbes for generating fuel-grade ethanol from a wide range of biomass. Different properties of *Saccharomyces cerevisiae* make them an attractive choice for ethanol generation over other organisms. These properties are as follows: (i) high sugar to alcohol conversion efficiency, (ii) tolerance to high ethanol concentration (Snoek et al. 2016), (iii) floc formation ability during fermentation, and (iv) nontoxic or safe (generally recognized as safe GRAS) organism (Lin and Tanaka 2006). Generally, *Saccharomyces cerevisiae* have the capability to secrete invertase enzyme that can hydrolyze sucrose-rich crop juices into fructose and glucose (Zabed et al. 2017). The another well-recognized organism for production of ethanol by fermentation of starch and lignocellulosic hydrolysate is gram-negative and facultative anaerobe *Zymomonas mobilis* (Cazetta et al. 2007). The *Zymomonas mobilis* has several better properties as compared to *Saccharomyces cerevisiae* such as higher ethanol tolerance and better glucose uptake with higher ethanol productivity and yield (Bai et al. 2008). One major disadvantage that limits *Zymomonas mobilis* from replacing the *Saccharomyces cerevisiae* as major bioethanol producer is its narrow substrate range. There are different microbes reported to help in ethanol production; however limitations are associated with these bio-agents to be used on commercial scale.

There are certain limitations associated with microorganism for ethanol generation. First, there is an inability of microorganisms to directly utilize the complex carbohydrate polymer for ethanol generation, as they require reducing sugars (glucose, fructose, and sucrose) for ethanol generation (Hahn-Hägerdal et al. 2006). Therefore, there is a need for an additional step converting naturally occurring carbohydrate polymers to simple sugars requiring costly enzymes which increase the overall cost and processing time. Second, these microbes are not capable to utilize both pentoses and hexoses generated after hydrolysis of hemicelluloses and cellulose, respectively. Third, the fermentation usually occurs at high temperature, so thermotolerant microbial strains are required. In addition, the enzyme production cost results in increase in overall ethanol generation cost. Therefore, keeping in view of these limitations, several researches have been attempted to develop different modified or recombinant strains using technological advancements in field of biotechnology. The different approaches used are altering the natural genetic makeup with the desired traits (Hahn-Hägerdal et al. 2007) or evolutionary microbial engineering techniques (Wisselink et al. 2009) for developing strains which can utilize the hexoses and pentoses or even capable of directly utilizing carbohydrate polymers.

A series of attempts have been made by different research groups for modifying yeast which can directly utilize starch for production of ethanol by cell surface engineering. The bioengineered yeast was capable of producing glucoamylase and α-amylase along with ethanol generation ability thus a single step conversion of starch to bioethanol can be performed (Shigechi et al. 2002, 2004; Aydemir et al. 2014). Several strains capable of tolerating high ethanol concentration have been isolated, or the already existing microbes are modified. Different thermotolerant yeasts have been isolated and adapted for the simultaneous saccharification and fermentation (Thammasittirong et al. 2013; Tikka et al. 2013).

7.3.6 Factors Affecting Bioethanol Production

Microorganism plays an important role in the generation of enzyme for hydrolysis and fermentation of the biomass to bioethanol. Therefore, the factors affecting the growth of microbes are physical (pH temperature, incubation time) and nutritional parameters (carbon and nitrogen source, metal ions, etc.), which are common for most of the microorganism (Ramirez et al. 2016). The major factors which are specific to ethanol generation are discussed below, i.e., initial solid load, microbial load, and accumulation of soluble by-product which sometime slow down the fermentation process due to end product inhibition.

7.3.6.1 Initial Solid/Substrate Load Concentration Affecting Bioethanol Production

The overall biochemical conversion efficiency of biomass to bioethanol is often affected by initial substrate (sugars, cellulose, starch, and hemicellulose) concentration (Modenbach and Nokes 2013). As discussed above, the excessive substrate load can inhibit the enzymes during enzymatic hydrolysis of the complex polymers, thus effecting the overall sugar yield due to incomplete conversion of carbohydrates (Mojović et al. 2006). Due to high substrate loading, gelatinization of starch occurs which increases the viscosity, leading to incomplete starch conversion because of poor mixing (Uthumporn et al. 2010). The initial solid loading of 12–38% is suggested for optimum hydrolysis (Foerster 2010). The increase in sugar concentration enhances the ethanol yield to a certain level after which high sugar concentration can negatively affect the fermentation due to osmotic pressure on cells affecting cell viability (Szymanowska-Powalowska et al. 2012). Therefore, there is a need to optimize the initial substrate concentration for efficient ethanol generation.

7.3.6.2 Microbial Load Concentration Affecting Bioethanol Production

Studies suggests that microbial load or initial inoculum concentration of fermenting microorganisms does not significantly affect the final concentration of bioethanol but can enhance consumption rate of sugar and fermentation time (Laopaiboon et al. 2007). The ethanol production is enhanced when initial inoculum concentration is increased from 10^4 to 10^7 cells/mL. However, further increase in cell concentration to 10^8 has no effect (Zabed et al. 2014). The increase in cell concentration at a certain level can speed up the growth of cells, which results in the decrease in fermentation time. Mojović et al. (2006) demonstrated that during fermentation of corn meal hydrolysate, increase in initial inoculum concentration from 1 to 2 % reduced fermentation time to 32 h from 48 h. Similarly, increase in yeast concentration from 3 to 6% can lead to decrement in fermentation time to 48 h from 72 h (Breisha 2010). Microbial load often increases by contamination of broth with other microbes which can compete with yeast for the substrate and thus negatively affect the ethanol generation. Thus, there is need to regulate process for preventing contamination by properly maintaining aseptic condition and use of antibiotics during fermentation process (Szymanowska-Powałowska et al. 2014).

7.3.6.3 Accumulation of By-products Affecting Bioethanol Production

The high ethanol concentration has negative effect on the ethanologenic strain due to end product inhibition, and dehydrating condition arises due to the presence of ethanol. Therefore, continuous recovery of ethanol during fermentation and application of alcohol-tolerant strains can minimize the effect of by-product accumulation. Metabolic activity of yeast and the contaminating bacteria can lead to generation of soluble inhibitory by-products such as lactic acid and acetic acid (Graves et al. 2006). Glycerol is a by-product of ethanol generation process by yeast or even by contamination of bacteria (Sarris and Papanikolaou 2016). The lactic or acetic acid is produced by contaminating bacteria. They cause accumulation of these by-products, which are undesirable as it can negatively affect the yeast growth and ethanol yield (Białas et al. 2010).

7.3.7 Bioethanol Production

7.3.7.1 First-Generation Feedstock-Based Bioethanol

7.3.7.1.1 Bioethanol Production Using Starch-Based Feedstock

The starch-based bioethanol is 60% of the total bioethanol production using first-generation feedstock with sugar-based feedstock contributing to 40% ethanol (Mussatto et al. 2010; Johnston and McAloon 2014). Starch-based bioethanol requires an additional step of conversion of starch to reducing sugar through utilizing amylases, whereas sugar-based crops only require direct extraction of fermentable sugar. Thus, the sugar-based ethanol is cheap as compared to starch-based technique. However, the limitation associated with the sugar-based ethanol generation is that the sugar crops require specific climate and soil type for their growth and thus cannot be cultivated around the globe (Barcelos et al. 2011).

Starch-based ethanol production from corn primarily utilizes dry grinding or wet milling method. Dry grinding method involves the following steps: (i) slurry preparation by mixing corn flour with water, (ii) cooking of the slurry, (iii) liquefaction using thermostable alpha amylase, (iv) saccharification using glucoamylase, (vi) ethanol fermentation, and (vii) distillation. After ethanol fermentation broth is subjected to distillation, leaving the solid fraction, thick and thin stillage. The thick stillage syrup mixed with the solid fraction is used as animal feed, whereas the thin stillage is recycled for water recovery. During the wet milling process, the biomass is first grounded followed by its separation from individual components, and the starch obtained is subjected to wet milling. The later stage of the process is similar to the dry grinding. Different co-products are generated in dry grinding and wet milling which can contribute to overall economy of the process. The dry grinding of corn gives ethanol yield of 0.395 L/kg which is slightly higher to 0.372 L/kg as obtained from wet milling method (Shapouri et al. 2002). Commercially, dry grinding method is more preferred because the wet milling method is costly and equipment is expensive. In the USA, 70–86% ethanol from corn is produced using dry grinding method (Mosier and Ileleji 2015). A summary of different pretreatment and fermentation methods used for bioethanol production using various starch-based feedstocks is summarized in Table 7.3.

7.3.7.1.2 Bioethanol Production Using Sugar-Based Feedstock

After the USA corn-based ethanol, Brazilian sugar-based ethanol is the second largest ethanol in the world. Sugarcane-based ethanol generation involves different steps: (i) extraction of juice from the sugarcane or beet sugar using roller press; (ii) purification of sugar using lime (calcium hydroxide) or calcium saccharate which reduces colorants and neutralize organic acids; (iii) filtration, that remove debris from juice which are collected as filter cake, (v) evaporation or condensation (14–18% sugar level, i.e., sugar tolerance capacity of microbes), and (vi) fermentation of the condensed syrup under sterilized condition at appropriate temperature and pH (Vohra et al. 2014). Additionally, nitrogenous source is also added to minimize osmotic stress due to high sugar content. The by-products majorly bagasse and filter cake obtained after the extraction of juice and filtration step can be used for different purposes. The generation of heat and electricity from bagasse can also help in minimizing the overall cost of production of ethanol. Bagasse can also be used as substrate for production of 2G bioethanol. The filter cakes are used as eco-friendly fertilizers (Zabed et al. 2017). A summary of different pretreatment and fermentation methods used for bioethanol production using various starch-based feedstocks is summarized in Table 7.3.

Table 7.3 Ethanol generation from different 1G and 2G feedstock using different pretreatment and fermentation method

Feedstock	Carbohydrate content	Pretreatment and hydrolysis (liquefaction and saccharification)	Fermentation condition	Ethanol yield	References
Corn varieties high sugary corn genotypes (HSG) parent field corn lines (PFCs)	HSG: starch 67.3%, total sugar 7.4%; PFC-starch 73.6%, total sugar 1.2%	Saccharification by STARGEN 002; enzyme load 2 Kg/MT; initial solid concentration 300 g/L	*S. cerevisiae ATCC* 96581 (2 mL inoculums /100mL media) incubated at 30°C, 4.2 pH for 96 h	Ethanol yield: HSG: 17.94%, 141.5 g/L; PFC: 16.85%, 130.5 g/L	Zabed et al. (2016b)
Cassava flour	Starch 76%–81%	High temperature liquefaction using Spezyme, amylase, and glucoamylase at 80 °C for 90 min	*S. cerevisiae* (inoculum 1.5×10^7 cells/mL) incubated at 30 °C, 5.5 pH for 72 h	Ethanol yield: At lab scale: 86.1% (17.2% v/v) theoretical yield. At pilot scale: 83.6% (16.5% v/v) of the theoretical yield	Shanavas et al. (2011); Nguyen et al. (2014)
Wild cassava	Carbohydrate content: 32–35% (fresh weight) 80–90% (dry matter)	Liquefaction using α-amylase and saccharification and β-glucanase	*Caloramator boliviensis* (inoculum load 50 mL/240 ml media) incubated at 60 °C, 7 pH for 48 h	Ethanol yield: 33.0 g/L, 85% of the theoretical ethanol yield	Montagnac et al. (2009); Moshi et al. (2015)
Sorghum juice	Not available	Not available	*S. cerevisiae* strain BY4741 (inoculum load 5×10^8 cells·mL^{-1}) incubated at 30 °C, pH 5.2 for 48 h	Ethanol yield: 115.2 g/L, 87.1% theoretical yield	(Sasaki et al. (2015))
Sugarcane molasses	Sucrose: 31%; Inverted sugar: 15%	Not available	Aule alcohol yeast (AY) and Aule baker's yeast (BY) (inoculum 1% w/v) incubated at RT, 4.3 pH for 72 h	Ethanol yield: AY-74.8 g/L, BY-102.9 g/L from 300 g/L sugar concentration	Mayzuhroh et al. (2016)

(continued)

Table 7.3 (continued)

Feedstock	Carbohydrate content	Pretreatment and hydrolysis (liquefaction and saccharification)	Fermentation condition	Ethanol yield	References
Sugar beet molasses	Molasses: 53% fermentable Sugar juice: 60% Fermentable sugar	Not available	Immobilized yeast (inoculum 1% w/v) at 30°C, pH 5.5, 144 h	Ethanol yield: molasses: 83.2 g/L, i.e., 96.8% of theoretical yield; thick juice: 132.4 g/L, i.e., 90.6% of theoretical yield	Razmovski and Vučurović (2012))
Cassava stem	Stem: cellulose 28.9%, hemicellulose 21.1%; peelings: cellulose 9.7%, hemicellulose 32.3%	Heating at 225 °C for 50 min followed by cellulase enzyme treatment	Inocolums load of 0.1 g dry biomass of *S. cerevisiae* or *Rhizopus* spp. per 10 ml hydrolysate	Ethanol yield: stem 0.052 g/g; peelings: 0.026 g/g	Nanssou et al. (2016)
Cassava cellulosic wastes	Carbohydrate 76.6% Starch 60.8%	Sequential enzymatic and acid hydrolysis using α-amylase and dilute HCl at 97–100 °C, saccharification using amyloglucosidase at 50 °C	*S. cerevisiae* (inoculum 0.002 g dry biomass/mL) incubated at 40 °C–50 °C, 4.6–5.5 pH for 8 h	Ethanol yield: 2.7 g/15 g cassava cellulosic waste, with concentration: 32.4% w/w	(Elemike et al. (2015))
Sugarcane bagasse	Cellulose 52% hemicellulose 20%	Delignification using 1 N sodium hydroxide for 2 h at reflux temperature	SSF using cellulase and *S. cerevisiae* incubated at RT for 5 days	Ethanol yield: 11.8 g/L	(Wahono et al. (2012), (2015))
Sugar beet pulp	Total carbohydrate 80%	Steam pretreatment at pressure of 4–8 bar and temperature of 152 °C–175.5 °C. Saccharification using commercial cellulase 50 °C for 24 h	*S. cerevisiae* incubated at 30 °C for 24 h	Ethanol yield: 0.5 g/g of glucose	Hamley-Bennett et al. 2016

Sweet sorghum bagasse	Cellulose: 45% hemicellulose: 27%	Mild alkali (NaOH) pretreatment Saccharification using commercial cellulase and xylanase	*Zymomonas mobilis* (inoculums load of 10% v/v) incubated at 32 °C, 6 pH for 30 h	Ethanol yield: 61.8% of the theoretical yield	Kim and Day (2011); Yu et al. (2016)
Sweet sorghum stover	Total carbohydrate ranges from 7 to 36%	Dilute sulfuric acid (0.37% v/v) treatment for 15 min at 150°C in high pressure reactor Saccharification by commercial cellulase (Zytex) at 50°C for 48 h	*S. cerevisiae (*inoculum 0.27 g/L) incubated at 30 °C for 48 h	Ethanol yield: 91.9 g ethanol/kg native sorghum	Akanksha et al. (2016); Sekhon et al. (2016)
Rice straw	Cellulose 47% hemicellulose 27 %	Aqueous ammonia (27% w/w) treatment at room temperature for 14 day Saccharification using Cellic Ctec2 (Cellulase) and Cellic Htec2 (xylanase)	Mixed culture of *S. cerevisiae* and *Candida tropicalis* (inoculum 1.2 g yeast/10 mL) incubated at 37°C for 72 h	Ethanol yield: 25.1 g/L	Phitsuwan et al. (2016)
Beechwood	Cellulose: 43.1% Hemicellulose: 20.2%	Organosolv with acetone and sulfuric acid (OAS) or Organosolv with ethanol and phosphoric acid (OEP)	High gravity SSF at High solid loading of 20% wt using 8.4 mg/g enzyme incubated for 144 h	OAS: Ethanol concentration: 8.0 % *w/v* Ethanol yield: 76.3 g/L OEP: ethanol yield: 80 g/L	Kalogiannis et al. (2018)
Eucalyptus dunnii bark	Glucans:37.1% Xylans:9.8%	1-Butyl-3-methylimidazolium chloride [BMIM]Cl: bark ratio of 5:1, at 140 °C, 8 h Saccharification using 50 FPU/g of pretreated material	*S. cerevisiae* (0.1 g cells/ ml)	Enhanced ethanol production by fourfold higher as compared with untreated bark Ethanol yield: 70%	(Reina et al. (2016)

(continued)

Table 7.3 (continued)

Feedstock	Carbohydrate content	Pretreatment and hydrolysis (liquefaction and saccharification)	Fermentation condition	Ethanol yield	References
Douglas-fir forest residue	Glucan: 40.97% Xylan: 5.70 Galactan: 2.0 Mannan: 9.67	Calcium bisulfite pretreatment temperature at 145 °C. Saccharification using commercial cellulase Cellic CTec3 enzyme (15 FPU/g glucan)	Yeast (inoculums of 0.4 mg cell/g)	Ethanol yield: 282 L/ ton, 70% theoretical yield with a 42 g/L titer	Zhu et al. (2015)

Table 7.4 Ethanol generation from different microalgae and macroalgae using different pretreatment and fermentation method

Algae	Pretreatment and hydrolysis (liquefaction and saccharification)	Fermentation	Yield	References
Microalgae				
Desmodesmus sp.	Solid load: 10% dry w/v, Dilute H_2SO_4 (2% v/v) for 20 min at 120°C	*S. cerevisiae* incubated at 28 °C, 120 rpm for 30 h	Reducing sugar: 137.2 g/L	Rizza et al. (2017)
			Ethanol titer: 61.2 g/L	
			Ethanol yield: 0.310g/g	
Chlorella vulgaris FSP-E	Dilute H_2SO_4 (1% w/v) for 20 min at 121 °C and pH 6.	*Z. mobilis* incubated at 30 °C for 24 h t	Reducing sugar: 0.477g/g	Ho et al. (2013)
			Ethanol titer: 11.7 g/L	
			Ethanol yield: 0.233g/g	
Nannochloropsis oculata	Dilute NaOH (0.75% w/v) for 10 min at RT	*S. cerevisiae* incubated at 30 °C, 50 rpm for 48 h	Reducing sugar: 1–2.4% (dw)	Reyimu and Ízšimen (2017)
			Ethanol yield: 0.037 g/g	
Chlamydomonas reinhardtii UTEX 90	Dilute H_2SO_4 (3% v/v) for 30 min at 110 °C	*S. cerevisiae* incubated at 30°C for 24 h	Reducing sugar: 0.58 g/g	Nguyen et al. (2009)
			Ethanol titer: 14.6 g/L	
			Ethanol yield: 0.292g/g	
Chlorococum sp.	Supercritical CO_2 assisted lipid extraction	SHF using *Saccharomyces bayanus* and for 60 h	Ethanol yield: 38.30 %	Harun et al. (2010)
			Ethanol concentration: 3.83 g/L from 10 g/L dry algal biomass	
Chlorococcum infusionum	Dilute NaOH pretreatment (0.75% w/v) for 30 min at 120°C	*Saccharomyces cerevisiae* for 72 h	Glucose yield: 350 mg/g,	Harun et al. (2011)
			Ethanol yield: 26.00%	
Chlamydomonas reinhardtii cw15	Sequential treatment of algae with 70 °C ethanol, 60 °C hexane, and 12 N H_2SO_4 at 121 °C and 2 atm	*Saccharomyces cerevisiae*, SHF, 48 h	Maximum ethanol concentration: 0.87 ± 0.01% (m/v)	Scholz et al. (2013)
			Ethanol generation rate: 14 ± 1 mL/g in 28 h.	
			Ethanol yield: 44.0%	

(continued)

Table 7.4 (continued)

Algae	Pretreatment and hydrolysis (liquefaction and saccharification)	Fermentation	Yield	References
Schizochytrium sp.	Solid load 25% (w/w) fractionation by hydrothermal treatment for 46.7 min at 115.5 °C	SSF using *Escherichia coli* KO11and incubating for 72 h	Ethanol yield: 1.18 g/L from 2.57 g/L of glucose;	Kim et al. (2012)
			Maximum theoretical ethanol yield: 89.8% (from glucan)	
Scenedesmus sp.	Ultrasound followed by acid/alkali mediated autoclave treatment for 10–40 min at 60–120 °C and 15 psi	Inoculum of *S. cerevisiae* (2.5 mL) incubating at 30 °C, 180 rpm for 72 h	Total sugar yield: 93%	Sivaramakrishnan and Incharoensakdi (2018)
			Theoretical ethanol: 86%	
			Practical ethanol: 93%	
Macroalgae				
Ulva lactuca	Sulfuric acid or sodium hydroxide treatment for 30 min at 121 °C	3 ml of inoculums of S. *cerevisiae* incubating at 35 °C for 48 h	Ethanol concentration: 12 ± 0.5 g/g	El-sayed (2016)
			Ethanol yield: 47.1%	
Sargassum spp.	Dilute H_2SO_4 (3.4–4.6% v/v) and enzymatic hydrolysis	SHF using *Saccharomyces cerevisiae,* incubating at 40 °C, pH of 4.5 for 48 h	Ethanol yield: 89%	Borines et al. (2013)
Gracilaria verrucosa	Cellulase- and β-glucosidase-based enzymatic hydrolysis	SHF using *Saccharomyces cerevisiae* (6.0% v/v) at 30 °C, pH 6.0 for 48 h	Reducing sugar yield: 0.87 g sugars/g cellulose	Kumar et al. (2013)
			Ethanol yield: 43 %	
Laminaria japonica	Mild acid treatment at 0.1 N HCl for 15 min at 121 °C followed by enzymatic hydrolysis using Celluclast, Viscozyme L	SSF using *Escherichia coli* KO11 at 30 °C, 150 rpm for 72 h	Sugar yield: mannitol 30.5% and glucose 6.98%	Kim et al. (2011)
			Ethanol yield: 16.1%	
Seaweed such as sea lettuce and chigaiso	Enzymatic treatment using Meicelase for 120 h at 50 °C	SHF using *Saccharomyces cerevisiae IAM 4178* at 30°C, 5.0 pH for 48 h under static condition	Sugar yield: sea lettuce 78.8 g/L; chigaiso 123 g/L	Yanagisawa et al. (2011)
			Ethanol yield: sea lettuce: 30.0 g/L; chigaiso: 34.4 g/L	

Gelidium elegans	Dilute H_2SO_4 (2.5% w/v) for 40 min at 120 °C	*Saccharomyces cerevisiae* incubated at 30 °C and 150 rpm	Reducing sugars 39.42% (galactose and glucose)	Hessami et al. (2019)
			Ethanol yield:13.27±0.47 g/L	
Ascophyllum nodosum and *Laminaria digitata*	Dilute H_2 SO_4 (0.2 M) for 20 min at 121°°C	Mixed culture of *Scheffersomyces (Pichia) stipitis* and *Kluyveromyces marxianus* incubated at 30°C, 100 rpm for 150 h	*L. digitata* a better choice for ethanol generation	Obata et al. (2016)
			L. digita hydrolysate with K. *marxianus* ethanol generation: 6 g/L	

7.3.7.2 Bioethanol Production Using Lignocellulosic Biomass

The conversion of the lignocellulosic biomass to bioethanol involves several steps: (i) particle size reduction, (ii) mixing of biomass with water to make slurry (solid load of 10–15%), (iii) pretreatment to break the recalcitrance of the biomass, i.e., removal of lignin and depolymerization of hemicelluloses and cellulose, (iv) chemical or enzymatic hydrolysis of the pretreated solid or liquid (based on carbohydrate composition) for breaking down of polymers to monomers, (v) fermentation of reducing sugars to ethanol, and (vi) distillation for ethanol recovery (Wyman 2018). The different by-products of 2G ethanol production process are usually lignin residues and wastewater generated during different steps of ethanol generation (Zabed et al. 2017). The lignin can be put to different applications, i.e., combustion for generation of heat and electricity for making overall process more economical. Recent trends show that lignin can be effectively used as raw material for production of different value-added products (Chaturvedi and Verma 2013). The wastewater generated can be recycled where a portion of wastewater can be recirculated as backset or the wastewater may be used for recovery of some organic compounds generated during the different ethanol generation steps, which is of great economical importance (Mathew et al. 2018). Due to morphological and physiochemical complexities of the lignocellulosic biomass and varying chemical constituents of the different lignocellulosic biomasses, the scientific community is struggling to come up with uniform conversion method or optimum production condition (Hassan et al. 2019). A summary of different pretreatment and fermentation methods used for bioethanol production using various lignocellulosic feedstocks is presented in Table 7.3.

7.3.7.3 Bioethanol Production Using Algal Biomass

The generation of ethanol from algal biomass involves harvesting, dehydration, pretreatment, hydrolysis, saccharification, and distillation. The first step involves harvesting of algal biomass from the open ponds or photo-bioreactors. The second step involves dehydration using sun drying or specialized drying (sprig, freeze, and fluidized bed drying) for up to 50% removal of water content (McKendry 2002; Grima et al. 2003). Different drying techniques have its own advantages and limitations, for example, sun drying is cheap but requires longer duration and larger surface area. Spig drying is used for costly product isolation; however, the process results in pigment loss and is usually a very expensive technique (Bibi et al. 2017). Freeze-drying facilities extraction of oils but are very expensive and difficult to upscale. The third step involved the extraction of by-product using different crushing or pretreatment techniques for the enhanced sugar and lipid yield resulting in better ethanol yield. The starch and cellulose are hydrolyzed using chemical or enzyme and fermented using yeast in container called fermentors (Singh et al. 2011). After alcoholic fermentation, the distillation is performed to recover the ethanol from fermentation broth by removing water and other components. The lipid part of algal biomass is used for generation of biodiesel. A summary of different pretreatment and fermentation methods used for bioethanol production using various micro -and macroalgae is presented in Table 7.4.

7.4 Current Global Scenario on Bioethanol Production

Developed nations such as the USA, Brazil, China, and Canada and several EU countries have already committed to enhance their reliance on bioethanol (Mohapatra et al. 2019). The trend of bioethanol production as transportation fuel from year 2007 to 2017 for the USA, Brazil, EU, China, Canada, and rest of the world is presented in Fig. 7.4. The figure clearly shows that the bioethanol production has enhanced gradually with major contribution from the USA and Brazil. The USA is the leading producer of bioethanol. The USA and Brazil have doubled their bioethanol generation capacity between 2007 and 2017, i.e., ~11 billion to 22 billion gallon (Fig. 7.4).

Figure 7.5 shows the comparison of ethanol production potential of the world and the USA from year 2007 till 2017. It clearly shows that the USA contributes approximately 50% of global production. Different sugar-based and grain-based feedstocks are used for ethanol generation such as in Brazil, where sugarcane is primarily used as ethanol feedstock, whereas the USA uses corn. Canada and EU

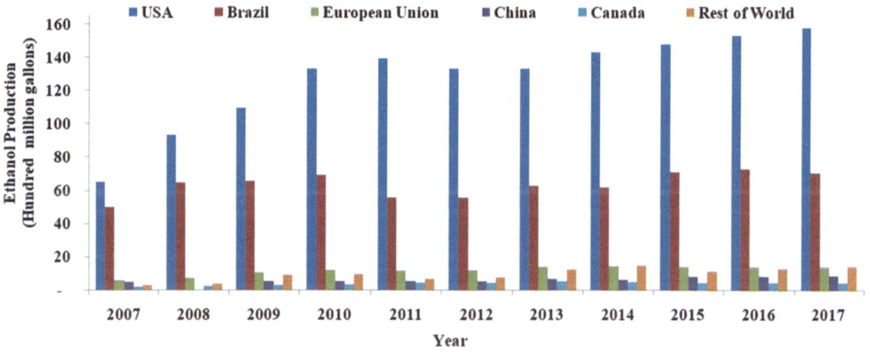

Fig. 7.4 Global ethanol production scenario from 2007 to 2017. (Data Retrieved from www.afdc.energy.gov/data. AFDC 2018)

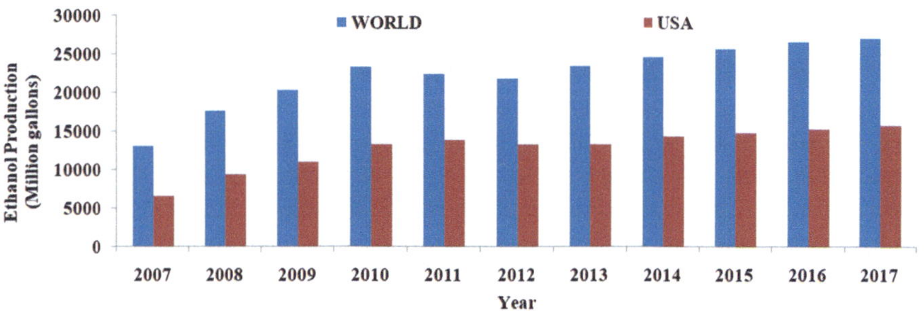

Fig. 7.5 Year-wise comparison of global ethanol production in the USA from 2007 to 2017. (Data Retrieved from www.afdc.energy.gov/data. AFDC 2018)

countries are majorly dependent on wheat and sorghum, respectively, for bioethanol production. Maize accounts for 67% of bioethanol production; however, in terms of biomass production, sugarcane is the leading contributor, as it requires the least amount of water as compared to other food crops (Gerbens-Leenes et al. 2012; Rulli et al. 2016). In the last two decades due to food feed competition and land-water utilization concern, lignocellulosic biomass has gained much interest and undergone several technological advancements. This can be supported by the fact that by 2016, around 67 commercial 2G bioethanol production facilities have been established around the world, among these about one third generate ethanol in tons (US Department of Energy, 2016). In the case of 2G bioethanol production, the USA is the leading 2G bioethanol producer with around 35% of the installed capacity. Currently, India is also moving toward commercialization of 2G ethanol generation based on techniques developed by DBT-ICT Centre for Energy Biosciences and commercialization led by Hindustan Petroleum and Bharat Petroleum (DBT 2017). For the first time in the context of Indian history, application of bioethanol as aircraft fuel was successfully demonstrated in India when a SpiceJet plane (72-seater aircraft) successfully completed its run from Dehradun to Delhi (TCI 2018). The processing cost involving pretreatment of the 2G feedstock due to its recalcitrance behavior and enzymes cost is a limiting factor for commercialization of 2G generation bioethanol (Behera and Ray 2015). National Renewable Energy Laboratory, USA, has demonstrated the application of low-cost hydrolyzing enzymes developed by private laboratories "Novozymes and Genencor," to make the overall ethanol production process cost-efficient (NREL 2010). Several ongoing researches for search of cost-efficient pretreatment strategies are under process as evident from the research article on biomass pretreatment. However, the literature survey demonstrates the fact that the economic feasibility of a pretreatment strategy is substrate dependent; therefore there is a need to understand the effect of different pretreatments on different feedstocks and develop a universal pretreatment strategy. Commercial production of the algal bioethanol is still in the laboratory or pilot scale due to several limitations of which one being the cost of the reactors to provide the controlled environment for enhanced and rapid bioethanol yield. The demand of bioethanol all over the world will keep on increasing and therefore create opportunities for algae and nonfood-based feedstocks for generation of ethanol.

7.5 Limitations and Future Prospects

The advantages and disadvantages of different ethanol generations have been tabulated in Table 7.5. The 1G bioethanol production is the most evolved technique covering most of the bioethanol yield. However, the major limitations associated with this generation are the food vs fuel conflict, which has forced the scientific community to focus toward the 2G feedstock. The shift from 1G to 2G is evident from the growing number of publications in the area of 2G biofuel accompanied by decrease in research publications in the area of first generation. However, it may be also because of the reason that the first-generation feedstock-based bioethanol

Table 7.5 Summary of different refinery-based concepts for generation of transportation fuel and essential chemicals: benefits and problems

Generation of fuel	Feedstocks	Products	Problems	Benefits
Petroleum refinery	Crude petroleum	CNG, LPG, petrol, kerosene and jet fuel,	Depletion of petroleum/coal reservoirs	Fully developed technology for production and purification
			Environmental and Ecological problems	Machines and other equipments developed suitably for such products
1G biorefinery	Corn, sugar	Biodiesel, corn ethanol, sugar ethanol	Food vs fuel competition	Environmental friendly
			Blending with conventional fuel	Economical
			Hydrolyzing enzyme cost	
2G biorefinery	Nonfood crops (grasses)	Lignocellulose-based bioethanol, butanol, and value-added chemicals	Technology not fully developed	No competition with the food
	Agricultural residues		Universal pretreatment method is needed	Environmental friendly
	Forest residues			
			Cost of enzyme	Waste management
3G biorefinery	Microalgae and macroalgae	Lipid-based biodiesel	Technology in lab scale	No completion with food
		Carbohydrate-based ethanol	Economical production of algal biofuel still not a reality	Environmental friendly
				Waste water can be used for algal growth
4G biorefinery	Modified organisms (microbes and plants)	Biodiesel, bioethanol, electricity	Limited to lab scale	No competition with food
			Ethical concern with recombinant and synthetic organism	
Integrated biorefinery	Biobased products	Ethanol, biodiesel, and value-added chemicals	Still a hypothesis	Will help to overcome the limitations of different generations of biofuel
	From all three generations			The overall process will be self-sustainable so less cost and less energy requirement

production technology is already developed and been implemented at the commercial level. Then also with the growing population, the food vs feed conflict is very serious, and therefore the shift to lignocellulosic biomass as feedstock is much needed. The lignocellulosic bioethanol is the most promising technology available at the current time but is limited to the recalcitrance nature of the biomass. Therefore, there is an urgent need to develop a universal pretreatment technology for biochemically and morphologically different wide range of potential 2G feedstock. Algal biomass has also grabbed the attention of the scientific community due to its potential to store lipids along with carbohydrate as building block of the algal biomass. The carbohydrate part can be used for bioethanol production, and lipid can be transesterified to biodiesel. The major limitations associated with this technique are the cost involved with the algal biomass production. The 3G approaches uses wastewater as substrate and value-added products are generated, thus compensating the cost associated with pretreatment and production of algal biomass. The biotechnological advances have led to the development of modified organism utilizing recombinant technology or developing an entirely new organism, and also the development of photovoltaic cells for generation of electricity has given rise to another approach called fourth-generation feedstock-based biofuel production.

The associated limitation of each approach can be overcome by developing an integrated biorefinery (Fig. 7.6) combining all the feedstock-based approaches for

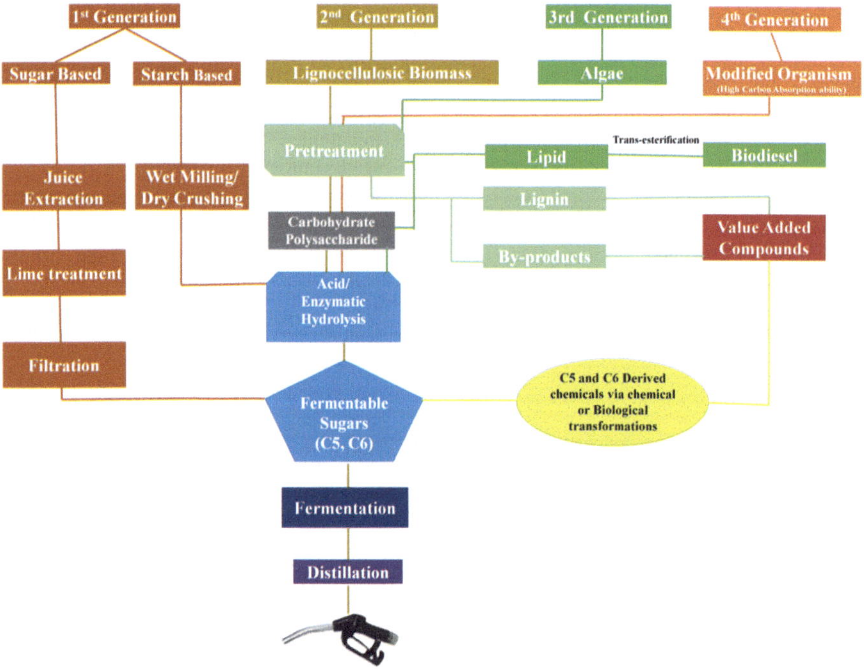

Fig. 7.6 Schematic diagram of integrated bio-based biorefinery

generation of bioethanol, biodiesel, electricity, and value-added compounds. Therefore, the integrated approach can help in meeting the growing energy and chemical demand replacing the overdependence on fossil fuels.

Acknowledgment Authors acknowledge the Department of Biotechnology, Government of India, for financially supporting the work (BT/PR7333/PBD/26/373/2012). B.K. acknowledges Jawaharlal Nehru Memorial Fund, New Delhi, CSIR-SRF for providing Doctoral Studies Scholarship. K.A. acknowledges the financial support provided by Central University of Rajasthan, Ajmer, India. N.B. gratefully acknowledges University Grants Commission, Government of India, for providing RGNF fellowship.

References

Aditiya HB, Mahlia TMI, Chong WT et al (2016) Second generation bioethanol production: a critical review. Renew Sust Energ Rev 66:631–653

AFDC (2016). https://afdc.energy.gov/data/10339

AFDC (2018) Alternative fuels data center: maps and data. https://afdc.energy.gov/data/

Agbor VB, Cicek N, Sparling R et al (2011) Biomass pretreatment: fundamentals toward application. Biotechnol Adv 29:675–685

Agência Nacional doPetróleo (2015) Oil, natural gas and biofuels statistical yearbook 2018. http://www.anp.gov.br

Aiyejagbara MO, Aderemi BO, Ameh AO et al (2016) Production of Bioethanol from Elephant Grass (*Pennisetum purpureum*) Stem. Int J Innov Math Stat Energy Policies 4:1–9

Akanksha K, Sukumaran RK, Pandey A et al (2016) Material balance studies for the conversion of sorghum stover to bioethanol. Biomass Bioenergy 85:48–52

Al-Hasan M (2003) Effect of ethanol unleaded gasoline blends on engine performance and exhaust emission. Energy Convers Manag 44:1547–1561

Alkasrawi M, Galbe M, Zacchi G (2002) Recirculation of process streams in fuel ethanol production from softwood based on simultaneous saccharification and fermentation. Biotechnol Fuels Chem 98–100:849–861

Alvira P, Tomás-Pejó E, Ballesteros MJ, Negro MJ (2010) Pretreatment technologies for an efficient bioethanol production process based on enzymatic hydrolysis: a review. Bioresour Technol 101:4851–4861

Aro E (2016) From first generation biofuels to advanced solar biofuels. Ambio 45:24–31. https://doi.org/10.1007/s13280-015-0730-0

Aydemir E, Demirci S, Dogan A et al (2014) Genetic modifications of *Saccharomyces cerevisiae* for ethanol production from starch fermentation: a review. J Bioprocess Biotech 4(7):1–8

Bai FW, Anderson WA, Moo-Young M (2008) Ethanol fermentation technologies from sugar and starch feedstocks. Biotechnol Adv 26:89–105

Balat M (2009) New biofuel production technologies. Energy Educ Sci Technol Part A 22:147–161

Balat M (2011) Production of bioethanol from lignocellulosic materials via the biochemical pathway: a review. Energy Convers Manag 52:858–875

Balat M, Balat H (2009) Recent trends in global production and utilization of bio-ethanol fuel. Appl Energy 86:2273–2282

Ban-Koffi L, Han Y (1990) Alcohol production from pineapple waste. World J Microbiol Biotechnol 6:281–284

Barcelos CA, Maeda RN, Betancur GJV, Pereira N Jr (2011) Ethanol production from sorghum grains [*Sorghum bicolor* (L.) *Moench*]: evaluation of the enzymatic hydrolysis and the hydrolysate fermentability. Braz J Chem Eng 28:597–604

Bayraktar H (2005) Experimental and theoretical investigation of using gasoline-ethanol blends in spark-ignition engines. Renew Energy 30:1733–1747

Bayrock DP, Ingledew WM (2001) Application of multistage continuous fermentation for production of fuel alcohol by very-high-gravity fermentation technology. J Ind Microbiol Biotechnol 27:87–93

Behera S, Ray RC (2015) Batch ethanol production from cassava (*Manihot esculenta Crantz.*) flour using *Saccharomyces cerevisiae* cells immobilized in calcium alginate. Ann Microbiol 65:779–783

Belboom S, Bodson B, Léonard A (2015) Does the production of Belgian bioethanol fit with European requirements on GHG emissions? Case of wheat. Biomass Bioenergy 74:58–65

Besada V, Andrade JM, Schultze F, González JJ (2009) Heavy metals in edible seaweeds commercialised for human consumption. J Mar Syst 75:305–313

Białas W, Szymanowska D, Grajek W (2010) Fuel ethanol production from granular corn starch using *Saccharomyces cerevisiae* in a long term repeated SSF process with full stillage recycling. Bioresour Technol 101:3126–3131

Bibi R, Ahmad Z, Imran M et al (2017) Algal bioethanol production technology: a trend towards sustainable development. Renew Sust Energ Rev 71:976–985

Borines MG, de Leon RL, Cuello JL (2013) Bioethanol production from the macroalgae *Sargassum* spp. Bioresour Technol 138:22–29

Breisha GZ (2010) Production of 16% ethanol from 35% sucrose. Biomass Bioenergy 34:1243–1249

Brosse N, Dufour A, Meng X et al (2012) *Miscanthus*: a fast-growing crop for biofuels and chemicals production. Biofuels Bioprod Biorefin 6:580–598

Buckow R, Weiss U, Heinz V, Knorr D (2007) Stability and catalytic activity of α-amylase from barley malt at different pressure--temperature conditions. Biotechnol Bioeng 97:1–11

Burphan T, Tatip S, Limcharoensuk T et al (2018) Enhancement of ethanol production in very high gravity fermentation by reducing fermentation-induced oxidative stress in *Saccharomyces cerevisiae*. Sci Rep 8(13069):1–11

Burton T, Lyons H, Lerat Y, et al (2009) A review of the potential of marine algae as a source of biofuel in Ireland, pp 1–88

Cardona CA, Quintero JA, Paz IC (2010) Production of bioethanol from sugarcane bagasse: status and perspectives. Bioresour Technol 101:4754–4766

Cazetta ML, Celligoi M, Buzato JB, Scarmino IS (2007) Fermentation of molasses by *Zymomonas mobilis*: effects of temperature and sugar concentration on ethanol production. Bioresour Technol 98:2824–2828

CBU (2007) Carbon capture and storage_the "fourth generation" Biofuels-crop biotech update, ISAAA. http://www.isaaa.org/kc/cropbiotechupdate/article/

Chandel AK, Chan ES, Rudravaram R et al (2007) Economics and environmental impact of bioethanol production technologies: an appraisal. Biotechnol Mol Biol Rev 2:14–32

Chandrasekaran M, Bahkali AH (2013) Valorization of date palm (*Phoenix dactylifera*) fruit processing by-products and wastes using bioprocess technology--Review. Saudi J Biol Sci 20:105–120

Chang J-W, Lin Y-H, Huang L-Y, Duan K-J (2011) The effect of fermentation configurations and FAN supplementation on ethanol production from sorghum grains under very-high-gravity conditions. J Taiwan Inst Chem Eng 42:1–4

Chaturvedi V, Verma P (2013) An overview of key pretreatment processes employed for bioconversion of lignocellulosic biomass into biofuels and value added products. 3 Biotech 3:415–431. https://doi.org/10.1007/s13205-013-0167-8

Chen M, Smith PM, Wolcott MP (2016) US biofuels industry: a critical review of the opportunities and challenges. Bioprod Bus 1:42–59

Cheng-Wu Z, Zmora O, Kopel R, Richmond A (2001) An industrial-size flat plate glass reactor for mass production of *Nannochloropsis* sp.(Eustigmatophyceae). Aquaculture 195:35–49

Cherney J, Small E (2016) Industrial hemp in North America: production, politics and potential. Agronomy 6(4):1–24

Chniti S, Djelal H, Hassouna M, Amrane A (2014) Residue of dates from the food industry as a new cheap feedstock for ethanol production. Biomass Bioenergy 69:66–70

Danquah M, Liu B, Harun R (2011) Analysis of process configurations for bioethanol production from microalgal biomass. In: Progress in biomass and bioenergy production. InTech

DBT (2017) http___www. http://www.dbtindia.nic.in/first-biomass-to-ethano

Demirbas A (2009) Political, economic and environmental impacts of biofuels: a review. Appl Energy 86:S108–S117

Demirbas A (2005) Bioethanol from cellulosic materials: a renewable motor fuel from biomass. Energy Sources 27:327–337

Demirbas MF (2011) Biofuels from algae for sustainable development. Appl Energy 88:3473–3480

Dien BS, Cotta MA, Jeffries TW (2003) Bacteria engineered for fuel ethanol production: current status. Appl Microbiol Biotechnol 63:258–266

Dogaris I, Vakontios G, Kalogeris E et al (2009) Induction of cellulases and hemicellulases from *Neurospora crassa* under solid-state cultivation for bioconversion of sorghum bagasse into ethanol. Ind Crop Prod 29:404–411

Duff SJB, Murray WD (1996) Bioconversion of forest products industry waste cellulosics to fuel ethanol: a review. Bioresour Technol 55:1–33

Duvernay WH, Chinn MS, Yencho GC (2013) Hydrolysis and fermentation of sweet potatoes for production of fermentable sugars and ethanol. Ind Crop Prod 42:527–537

Dziugan P, Balcerek M, Pielech-Przybylska K, Patelski P (2013) Evaluation of the fermentation of high gravity thick sugar beet juice worts for efficient bioethanol production. Biotechnol Biofuels 6(158):1–10

Ekefre DE, Mahapatra AK, Latimore M Jr et al (2017) Evaluation of three cultivars of sweet sorghum as feedstocks for ethanol production in the Southeast United States. Heliyon 3 (e00490):1–18

El-sayed WMM (2016) Evaluation of bioethanol production from Ulva lactuca by *Saccharomyces cerevisiae*. Journal of Biotechnol Biomater 6(2):1–10. https://doi.org/10.4172/2155-952X.1000226

Elemike EE, Oseghale OC, Okoye AC (2015) Utilization of cellulosic cassava waste for bio-ethanol production. J Environ Chem Eng 3:2797–2800

Enquist-Newman M, Faust AME, Bravo DD et al (2014) Efficient ethanol production from brown macroalgae sugars by a synthetic yeast platform. Nature 505(7482):239–243

Eom I-Y, Yu J-H, Jung C-D, Hong K-S (2015) Efficient ethanol production from dried oil palm trunk treated by hydrothermolysis and subsequent enzymatic hydrolysis. Biotechnol Biofuels 8 (83):1–11

Fengel D, Wegener G (2011) Wood: chemistry, ultrastructure, reactions. Walter de Gruyter, Berlin, pp 1–613

Foerster H (2010) Granular starch hydrolysis (GSHE) for conversion of grains to ethanol. In: Prezentacja: Near-term opportunities for biorefineries symposium, Champaign, Il, p 2010

Furlan FF, Tonon Filho R, Pinto FHPB et al (2013) Bioelectricity versus bioethanol from sugarcane bagasse: is it worth being flexible? Biotechnol Biofuels 6:142

Gerbens-Leenes PW, Van Lienden AR, Hoekstra AY, der Meer TH (2012) Biofuel scenarios in a water perspective: the global blue and green water footprint of road transport in 2030. Glob Environ Chang 22:764–775

Girio FM, Fonseca C, Carvalheiro F et al (2010) Hemicelluloses for fuel ethanol: a review. Bioresour Technol 101:4775–4800

Graves T, Narendranath NV, Dawson K, Power R (2006) Effect of pH and lactic or acetic acid on ethanol productivity by *Saccharomyces cerevisiae* in corn mash. J Ind Microbiol Biotechnol 33 (6):469–474

Grima EM, Belarbi E-H, Fernández FGA et al (2003) Recovery of microalgal biomass and metabolites: process options and economics. Biotechnol Adv 20:491–515

Gris LRS, de Paim AC, Farenzena M, Trierweiler JO (2013) Laboratory apparatus to evaluate microalgae production. Braz J Chem Eng 30:487–497

Gryta M (2012) Effectiveness of water desalination by membrane distillation process. Membranes (Basel) 2:415–429

Hadar Y (2013) Sources for lignocellulosic raw materials for the production of ethanol. In: Lignocellulose conversion. Springer, Berlin, pp 21–38

Hahn-Hägerdal B, Galbe M, Gorwa-Grauslund MF et al (2006) Bio-ethanol – the fuel of tomorrow from the residues of today. Trends Biotechnol 24:549–556. https://doi.org/10.1016/j.tibtech.2006.10.004

Hahn-Hägerdal B, Karhumaa K, Jeppsson M, Gorwa-Grauslund MF (2007) Metabolic engineering for pentose utilization in *Saccharomyces cerevisiae*. In: Biofuels. Springer, Berlin, pp 147–177

Hamelinck CN, Van Hooijdonk G, Faaij APC (2005) Ethanol from lignocellulosic biomass: techno-economic performance in short-, middle-and long-term. Biomass Bioenergy 28:384–410

Hamley-Bennett C, Lye GJ, Leak DJ (2016) Selective fractionation of sugar beet pulp for release of fermentation and chemical feedstocks; optimisation of thermo-chemical pre-treatment. Bioresour Technol 209:259–264

Han B, Wang L, Li S et al (2010) Ethanol production from sweet sorghum stalks by advanced solid state fermentation (ASSF) technology. Chin J Biotechnol 26:966–973

Hara T, Tanoue K (2006) Laminar flame speed of ethanol, n-heptane, iso-octane air mixtures. JSAE Pap 20068518

Hargreaves PI, Barcelos CA, da Costa ACA, Pereira N Jr (2013) Production of ethanol 3G from *Kappaphycus alvarezii*: evaluation of different process strategies. Bioresour Technol 134:257–263

Harun R, Danquah MK, Forde GM (2010) Microalgal biomass as a fermentation feedstock for bioethanol production. J Chem Technol Biotechnol 85:199–203

Harun R, Jason WSY, Cherrington T, Danquah MK (2011) Exploring alkaline pre-treatment of microalgal biomass for bioethanol production. Appl Energy 88:3464–3467

Hassan SS, Williams GA, Jaiswal AK (2019) Moving towards the second generation of lignocellulosic biorefineries in the EU: drivers, challenges, and opportunities. Renew Sust Energ Rev 101:590–599

Hessami MJ, Cheng SF, Ambati RR et al (2019) Bioethanol production from agarophyte red seaweed, *Gelidium elegans*, using a novel sample preparation method for analysing bioethanol content by gas chromatography. 3 Biotech 9(1):25

Ho S-H, Huang S-W, Chen C-Y et al (2013) Bioethanol production using carbohydrate-rich microalgae biomass as feedstock. Bioresour Technol 135:191–198

Hoover R (2001) Composition, molecular structure, and physicochemical properties of tuber and root starches: a review. Carbohydr Polym 45:253–267

Humbird D, Aden A (2009) Biochemical production of ethanol from corn stover: 2008 state of technology model. NREL/TP-510-46214, pp 1–16

Izmirlioglu G, Demirci A (2012) Ethanol production from waste potato mash by using *Saccharomyces cerevisiae*. Appl Sci 2:738–753

John RP, Anisha GS, Nampoothiri KM, Pandey A (2011) Micro and macroalgal biomass: a renewable source for bioethanol. Bioresour Technol 102:186–193

Johnston DB, McAloon AJ (2014) Protease increases fermentation rate and ethanol yield in dry-grind ethanol production. Bioresour Technol 154:18–25

Jung KA, Lim S-R, Kim Y, Park JM (2013) Potentials of macroalgae as feedstocks for biorefinery. Bioresour Technol 135:182–190

Kagan J (2015) Third and fourth generation biofuels_technologies, markets and economics through 2015. https://www.woodmac.com/our-expertise/focus/Power-

Kalogiannis K, Matsakas L, Aspden J et al (2018) Acid assisted organosolv delignification of beechwood and pulp conversion towards high concentrated cellulosic ethanol via high gravity enzymatic hydrolysis and fermentation. Molecules 23:1647

Kar Y, Deveci H (2006) Importance of P-series fuels for flexible-fuel vehicles (FFVs) and alternative fuels. Energy Sources Part A 28:909–921

Kaur S, Kaushal N, Oberoi HS, Phutela R (2018) Comparative study of simultaneous saccharification and fermentation (SSF) and separate hydrolysis and fermentation (SHF) for rice wine production by *Pichia kudriavezii*. Int J Food Ferment Technol 8:45–50

Kawa-Rygielska J, Pietrzak W (2014) Ethanol fermentation of very high gravity (VHG) maize mashes by *Saccharomyces cerevisiae* with spent brewer's yeast supplementation. Biomass Bioenergy 60:50–57

Keshwani DR, Cheng JJ (2009) Switchgrass for bioethanol and other value-added applications: a review. Bioresour Technol 100:1515–1523

Kim H, Ra CH, Kim S-K (2013) Ethanol production from seaweed (*Undaria pinnatifida*) using yeast acclimated to specific sugars. Biotechnol Bioprocess Eng 18:533–537

Kim JK, Um B-H, Kim TH (2012) Bioethanol production from micro-algae, *Schizocytrium* sp., using hydrothermal treatment and biological conversion. Korean J Chem Eng 29:209–214

Kim M, Day DF (2011) Composition of sugar cane, energy cane, and sweet sorghum suitable for ethanol production at Louisiana sugar mills. J Ind Microbiol Biotechnol 38:803–807

Kim N-J, Li H, Jung K et al (2011) Ethanol production from marine algal hydrolysates using *Escherichia coli* KO11. Bioresour Technol 102:7466–7469

Kim S, Dale BE (2004) Global potential bioethanol production from wasted crops and crop residues. Biomass Bioenergy 26:361–375

Kuhad RC, Singh A (1993) Lignocellulose biotechnology: current and future prospects. Crit Rev Biotechnol 13:151–172

Kumar S, Gupta R, Kumar G et al (2013) Bioethanol production from *Gracilaria verrucosa*, a red alga, in a biorefinery approach. Bioresour Technol 135:150–156

Kumar S, Singh N, Prasad R (2010) Anhydrous ethanol: a renewable source of energy. Renew Sust Energ Rev 14:1830–1844

Kumari D, Singh R (2018a) Pretreatment of lignocellulosic wastes for biofuel production: a critical review. Renew Sust Energ Rev 90:877–891. https://doi.org/10.1016/j.rser.2018.03.111

Kumari D, Singh R (2018b) Pretreatment of lignocellulosic wastes for biofuel production: a critical review. Renew Sust Energ Rev 90:877–891

Lamsal BP, Wang H, Johnson LA (2011) Effect of corn preparation methods on dry-grind ethanol production by granular starch hydrolysis and partitioning of spent beer solids. Bioresour Technol 102:6680–6686

Laopaiboon L, Nuanpeng S, Srinophakun P et al (2009) Ethanol production from sweet sorghum juice using very high gravity technology: effects of carbon and nitrogen supplementations. Bioresour Technol 100:4176–4182

Laopaiboon L, Thanonkeo P, Jaisil P, Laopaiboon P (2007) Ethanol production from sweet sorghum juice in batch and fed-batch fermentations by *Saccharomyces cerevisiae*. World J Microbiol Biotechnol 23:1497–1501

Lareo C, Ferrari MD, Guigou M et al (2013) Evaluation of sweet potato for fuel bioethanol production: hydrolysis and fermentation. Springerplus 2(493):1–11

Lee JY, Kim YS, Um BH, Oh K (2013) Pretreatment of *Laminaria japonica* for bioethanol production with extremely low acid concentration. Renew Energy 54:196–200

Lee OK, Lee EY (2016) Sustainable production of bioethanol from renewable brown algae biomass. Biomass Bioenergy 92:70–75

Lewandowski I, Scurlock JMO, Lindvall E, Christou M (2003) The development and current status of perennial rhizomatous grasses as energy crops in the US and Europe. Biomass Bioenergy 25:335–361

Li S, Li G, Zhang L et al (2013) A demonstration study of ethanol production from sweet sorghum stems with advanced solid state fermentation technology. Appl Energy 102:260–265

Liang L, Zhang Y, Zhang L et al (2008) Study of sugarcane pieces as yeast supports for ethanol production from sugarcane juice and molasses. J Ind Microbiol Biotechnol 35:1605–1613

Liew WH, Hassim MH, Ng DKS (2014) Review of evolution, technology and sustainability assessments of biofuel production. J Clean Prod 71:11–29

Lim Y, Jang Y, Kim K (2013) Production of a high concentration of ethanol from potato tuber by high gravity fermentation. Food Sci Biotechnol 22:441–448
Lin Y, Tanaka S (2006) Ethanol fermentation from biomass resources: current state and prospects. Appl Microbiol Biotechnol 69:627–642
Liu Z, Ho S-H, Hasunuma T et al (2016) Recent advances in yeast cell-surface display technologies for waste biorefineries. Bioresour Technol 215:324–333
Loow Y-L, Wu TY, Jahim JM et al (2016) Typical conversion of lignocellulosic biomass into reducing sugars using dilute acid hydrolysis and alkaline pretreatment. Cellulose 23:1491–1520
Lynd LR (1996) Overview and evaluation of fuel ethanol from cellulosic biomass: technology, economics, the environment, and policy. Annu Rev Energy Environ 21:403–465
Mahdy A, Mendez L, Ballesteros M, González-Fernández C (2014) Enhanced methane production of *Chlorella vulgaris* and *Chlamydomonas reinhardtii* by hydrolytic enzymes addition. Energy Convers Manag 85:551–557
Marin D, Posadas E, Cano P et al (2018) Seasonal variation of biogas upgrading coupled with digestate treatment in an outdoors pilot scale algal-bacterial photobioreactor. Bioresour Technol 263:58–66
Marx S, Brandling J, van der Gryp P (2012) Ethanol production from tropical sugar beet juice. Afr J Biotechnol 11:11709–11720
Maryana R, Ma'rifatun D, Wheni AI et al (2014) Alkaline pretreatment on sugarcane bagasse for bioethanol production. Energy Procedia 47:250–254
Mata TM, Martins AA, Caetano NS (2010) Microalgae for biodiesel production and other applications: a review. Renew Sust Energ Rev 14:217–232
Mathew AK, Abraham A, Mallapureddy KK, Sukumaran RK (2018) Lignocellulosic biorefinery wastes, or resources? In: Waste biorefinery. Elsevier, pp 267–297
Mayzuhroh A, Arindhani S, Caroenchai C, others (2016) Studies on bioethanol production of commercial baker's and alcohol yeast under aerated culture using sugarcane molasses as the media. Agric Agric Sci Procedia 9:493–499
McBride J, Brevnova E, Mellon M, et al (2018) Yeast expressing cellulases for simultaneous saccharification and fermentation using cellulose. US2012129229
McKendry P (2002) Energy production from biomass (part 2): conversion technologies. Bioresour Technol 83:47–54
McLeod JG, May WE, Salmon DF et al (2010) Changes in ethanol production potential due to species, cultivar and location on the Canadian prairie. Can J Plant Sci 90:163–171
Meinita MDN, Marhaeni B, Winanto T et al (2013) Comparison of agarophytes (*Gelidium*, *Gracilaria*, and *Gracilariopsis*) as potential resources for bioethanol production. J Appl Phycol 25:1957–1961
Mishra V, Dubey A, Prajapti SK (2017) Algal biomass pretreatment for improved biofuel production. In: Algal biofuels. Springer, Cham, pp 259–280
Modenbach AA, Nokes SE (2013) Enzymatic hydrolysis of biomass at high-solids loadings-a review. Biomass Bioenergy 56:526–544
Mohapatra S, Ray RC, Ramachandran S (2019) Bioethanol from bio-renewable feedstocks: technology, economics, and challenges. In: Bioethanol production from food crops. Elsevier, London, pp 3–27
Mojović L, Nikolić S, Rakin M, Vukasinović M (2006) Production of bioethanol from corn meal hydrolyzates. Fuel 85:1750–1755
Montagnac JA, Davis CR, Tanumihardjo SA (2009) Nutritional value of cassava for use as a staple food and recent advances for improvement. Compr Rev Food Sci Food Saf 8:181–194
Moshi AP, Hosea KMM, Elisante E et al (2015) High temperature simultaneous saccharification and fermentation of starch from inedible wild cassava (*Manihot glaziovii*) to bioethanol using *Caloramator boliviensis*. Bioresour Technol 180:128–136
Mosier NS, Ileleji KE (2015) How fuel ethanol is made from corn. In: Bioenergy, pp 379–384
Muktham R, Bhargava SK, Bankupalli S, Ball AS (2016) A review on 1st and 2nd generation bioethanol production-recent progress. J Sustain Bioenergy Syst 6:72–92

Mussatto SI, Dragone G, Guimarães PMR et al (2010) Technological trends, global market, and challenges of bio-ethanol production. Biotechnol Adv 28:817–830

Naik SN, Goud VV, Rout PK, Dalai AK (2010) Production of first and second generation biofuels: a comprehensive review. Renew Sust Energ Rev 14:578–597

Nanssou PAK, Nono YJ, Kapseu C (2016) Pretreatment of cassava stems and peelings by thermohydrolysis to enhance hydrolysis yield of cellulose in bioethanol production process. Renew Energy 97:252–265

Nguyen C-N, Le T-M, Chu-Ky S (2014) Pilot scale simultaneous saccharification and fermentation at very high gravity of cassava flour for ethanol production. Ind Crop Prod 56:160–165

Nguyen MT, Choi SP, Lee J et al (2009) Hydrothermal acid pretreatment of *Chlamydomonas reinhardtii* biomass for ethanol production. J Microbiol Biotechnol 19:161–166

Nigam PS, Singh A (2011) Production of liquid biofuels from renewable resources. Prog Energy Combust Sci 37:52–68

Nikolić S, Mojović L, Pejin D et al (2010) Production of bioethanol from corn meal hydrolyzates by free and immobilized cells of *Saccharomyces cerevisiae* var. ellipsoideus. Biomass Bioenergy 34:1449–1456

Njoku SI, Uellendahl H, Ahring BK (2013) Comparing oxidative and dilute acid wet explosion pretreatment of Cocksfoot grass at high dry matter concentration for cellulosic ethanol production. Energy Sci Eng 1:89–98

NREL (2010) Bioenergy reducing enzyme costs increases the market potential of biofuels. https://www.nrel.gov/docs/fy10osti/47572.pdf

Obata O, Akunna J, Bockhorn H, Walker G (2016) Ethanol production from brown seaweed using non-conventional yeasts. Bioethanol 2:134–145

Pandey A, Soccol CR, Mitchell D (2000) New developments in solid state fermentation: I-bioprocesses and products. Process Biochem 35:1153–1169

Paridah MT, Basher AB, SaifulAzry S, Ahmed Z (2011) Retting process of some bast plant fibres and its effect on fibre quality: a review. Bioresources 6:5260–5281

Peng F, Ren J-L, Xu F et al (2009) Comparative study of hemicelluloses obtained by graded ethanol precipitation from sugarcane bagasse. J Agric Food Chem 57:6305–6317

Phitsuwan P, Permsriburasuk C, Waeonukul R et al (2016) Evaluation of fuel ethanol production from aqueous ammonia-treated rice straw via simultaneous saccharification and fermentation. Biomass Bioenergy 93:150–157

Pickett J, Anderson D, Bowles D et al (2008) Sustainable bio-fuels: prospects and challenges. The Royal Society, London, pp 1–90

Pirwitz K, Rihko-Struckmann L, Sundmacher K (2016) Valorization of the aqueous phase obtained from hydrothermally treated *Dunaliella salina remnant* biomass. Bioresour Technol 219:64–71

Plumier BM, Danao M-GC, Rausch KD, Singh V (2015) Changes in unreacted starch content in corn during storage. J Stored Prod Res 61:85–89

Pradeep P, Reddy OVS (2010) High gravity fermentation of sugarcane molasses to produce ethanol: effect of nutrients. Indian J Microbiol 50:82–87

Puligundla P, Smogrovicova D, Obulam VSR, Ko S (2011) Very high gravity (VHG) ethanolic brewing and fermentation: a research update. J Ind Microbiol Biotechnol 38:1133–1144

Rabemanolontsoa H, Saka S (2016) Various pretreatments of lignocellulosics. Bioresour Technol 199:83–91

Rajandran K (2013) Bioethanol production from cassava by fermentation process using *Saccharomyces Cerevisiae*. Universiti Malaysia Pahang, pp 1–24.

Rakin M, Mojovic L, Nikolic S et al (2009) Bioethanol production by immobilized *Sacharomyces cerevisiae* var. ellipsoideus cells. Afr J Biotechnol 8(3):464–471

Ramirez MB, Ferrari MD, Lareo C (2016) Fuel ethanol production from commercial grain sorghum cultivars with different tannin content. J Cereal Sci 69:125–131

Ravagnani M, Reis MHM, Maciel Filho R, Wolf-Maciel MR (2010) Anhydrous ethanol production by extractive distillation: a solvent case study. Process Saf Environ Prot 88:67–73

Ray RC, Swain MR (2011) Bioethanol, bioplastics and other fermented industrial products from cassava starch and flour. In: Cassava farming, uses econ impacts. Nova, Hauppauge, pp 1–32
Razmovski R, Vučurović V (2012) Bioethanol production from sugar beet molasses and thick juice using *Saccharomyces cerevisiae* immobilized on maize stem ground tissue. Fuel 92:1–8
Reina L, Botto E, Mantero C et al (2016) Production of second generation ethanol using Eucalyptus dunnii bark residues and ionic liquid pretreatment. Biomass Bioenergy 93:116–121
Reyimu Z, Ízšimen D (2017) Batch cultivation of marine microalgae *Nannochloropsis oculata* and *Tetraselmis suecica* in treated municipal wastewater toward bioethanol production. J Clean Prod 150:40–46
RFA (2015) Going Global 2015 Bioethanol Industry Outlook. Renew Fuel Assoc
Rizza LS, Smachetti MES, Do Nascimento M et al (2017) Bioprospecting for native microalgae as an alternative source of sugars for the production of bioethanol. Algal Res 22:140–147
Robert D, Perlack LLW, Turhollow AF, et al (2005) Biomass as feedstock for a bioenergy and bioproducts industry: the technical feasibility of a billion-ton annual supply. Oak Ridge Natl Lab
Rodriguez LA, Toro ME, Vazquez F et al (2010) Bioethanol production from grape and sugar beet pomaces by solid-state fermentation. Int J Hydrog Energy 35:5914–5917
Romani A, Garrote G, López F, Parajó JC (2011) *Eucalyptus globulus* wood fractionation by autohydrolysis and organosolv delignification. Bioresour Technol 102:5896–5904
Rubens C (2008) Fourth-generation biofuels. https://gigaom.com/2008/03/04/wtf-are-fourth-gener
Rulli MC, Bellomi D, Cazzoli A et al (2016) The water-land-food nexus of first-generation biofuels. Sci Rep 6(22521):1–10
Saini JK, Saini R, Tewari L (2015) Lignocellulosic agriculture wastes as biomass feedstocks for second-generation bioethanol production: concepts and recent developments. 3 Biotech 5:337–353
Salazar-Ordóñez M, Pérez-Hernández PP, Martin-Lozano JM (2013) Sugar beet for bioethanol production: an approach based on environmental agricultural outputs. Energy Policy 55:662–668
Sánchez C (2009) Lignocellulosic residues: biodegradation and bioconversion by fungi. Biotechnol Adv 27:185–194
Sanchez OJ, Cardona CA (2008) Trends in biotechnological production of fuel ethanol from different feedstocks. Bioresour Technol 99:5270–5295
Sangodoyin AY, Amori AA (2013) Aerobic composting of cassava peels using cowdung, sewage sludge and poultry manure as supplements. Eur Int J Sci Technol 2:22–34
Sankh SN, Deshpande PS, Arvindekar AU (2011) Improvement of ethanol production using *Saccharomyces cerevisiae* by enhancement of biomass and nutrient supplementation. Appl Biochem Biotechnol 164:1237–1245
Santana ÁL, Meireles MAA (2014) New starches are the trend for industry applications: a review. Food Public Heal 4:229–241
Sarkar N, Ghosh SK, Bannerjee S, Aikat K (2012) Bioethanol production from agricultural wastes: an overview. Renew Energy 37:19–27
Sarris D, Papanikolaou S (2016) Biotechnological production of ethanol: biochemistry, processes and technologies. Eng Life Sci 16:307–329
Sasaki K, Tsuge Y, Sasaki D et al (2015) Repeated ethanol production from sweet sorghum juice concentrated by membrane separation. Bioresour Technol 186:351–355
Saunders J, Izydorczyk M, Levin DB (2011) Limitations and challenges for wheat-based bioethanol production. In: Economic effects of biofuel production. InTech, pp 1–27
Scholz MJ, Riley MR, Cuello JL (2013) Acid hydrolysis and fermentation of microalgal starches to ethanol by the yeast *Saccharomyces cerevisiae*. Biomass Bioenergy 48:59–65
Sekhon RS, Breitzman MW, Silva RR et al (2016) Stover composition in maize and sorghum reveals remarkable genetic variation and plasticity for carbohydrate accumulation. Front Plant Sci 7(822):1–10

Semhaoui I, Zarguili I, Rezzoug SA et al (2017) Bioconversion of Moroccan Alfa (*Stipa Tenacissima*) by thermomechanical pretreatment combined to acid or alkali spraying for ethanol production. J Mater 8:2619–2631
Shanavas S, Padmaja G, Moorthy SN et al (2011) Process optimization for bioethanol production from cassava starch using novel eco-friendly enzymes. Biomass Bioenergy 35:901–909
Shapouri H, Duffield JA, Wang MQ, others (2002) The energy balance of corn ethanol: an update. U.S. Department of Agriculture, Office of the Chief Economist, Office of Energy Policy and New Uses. Agricultural Economic Report No. 814, pp 1–19
Shigechi H, Fujita Y, Koh J et al (2004) Energy-saving direct ethanol production from low-temperature-cooked corn starch using a cell-surface engineered yeast strain co-displaying glucoamylase and α-amylase. Biochem Eng J 18:149–153
Shigechi H, Uyama K, Fujita Y et al (2002) Efficient ethanol production from starch through development of novel flocculent yeast strains displaying glucoamylase and co-displaying or secreting α-amylase. J Mol Catal B Enzym 17:179–187
Shikida PFA, Finco A, Cardoso BF et al (2014) A comparison between ethanol and biodiesel production: the Brazilian and European experiences. In: Liquid biofuels: emergence, development and prospects. Springer, London, pp 25–53
Sindhu R, Gnansounou E, Binod P, Pandey A (2016) Bioconversion of sugarcane crop residue for value added products – an overview. Renew Energy 98:203–215
Singh A, Nigam PS, Murphy JD (2011) Renewable fuels from algae: an answer to debatable land based fuels. Bioresour Technol 102:10–16
Singh V (2012) Effect of corn quality on bioethanol production. Biocatal Agric Biotechnol 1:353–355
Sivaramakrishnan R, Incharoensakdi A (2018) Utilization of microalgae feedstock for concomitant production of bioethanol and biodiesel. Fuel 217:458–466
Snoek T, Verstrepen KJ, Voordeckers K (2016) How do yeast cells become tolerant to high ethanol concentrations? Curr Genet 62:475–480
Sørensen A, Teller PJ, Hilstrøm T, Ahring BK (2008) Hydrolysis of *Miscanthus* for bioethanol production using dilute acid presoaking combined with wet explosion pre-treatment and enzymatic treatment. Bioresour Technol 99:6602–6607
Spolaore P, Joannis-Cassan C, Duran E, Isambert A (2006) Commercial applications of microalgae. J Biosci Bioeng 101:87–96
Srichuwong S, Fujiwara M, Wang X et al (2009) Simultaneous saccharification and fermentation (SSF) of very high gravity (VHG) potato mash for the production of ethanol. Biomass Bioenergy 33:890–898
SugarCane.org Ethanol – SugarCane (2019). Accessed on 13/01/2019
Szymanowska-Powalowska D, Lewandowicz G, Blaszczak W, Szwengiel A (2012) Structural changes of corn starch during fuel ethanol production from corn flour. Biotechnol J Biotechnol Comput Biol Bionanotechnol 93:333–341
Szymanowska-Powałowska D, Lewandowicz G, Kubiak P, Błaszczak W (2014) Stability of the process of simultaneous saccharification and fermentation of corn flour. The effect of structural changes of starch by stillage recycling and scaling up of the process. Fuel 119:328–334
Takara D, Khanal SK (2015) Characterizing compositional changes of Napier grass at different stages of growth for biofuel and biobased products potential. Bioresour Technol 188:103–108
TCI (2018) In a first, SpiceJet operates India's first biojet fuel flight. https://yourstory.com/2018/08/indian-farmers-bio-f
Thammasittirong SN-R, Thirasaktana T, Thammasittirong A, Srisodsuk M (2013) Improvement of ethanol production by ethanol-tolerant *Saccharomyces cerevisiae* UVNR56. Springerplus 2 (583):1–5
Thatoi H, Dash PK, Mohapatra S, Swain MR (2016) Bioethanol production from tuber crops using fermentation technology: a review. Int J Sustain Energy 35:443–468
Tikka C, Osuru HP, Atluri N et al (2013) Isolation and characterization of ethanol tolerant yeast strains. Bioinformation 9(8):421–425

Trivedi N, Gupta V, Reddy CRK, Jha B (2013) Enzymatic hydrolysis and production of bioethanol from common macrophytic green alga *Ulva fasciata Delile*. Bioresour Technol 150:106–112

Tye YY, Lee KT, Abdullah WNW, Leh CP (2016) The world availability of non-wood lignocellulosic biomass for the production of cellulosic ethanol and potential pretreatments for the enhancement of enzymatic saccharification. Renew Sust Energ Rev 60:155–172

Upadhyay A, Lama JP, Tawata S (2010) Utilization of pineapple waste: a review. J Food Sci Technol Nepal 6:10–18

Uthumporn U, Zaidul IS, Karim AA (2010) Hydrolysis of granular starch at sub-gelatinization temperature using a mixture of amylolytic enzymes. Food Bioprod Process 88:47–54

Vanhala P, Bergström I, Haaspuro T et al (2016) Boreal forests can have a remarkable role in reducing greenhouse gas emissions locally: land use-related and anthropogenic greenhouse gas emissions and sinks at the municipal level. Sci Total Environ 557:51–57

Verma P, Watanabe T, Honda Y, Watanabe T (2011) Microwave-assisted pretreatment of woody biomass with ammonium molybdate activated by H_2O_2. Bioresour Technol 102:3941–3945

Vohra M, Manwar J, Manmode R et al (2014) Bioethanol production: feedstock and current technologies. J Environ Chem Eng 2:573–584

Wahono SK, Rosyida VT, Afrizal A, Kismurtono M (2012) Residual sugar reduction concentration parameter at the product of simultaneous saccharification and fermentation process of second-generation bioethanol from bagasse cane. In: Proceeding of international conference on sustainable engineering and application. Yogyakarta, pp 27–30

Wahono SK, Rosyida VT, Darsih C et al (2015) Optimization of simultaneous saccharification and fermentation incubation time using cellulose enzyme for sugarcane bagasse on the second-generation bioethanol production technology. Energy Procedia 65:331–336

Wang E-Q, Li S-Z, Tao L et al (2010) Modeling of rotating drum bioreactor for anaerobic solid-state fermentation. Appl Energy 87:2839–2845

Wang P, Chung T-S (2015) Recent advances in membrane distillation processes: membrane development, configuration design and application exploring. J Membr Sci 474:39–56

Wargacki AJ, Leonard E, Win MN et al (2012) An engineered microbial platform for direct biofuel production from brown macroalgae. Science (80-) 335:308–313

Wen Z-Y, Chen F (2003) Heterotrophic production of eicosapentaenoic acid by microalgae. Biotechnol Adv 21:273–294

Wisselink HW, Toirkens MJ, Wu Q et al (2009) Novel evolutionary engineering approach for accelerated utilization of glucose, xylose, and arabinose mixtures by engineered *Saccharomyces cerevisiae* strains. Appl Environ Microbiol 75:907–914

Wu F-C, Wu J-Y, Liao Y-J et al (2014) Sequential acid and enzymatic hydrolysis in situ and bioethanol production from *Gracilaria* biomass. Bioresour Technol 156:123–131

Wyman CE (2018) Ethanol production from lignocellulosic biomass: overview. In: Handbook on bioethanol. Routledge, pp 1–18

Xiao D, Wu S, Zhu X et al (2010) Effects of soya fatty acids on cassava ethanol fermentation. Appl Biochem Biotechnol 160:410–420

Yanagisawa M, Nakamura K, Ariga O, Nakasaki K (2011) Production of high concentrations of bioethanol from seaweeds that contain easily hydrolyzable polysaccharides. Process Biochem 46:2111–2116

Yang L, Liu Y, Ruan R et al (2011) Advances in production of 5-hydroxymethylfurfural from starch. Mod Chem Ind 31:332–337

Yao C, Yang X, Roy Raine R et al (2009) The effects of MTBE/ethanol additives on toxic species concentration in gasoline flame. Energy Fuel 23:3543–3548

Yu J, Tan T, others (2008) Ethanol production by solid state fermentation of sweet sorghum using thermotolerant yeast strain. Fuel Process Technol 89:1056–1059

Yu M, Li J, Chang S et al (2016) Bioethanol production using the sodium hydroxide pretreated sweet sorghum bagasse without washing. Fuel 175:20–25

Zabed H, Faruq G, Boyce AN et al (2016a) Evaluation of high sugar containing corn genotypes as viable feedstocks for decreasing enzyme consumption during dry-grind ethanol production. J Taiwan Inst Chem Eng 58:467–475

Zabed H, Faruq G, Sahu JN et al (2016b) A comparative study on normal and high sugary corn genotypes for evaluating enzyme consumption during dry-grind ethanol production. Chem Eng J 287:691–703

Zabed H, Faruq G, Sahu JN et al (2014) Bioethanol production from fermentable sugar juice. Sci World J 2014(957102):1–11

Zabed H, Sahu JN, Suely A et al (2017) Bioethanol production from renewable sources: current perspectives and technological progress. Renew Sust Energ Rev 71:475–501

Zafar S (2018) Ethanol from lignocellulosic biomass. BioEnergy: https://www.bioenergyconsult.com/production-cellul

Zhang Y-HP, Ding S-Y, Mielenz JR et al (2007) Fractionating recalcitrant lignocellulose at modest reaction conditions. Biotechnol Bioeng 97:214–223

Zhu JY, Chandra MS, Gu F et al (2015) Using sulfite chemistry for robust bioconversion of douglas-fir forest residue to bioethanol at high titer and lignosulfonate: a pilot-scale evaluation. Bioresour Technol 179:390–397

Zhu JY, Pan X, Zalesny RS (2010) Pretreatment of woody biomass for biofuel production: energy efficiency, technologies, and recalcitrance. Appl Microbiol Biotechnol 87:847–857

Algal Biomass: Potential Renewable Feedstock for Biofuels Production – Part I

8

Komal Agrawal, Ankita Bhatt, Nisha Bhardwaj, Bikash Kumar, and Pradeep Verma

Abstract

The present age has seen technological development at a large scale and has imposed tremendous pressure on the natural resources. The usage of fossil fuel has resulted in the release of greenhouse gases, thereby promoting global warming and increased environmental concern among the researchers. Therefore, the quest to find "clean energy" has become the chief concern of the environmentalist. In order to address the issue, third-generation biofuel involving microalgae has been regarded as one of the most effective biological sources as it only requires sunlight, carbon dioxide, and nutrition for its growth. The biofuel derived from the living organism has numerous advantages as it effectively decreases the concentration of emitted greenhouse gases. The microalgae have numerous applications and can be effectively used in biofuels, cosmetics, and pharmaceuticals and as human and animal nutritional sources. Thus, the present chapter would focus on microalgae production processes, advantages and disadvantages of natural and artificial cultivation system, various harvesting techniques followed by its application in various sectors, and lastly the limitations and its future prospects.

Keywords

Cultivation · Harvesting · Extraction · Biofuel

Authors "Komal Agrawal" and "Ankita Bhatt" contributed equally to this work.

K. Agrawal · A. Bhatt · N. Bhardwaj · B. Kumar · P. Verma (✉)
Bioprocess and Bioenergy Laboratory, Department of Microbiology, Central University of Rajasthan, Ajmer, Rajasthan, India
e-mail: pradeepverma@curaj.ac.in

N. Srivastava et al. (eds.), *Biofuel Production Technologies: Critical Analysis for Sustainability*, Clean Energy Production Technologies,
https://doi.org/10.1007/978-981-13-8637-4_8

List of Abbreviation

AD	Anaerobic digestion
ALPBRs	Airlift photobioreactors
BCPBRs	Bubble column photobioreactors
CPBRS	Closed photobioreactor system
CPBRs	Closed photobioreactors
FAME	Fatty acid methyl ester
FPBRs	Flat plate photobioreactors
F^{st}GB	First-generation biofuel
GHGE	Greenhouse gas emission
HTC	Hydrothermal carbonization
HTL	Hydrothermal liquefaction
HTPBRs	Helical type photobioreactors
HTSCS	Hybrid two-stage cultivation system
ICS	Indoor cultivation system
MAOSE	Microwave-assisted organic solvent extraction
OPCS	Open pond cultivation system
OS	Outdoor system
PBPBRs	Plastic bag photobioreactors
PBR	Photobioreactor
PRPBRs	Penthouse-roof photobioreactors
PTC	Phototrophic cultures
PUFAs	Polyunsaturated fatty acids
SCFE	Supercritical fluid extraction
S^{nd}GB	Second-generation biofuel
STPBRs	Stirred tank photobioreactors
TPBRs	Tubular photobioreactors
T^{rd}GB	Third-generation biofuel

8.1 Introduction

The present age has seen high technological developments in both industrial and agricultural sectors. The advancement in the process has led to environmental pollution in the form of greenhouse gas emission (GHGE) and depletion of natural resources (i.e., deforestation). The exhaustion of nonrenewable natural sources to produce fuel has generated huge concern among the researchers worldwide. Globally, the community is now focused on building an energy-efficient future, without having any reliability on the nonrenewable natural resources and being carbon neutral as well. Thus, biofuel from renewable feedstock is developed and can be classified into four types depending on the sources used (Mathimani et al. 2018) (Fig. 8.1).

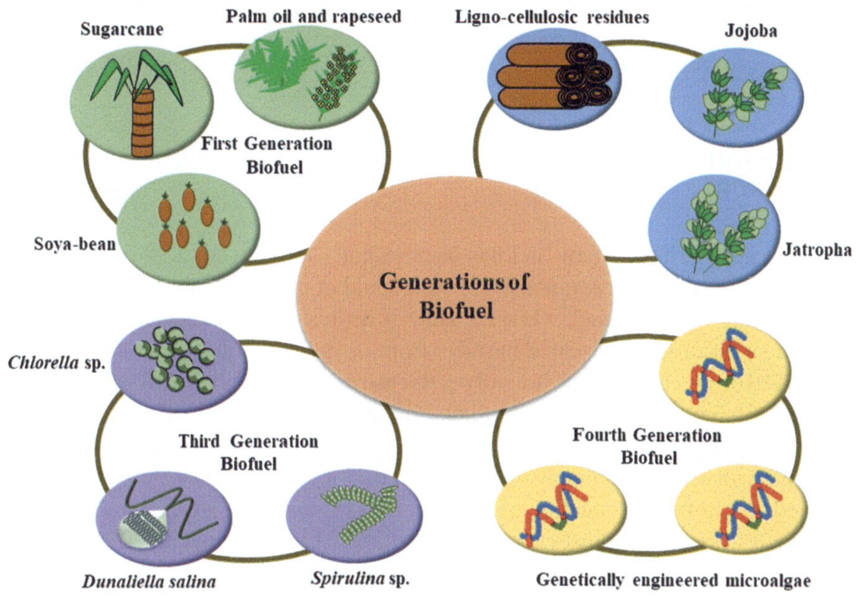

Fig. 8.1 The representation of generations of biofuels and its example

8.1.1 First-Generation Biofuel (FstGB)

The production of biofuel relies on food crops which depends on the farmland and thus promotes the conversion of forest areas into agricultural land and competes in the "food vs. fuel" controversy (Maity et al. 2014).

8.1.2 Second-Generation Biofuel (SndGB)

In order to overcome the above limitation, SndGB was developed based entirely on nonedible sources (e.g., lignocellulosic sources and non-food have been considered under SndGB). However, it also requires larger surface area for the cultivation, the expense associated with the equipment is very high, and the biofuel generated cannot replace the fossil fuel completely (Mathimani et al. 2018).

8.1.3 Third-Generation Biofuel (TrdGB)

Thus, to overcome the limitations, the concept of microalgae gained interest among the researchers. They are photosynthetically more competent, effective fixers of carbon dioxide and miniature biochemical factories (Demirbas 2011). In comparison to the FstGB and SndGB, the cost associated with the processing is low, and it can directly generate 10–20 times more biofuel end products (Chisti 2007; Gouveia and Oliveira 2009; Posten and Schaub 2009).

8.1.4 Fourth-Generation Biofuel (FthGB) Photosynthetic Biofuel (PCB)

The concept of FthGB is new and has been recently introduced; TrdGB and FthGB have photosynthetic microorganisms in common, but the former uses algae biomass for the production of biofuel, while the latter is dedicated to the metabolic engineering of algae in order to produce biofuels from oxygenic photosynthetic organisms (Kagan 2010; Lu et al. 2011). It involves the use of recombinant DNA and bioengineering techniques to enhance biofuel production (Lu et al. 2011). The main advantage associated with FthGB is that the product will be secreted out of the cells and will thus completely reduce the cost of fermentation/processing steps involved in biofuel production (Lu et al. 2011), and it also enhances biofuel production from algal strains (Anandarajah et al. 2012; Daroch et al. 2013).

Therefore, the present chapter discusses various methods for the cultivation, harvesting, and utilization of algal biosystem as an effective feedstock to produce biofuel. Further, the application, limitations, and future prospect of the algal system are discussed in detail in the chapter.

8.2 The Cultivation System of Microalgae

Three basic mechanisms are involved in the cultivation of microalgae, namely, photoautotrophic, heterotrophic, and mixotrophic. The photoautotrophic production relies on autotrophic photosynthesis, while the heterotrophic production relies on the presence of organic substances, and the combination of the two is a mixotrophic process (Brennan and Owende 2010) (Fig. 8.2).

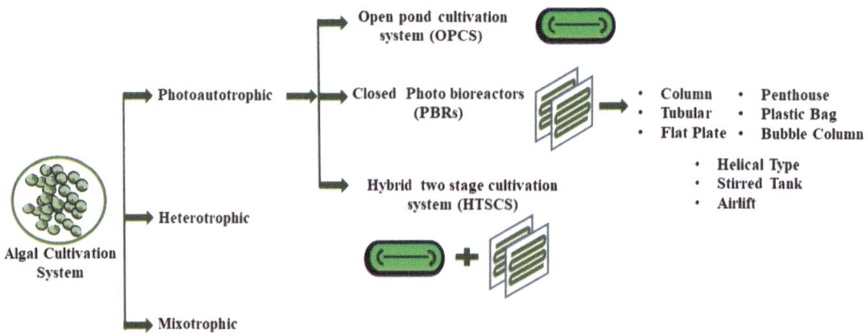

Fig. 8.2 Representation of various cultivation systems for algal biomass

Table 8.1 Comparative analysis of various types of open pond cultivation system (OPCS)

OPCS	Operation	Properties	Productivity	Tested species
Raceway	Employs paddle wheel to ensure continuous flow of water	Exhibits efficient mixing, low hydrodynamic stress, and appropriate light distribution, but the gas transfer is inappropriate	14–50 $g/m^2/d$	*Chlorella* sp. *Spirulina* sp. *Dunaliella* sp.
Circular	Consists of paddle wheel, often used for wastewater treatment	Mixing and gas transfer are poor, low hydrodynamic stress, light distribution is improper	21 $g/m^2/d$	*Spirulina* sp. *Chlorella* sp.
Inclined	Operation is either aided by a pump or based on the principle of gravity	Mixing and gas transfer are poor, low hydrodynamic stress	31 $g/m^2/d$	*Chlorella* sp. *Haematococcus* sp. *Spirulina* sp.
Unmixed	Treatment of wastewater	Mixing and gas transfer are poor, low hydrodynamic stress	< 1 $g/m^2/d$	*Dunaliella* sp. *Spirulina* sp.

8.2.1 Photoautotrophic Cultivation System of Microalgae

The phototrophic cultures (PTC) store the energy obtained from light in the form of ATP or NADPH, which further enter the Calvin cycle for glucose production. However, this process is restricted by the amount of available light and the carbon dioxide supply (Suali and Sarbatly 2012). The reliability of the phototrophic cultures depends on light and results in less lipid production in comparison to the heterotrophic cultures (Brennan and Owende 2010). However, the PTC depending on the algae strain, land and water cost, and climatic conditions can be cultivated by three different methods (i.e., natural, artificial, and hybrid).

8.2.1.1 Natural System

Natural system is also called as the outdoor system (OS) or the open pond cultivation system (OPCS) (Table 8.1). It is the oldest method utilized for large-scale cultivation of algae (Junying et al. 2013; Mathimani et al. 2018). This method can be combined with wastewater treatment, thereby offering double benefit as the microalgal cells can utilize nitrogen and phosphorus present in wastewater along with its treatment (Manninen et al. 2016; Mathimani et al. 2018). This system can have numerous advantages and disadvantages, which have been presented in Fig. 8.3. However, the OPCS can be used to cultivate monocultures if extreme conditions are maintained, e.g., *Chlorella*, *D. salina*, and *Spirulina* have the potential to endure extreme environmental conditions (Brennan and Owende 2010).

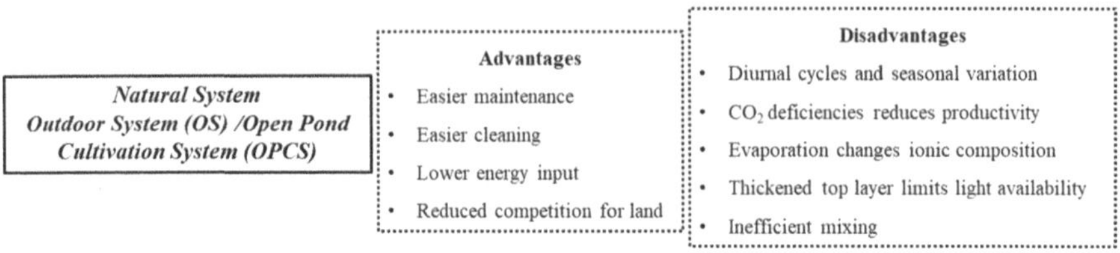

Fig. 8.3 Various advantages and disadvantages associated with the open pond cultivation system (OPCS)

8.2.1.2 Artificial System

Artificial system is also called indoor cultivation system (ICS) or closed photobioreactor system (CPBRS) and is the most stringently controlled and monitored method of algal biomass cultivation. It was designed to address the limitations of the OPCS such as pollution and contamination risks that impede their use in pharmaceuticals and cosmetics industries. In addition, CBPRS allows monoculture cultivation of algal species for longer durations along with a lower contamination risk (Chisti 2007; Brennan and Owende 2010; Ugwu et al. 2008). As the artificial means of production employs fluorescent lamps for continuous production, it requires higher-energy input and increased carbon footprint (Brennan and Owende 2010). On the basis of duration and method employed, three types of cultivation system were developed (i.e., batch, continuous, and semicontinuous) (Chisti 2013). In batch system, the target culture is inoculated in the growth media and allowed to grow until maximum population density is attained (Mathimani et al. 2018), and the products are harvested in a single step (Suali and Sarbatly 2012). It is then followed by a new cycle of cultivation with fresh inoculum of the microalgal cells and the growth medium. In the continuous system, continuous inflow of the culture medium with a simultaneous growth of biomass is maintained (Mathimani et al. 2018). Despite the wide use of the batch culture method, the production costs were reduced by approximately 40% in continuous culture systems (Suali and Sarbatly 2012). Lastly, in the semicontinuous system, culture medium is added continuously, but only a partial amount of biomass is harvested to attain a constant growth rate of the culture (Mathimani et al. 2018). These systems can be operated as various types of photobioreactors (PBRs), namely, the column, tubular, flat plate, penthouse-roof, plastic bag, bubble column, helical type, stirred tank, and airlift (Bahadar and Khan 2013).

8.2.1.2.1 Column Photobioreactors (CPBRs)

CPBRs are easy to operate, compact, and cheap. The CPBRs are composed of a series of straight glass or plastic tubes arranged horizontally, vertically, inclined, or helically, which is responsible for capturing sunlight. A mechanical pump/airlift system performs the recirculation of the algal culture and allows the exchange of carbon dioxide and oxygen between the aeration gas and the growth medium (Ugwu et al. 2002, 2008; Eriksen 2008; Brennan and Owende 2010).

8.2.1.2.2 Tubular Photobioreactors (TPBRs)

TPBRs are mainly used for outdoor cultivation processes as they have high surface-to-volume ratio exposed to sunlight and high photosynthetic efficacy (Brennan and Owende 2010; Rastogi et al. 2018). Further, they provide an excellent temperature control and have lower risks of photoinhibition and contamination, thereby allowing monoalgal culture cultivation (Rastogi et al. 2018). These TPBRs further have a drawback of possible fouling accompanied by algal growth along the walls (Bahadar and Khan 2013), possibility of high dissolved oxygen levels, cell damage by shear stress of pumping, and large space requirement for setup (Rastogi et al. 2018).

8.2.1.2.3 Flat Plate Photobioreactors(FPBRs)

The algal culture proceeds across the flat plate in the form of a thin and dense layer, thereby resulting in absorption of radiation within the initial few millimeters of thickness (Richmond et al. 2003; Brennan and Owende 2010). Contrary to the tubular versions, the flat plate PBRs involve lower dissolved oxygen accumulation and lower chances of contamination (Rastogi et al. 2018), it provides higher photosynthetic efficiencies, and are thus widely used for algal mass cultivation (Richmond 2000; Brennan and Owende 2010). Despite various advantages, its limitations include difficulties in scale-up, low surface-to-volume ratios, poor control over temperature, hydrodynamic stress, and wall growth (Rastogi et al. 2018).

8.2.1.2.4 Penthouse-Roof Photobioreactors (PRPBRs)

PRPBRs are mostly used in regions of temperate climatic conditions and consist of both indoor and outdoor units. The various parameters of cultivation such as flow rate, temperature, and oxygen levels can be easily maintained, and collectors are used to focus and direct the light (Bahadar and Khan 2013).

8.2.1.2.5 Plastic Bag Photobioreactors (PBPBRs)

PBPBRs are attractive and are used commercially due to their lower production costs. However, the disadvantages are photo-limitation, insufficient mixing, possibility of leakage, shorter life span, etc. Further, the major hurdle in the use of these systems is the disposal of large amounts of plastic bags (Wang et al. 2012; Huang et al. 2017).

8.2.1.2.6 Bubble Column Photobioreactors (BCPBRs)

BCPBRs are cylindrical vessels with a height that is more than twice the diameter, where gas mixture is bubbled through the sparger and an external light source is used to determine its photosynthetic efficiency (Singh and Sharma 2012; Janssen et al. 2002).

8.2.1.2.7 Helical-Type Photobioreactors (HTPBRs)

HTPBRs are comprised of coiled tubes that are flexible and transparent and have a small diameter. The degassing unit is either separate or attached, and the culture is made to traverse through the long tube toward the degassing unit by a centrifugal pump. Better photosynthetic efficiency can be obtained when carbon dioxide mixture and the culture medium are introduced at the bottom of the bioreactor (Morita et al. 2001; Singh and Sharma 2012).

8.2.1.2.8 Stirred Tank Photobioreactors (STPBRs)

STPBRs involve mechanical agitation provided by various kinds of impellers and help algae to obtain the carbon needed for its growth from the carbon dioxide-enriched air which is bubbled in the system from the bottom of the reactor. This type of PBR employs optical fibers or fluorescent lamps for illumination in the system (Singh and Sharma 2012).

8.2.1.2.9 Airlift Photobioreactors (ALPBRs)

These photobioreactors are comprised of two distinct interconnected zones, namely, the riser and the downcomer. The gas mixture is sparged in the riser, while no gas is received by the downcomer. Two main forms of airlift bioreactors consist of the internal loop and the external loop structure. A draft tube/split cylinder separates the riser and the downcomer in the internal loop ALPBR. Moreover, the external loop ALPBR consists of two tubes that physically separate the riser and downcomer (Singh and Sharma 2012).

8.2.1.3 Hybrid Two-Stage Cultivation System (HTSCS)

HTSCS combines CPBRS and OPCS. The first stage comprises of CPBRS in which the contamination and pollution risks are significantly reduced owing to the stringently controlled culture conditions, and continuous cell division is favored by the CPBRS. The second production stage is focused at enhancing the synthesis of the desired lipid product which is achieved by subjecting the microalgal cells to nutrient stresses (Rodolfi et al. 2009; Brennan and Owende 2010). OPCS are convenient for the second stage as transfer of microalgal cultures from the CPBRS to the OPCS results in the generation of environmental stresses that enhance production (Brennan and Owende 2010). On the basis of the comparative analysis of natural and artificial algal cultivation system, various advantages and disadvantages are associated which have been represented in Table 8.2.

8.2.2 Heterotrophic Cultivation System of Microalgae

This method is extensively used for the production of algal biomass and metabolites (Miao and Wu 2006; Brennan and Owende 2010). This type of method employs fermenters or stirred tank bioreactors for the cultivation of microalgae on organic carbon substrates such as glucose glycerol and sweet sorghum. The type and concentrations of the source of carbon determine the content of lipid obtained and the yield of biomass (Suali and Sarbatly 2012). The scale-up of the systems is easy as the algal growth is lightly independent, thereby allowing smaller surface area-to-volume ratios (Eriksen 2008). Advantages of these systems include higher biomass productivities and high degree of control over the growth of microalgal cells (Chen and Chen 2006; Brennan and Owende 2010). One limitation of the heterotrophic

Table 8.2 Advantages and disadvantages associated with OPCS and PBRs

Parameters	Open pond cultivation system (OPCS)	Closed photobioreactor system (CPBRS)
Biomass productivity	Low	High
Contamination	Highly susceptible	Less susceptible
Growth parameters (pH, temperature, mixing, carbon dioxide, oxygen)	Difficult to monitor and control	Easily controlled
Maintenance	Easy to operate and maintain	Difficult to operate due to technicalities
Agitation and flow	Paddle wheel, water jet, air pumps	Compressible circulators, air pumps, spargers
Building and operating costs	Low	High
Scale-up	Easy	Difficult
Drawbacks	Cell damage due to shear stress, overtime deterioration of materials	Overheating, biofouling, oxygen accumulation

system that precludes its use is the cost of the carbon source. However, the production cost can be reduced by using industrial waste or coproduct of refinery plants as carbon source for the heterotrophic cultures. For instance, the microalgal productivity increases upon utilization of glycerol (crude) as the carbon source which is obtained as a coproduct in biodiesel refinery plant (Suali and Sarbatly 2012). Relative to the photosynthetic production, heterotrophic cultivation system consumes more energy owing to the photosynthetic production of the initial organic carbon source (Chisti 2007; Brennan and Owende 2010).

8.2.3 Mixotrophic Cultivation System of Microalgae

Many algal species can employ either phototrophic or autotrophic method of growth. This implies that any metabolism process can be utilized as they possess photosynthetic ability and can also consume prey or exploit organic resources for growth (Graham et al. 2009). Either light or organic carbon source can assist in microalgal growth, and it is not stringently determined by photosynthesis, thereby exempting light from being a growth limiting factor (Andrade and Costa 2007). The green algae *Spirulina platensis* and *Chlamydomonas reinhardtii* are the representative organisms displaying mixotrophic metabolism (Chen et al. 1996). In the mixotrophic type of production, lesser amount of biomass is lost during the dark phase because the microalgal growth is affected by media supplements along with glucose in both the light and dark phases (Andrade and Costa 2007). As compared with the CPBRs and OPCS of photoautotrophic microorganisms, the mixotrophic microalgal growth rates are comparable with respect to the CPBR and higher than OPCS, but these rates are significantly lower contrary to the heterotrophic production. Mixotrophic method serves as a vital part of the microalgal biofuel production owing to its potential features such as lower biomass loss during the respiratory phase (dark phase) and reduced utilization of organic substrates during growth phase of the microalgal cells (Brennan and Owende 2010).

8.3 Harvesting of Algal Biomass for Efficient Production of Biofuel

Harvesting accounts for >30% of the total production cost in the open pond systems (Zittelli et al. 2006; Shuba and Kifle 2018; Mathimani and Mallick 2018). The harvesting of microalgae involves a two-stage process (i.e., bulk harvesting and thickening). Biomass is separated from the bulk of the suspension in the process of bulk harvesting, while thickening is characterized as the process of concentrating the slurry (Brennan and Owende 2010; Shuba and Kifle 2018). As the microalgal cells are small in size and have low density, this step incurs additional costs in the production process. Thus, cost-effective processes for dewatering and harvesting need to be chosen to make the entire process economically viable (Shuba and Kifle 2018). The following are the widely used methods for the biomass harvesting and recovery (Fig. 8.4):

8.3.1 Flocculation for Harvesting Microalgae

Flocculation is characterized as the process of aggregate formation. It is used as a pretreatment for the increment of cell density by physical, chemical, or natural means (Bhatt et al. 2014). Flocculation is induced by flocculants which may be organic (starch or chitosan) or inorganic (Al^{3+}, Zn^{2+}, Fe^{3+}) (Vandamme et al. 2009; Morales et al. 1985; Knuckey et al. 2006). The adsorption of ions from the growth medium and the functional groups on the cell wall of microalgae generally make the algal surface negatively charged. These negatively charged surfaces are neutralized by the application of cationic polymers and electrodes having positive charge,

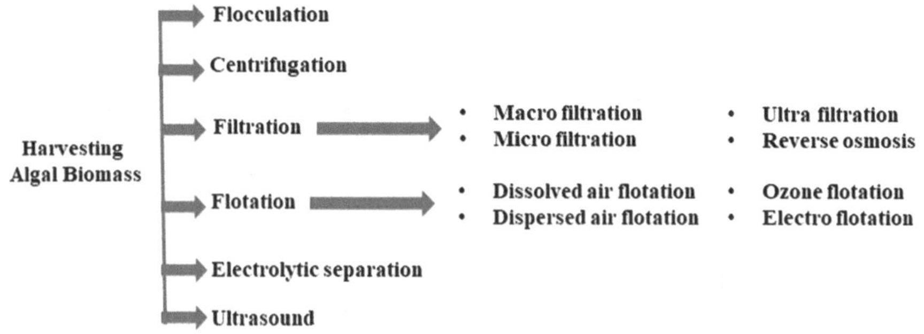

Fig. 8.4 Schematic representation of various methods for harvesting algal biomass

thereby helping in the process of flocculation (Shuba and Kifle 2018). Synthetic polymers are used which are highly efficient and produce stable flocs but are associated with various disadvantages such as hazardous nature that result in the production of a low-quality harvest product. The alternative to the synthetic polymers is the natural polymer (i.e., chitosan), which is a product of chitin deacetylation and can also be produced from fungi under anaerobic conditions (Rinaudo 2006; Rashid et al. 2013, 2014). Flocculation can be of two types: bio-flocculation and auto-flocculation. Bio-flocculants employ the activities of a single flocculating microalga which further lead to concentration of the desired non-flocculating microalgal biomass (Shuba and Kifle 2018; Salim et al. 2010). The availability of nutrient (mainly nitrogen and phosphorous) and physical parameters is mainly responsible for bio-flocculation. Exopolysaccharide (EPS) is produced by the microalgal cells in conditions of nutrition deficiency, and the EPS further helps in the process of bio-flocculation (Rashid et al. 2014). In case of auto-flocculation, the mature microalgae are exposed to sunlight for a longer period with low concentrations of carbon dioxide. This process is beneficial for large-scale harvesting as it leads to substantial reductions in the production costs and is governed by the presence of light (Gouveia 2011; Milano et al. 2016).

8.3.2 Harvesting of Microalgae Using Centrifugation

Centrifugation involves separation of the particles on the basis of size and density. The microalgae are subjected to higher centripetal acceleration which results in the separation of the cells into a greater density and low-density area (Milano et al. 2016). For the size and type of the particles, centrifugation techniques can be of various types such as imperforate basket, decanter, tubular, nozzle type, multichamber, solid-ejecting type disk, and solid-retaining disk (Shelef et al. 1984; Shuba and Kifle 2018). The size of the desired algal species determines the efficiency of the process. This method of algal separation cannot be employed on a large scale as it is a relatively expensive (Uduman et al. 2010) and energy-consuming technique (Heasman et al. 2000; Shuba and Kifle 2018).

8.3.3 Harvesting of Microalgae Using Filtration

Filtration is the process in which the solution containing the microalgal cells is passed through a filter, thereby trapping the cells and allowing only the medium to pass through. Filtration techniques can be divided into four types (i.e., macro-filtration (>10 μm), microfiltration (size of the pores ranges from 0.1 to 10 μm), ultrafiltration (0.02–2 μm), and reverse osmosis (<0.001 μm) (Harun et al. 2010; Shuba and Kifle 2018). Filtration techniques are limited by the high operational costs involved and longer duration of processing (Shuba and Kifle 2018). Microalgae production mainly employs rotary filters and micro-strainers as these are cost effective and easy to operate owing to the fine mesh containing micro-strainers

(Grima et al. 2003; Patel et al. 2017). The secondary alternatives such as ultrafiltration and membrane filtration techniques can also be employed, but they are expensive due to the presence of many filters along with the primary filter (Patel et al. 2017). The surface charge of the microalgal cells, culture age, size, temperature, contact angle, hydrophilic or hydrophobic nature of the membrane, and concentration of the microalgal cells are the factors that influence the process of filtration. Fouling acts as a major bottleneck in the filtration process and demands frequent membrane replacements and backwashing that increases the production costs involved. Fouling can be reduced to some extent by application of pressure on the filter of the system (Rashid et al. 2014).

8.3.4 Harvesting of Microalgae Using Flotation

This method depends on the mechanism of interaction in between the negatively charged microalgal surfaces that are hydrophilic in nature (Patel et al. 2017). The size of the bubbles (microbubbles/nanobubbles/fine bubbles) is responsible for the determination of the efficiency of the harvesting process (Shuba and Kifle 2018). The technique of flotation offers many benefits (e.g., inexpensive, easy to operate, and involves less processing time), but it is also associated with various drawbacks (e.g., difficulties in scale-up operations and higher-energy consumption) (Rashid et al. 2014); however, various methods for performing flotation were designed which includes the following:

8.3.4.1 Dissolved Air Flotation

In this method, liquid stream is injected via a nozzle in the microalgal suspension. This stream saturated with air and the generated air bubbles from the nozzle rise to the surface after attachment with the microalgal cells (Pragya et al. 2013).

8.3.4.2 Dispersed Air Flotation

The injection of unpressurized air results in the generation of larger bubbles, thereby resulting in relatively lower efficiency (Laamanen et al. 2016).

8.3.4.3 Ozone Flotation

In this process, the proteins are released after the disruption of cell walls of microalgae by applying ozone, and this protein further acts as bio-flocculant (Singh and Patidar 2018).

8.3.4.4 Electro Flotation

In this process, the electrolysis of water leads to production of hydrogen gas bubbles that further carry the microalgal cells to the surface for skimming (Uduman et al. 2010; Rashid et al. 2014).

8.3.5 Electrolytic Separation of Algal Biomass

Microalgal biomass are extracted from the culture medium using electric fields. In the process, hydrogen ions produced by water electrolysis get attached to the algal flocks that make the biomass move toward the surface. This method is highly efficient, economically feasible, and environment friendly (Chen et al. 2011; Patel et al. 2017). Electrolysis can be categorized as electrolytic flocculation and electrolytic coagulation. The positively charged coagulants at the sacrificial anode get complexed with the hydroxide ions and react with the microalgal cells (Lee et al. 2012).

8.3.6 The Use of Ultrasound for Harvesting Algal Biomass

In this process, the microalgal biomass is harvested at low amplitude and low frequency using ultrasonication. The microalgal cells are disrupted upon sonication, and their buoyancy is decreased; thus, they become stable and settle down, resulting in increased efficiency of harvesting process. Cases of high-frequency and high-amplitude sonification lead to disruption of the microalgal cells and subsequent lipid release in the aqueous medium (Adam et al. 2012; Rashid et al. 2014). As sonification is deleterious, it is not employed for the harvesting process, and commercialization has not been feasible due to large input of energy and high cost (Milledge and Heaven 2012; Rashid et al. 2014).

8.4 Extraction of Algal Oil for Production of Biofuel

The production of biofuels from the algal biomass involves the removal of water from the biomass, and lipid is extracted from the cells, and further its recovery or lipid concentration constitutes the process of extraction (Patel et al. 2017). The oil extraction acts as a bottleneck for the production of biofuel from potential microalgal species due to the high demand of effort and cost (Shuba and Kifle 2018). Hence, an appropriate balance between the cost involved and drying efficiency is vital to obtain the maximum possible energy output from biofuels (Li et al. 2008; Brennan and Owende 2010). Several methods used for the extraction of oil from biomass include physical, chemical (with organic solvents), and supercritical methods (Fig. 8.5); among them, organic solvent extraction is the most widely employed method (Borowitzka and Moheimani 2013; Milano et al. 2016).

8.4.1 Extraction of Algal Oil by Physical Method

In the physical method, cells are disrupted using methods such as bed mills, ultrasound (Hosikian et al. 2010), and cell homogenizers and autoclaved to release intracellular lipids from microalgal cells (Patel et al. 2017). In this method, oil is

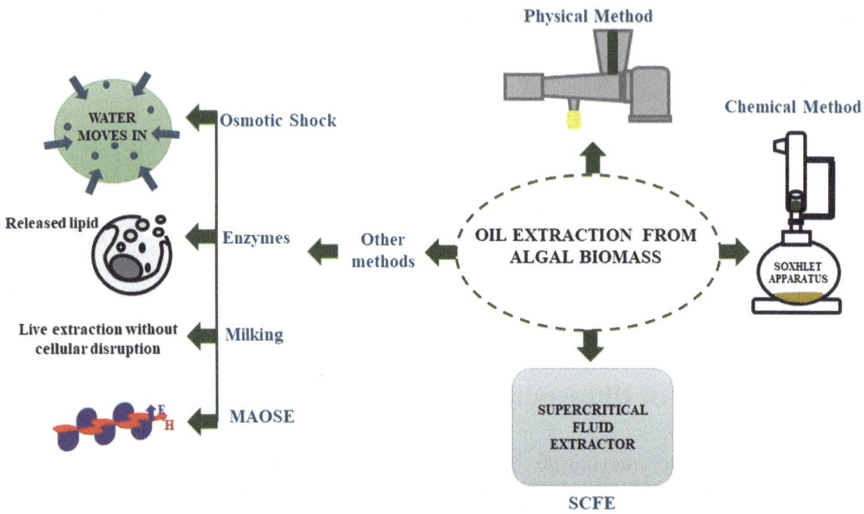

Fig. 8.5 Representation of various methods employed for the extraction of oil from algal biomass

extracted via mechanical crushing to separate oil from the non-oil biomass. It can be achieved by drying the algal biomass followed by pressing it in an oil press. The disruption vessels are used for commercial scale extraction process (Richmond 2004; Shuba and Kifle 2018). These vessels are designed in such a way that they consist of a stream of target materials which are in a continuous flow (Shuba and Kifle 2018). The method employed for the extraction of algal biomass relies on the algal strain (cell wall type) and the nature of target product (Patel et al. 2017).

8.4.2 Extraction of Algal Oil by Chemical Method

In this process, after mechanically disrupting the algal cells, chemical extraction of oil follows, employing various solvents such as hexane, benzene, ether, hexane–ethanol, methanol–chloroform, etc. The solvent to be utilized depends on the algal biomass type and the purity of the final product (Richmond 2004; Patel et al. 2017; Shuba and Kifle 2018). Extraction solvents should be cheap, nonpolar, poor extractor of undesired cellular components, and volatile in nature. The basic principle underlying the chemical extraction process using organic solvents depends on "like dissolves like" (polar molecules dissolves in polar solvents whereas non-polar dissolves in non-polar solvents). (Geciova et al. 2002; Bahadar and Khan 2013; Patel et al. 2017). Five basic steps are followed in the extraction process: first, the microalgal cells are subjected to organic solvents, followed by penetration of cell membranes by the solvents and their subsequent entry into the cytoplasm of the cells. A solvent–lipid complex is produced due to the interaction between the solvents and the neutral lipids through the van der Waals forces. This phenomenon is followed by the diffusion of solvent–lipid complex across the cell membrane which further results in the organic phase containing the neutral lipids, while the aqueous phase contains the water and solvent–carbohydrate or solvent–protein complexes. The biofuel is further produced after the separation of organic phase and subsequent transesterification of the crude lipids (Suali and Sabartly 2012; Bahadar and Khan 2013). Use of solvent extraction on a larger scale is associated with various disadvantages such as the high-energy consumption and possible contamination of the algal solids which delimits its use at a commercial scale. Further, serious health issues can result from contact between a solvent and body surface or exposure to vapors, thereby restricting the use of this method for oil extraction (Bahadar and Khan 2013).

8.4.3 Supercritical Fluid Extraction (SCFE)

The traditional solvent extraction process is replaced by the SCFE method as the latter is less toxic and supports the mass transfer equilibrium favorably (Patel et al. 2017). It relies on the principle that a fluid behaves both as a gas and a liquid when subjected to above critical point temperature and pressure conditions. This green technology is efficient for lipid extraction as solvent-free crude lipid products are obtained; higher yield of lipid is achieved owing to the rapid penetration of the solvent in the algal cells. The fluid density determines the solvent power, and the former can be adjusted by changing the temperature–pressure conditions, thereby facilitating the production of neutral lipids. Various features of SCFEs are lower toxicity, inert nature, and non-corrosiveness, and inflammable characteristic promotes its wide usability (Sahena et al. 2009; Bahadar and Khan 2013).

8.4.4 Other Methods Employed for Oil Extraction via Algal Biomass

Other methods which have been effectively used for oil extraction via algal biomass are as follows:

8.4.4.1 Osmotic Shock

The oil is extracted using osmotic shocks where pressure is developed across the cell wall that results in disruption of the microalgal cells (Kim et al. 2013; Rashid et al. 2014). Hyper-osmotic stress is developed on the cell wall due to higher concentration of salt in the liquid medium and causes the microalgal cells to bulge in the outward direction. A higher salt concentration inside the cells relative to the outside medium also results in cell disruption (Rashid et al. 2014; Kumar et al. 2015).

8.4.4.2 Enzymes

The most commonly employed enzymes for extraction are cellulase, pectinase, and neutrase. Sporopollenin layer is destroyed by the enzyme without causing any harm to the structure of the whole cell. As opposed to the chemical methods, enzymes are inexpensive, serve as a highly efficient extraction method, and do not hinder with the fatty acids in the cell (Taher et al. 2014; Rashid et al. 2014).

8.4.4.3 Milking

A more recent technique developed for oil extraction is the process of milking which consists of direct extraction of oil from the live cells with the help of various organic solvents, provided these are not toxic to microalgal cells (Rashid et al. 2014; Jackson et al. 2017).

8.4.4.4 Microwave-Assisted Organic Solvent Extraction

Electromagnetic radiations of a specific frequency are used to heat the cells, resulting in an increased internal pressure. This leads to rupturing of the cell and subsequent release of all the cellular constituents. Hence, the lipids get diffused into the organic solvent by this combined method of microwaves and organic solvents (Sostaric et al. 2012; Bahadar and Khan 2013).

8.5 Techniques Implemented for the Conversion of Algal Feedstock to Biofuel

Thermochemical and biochemical methods are the two categories for the conversion of algal feedstock to biofuel (Fig. 8.6). The type of conversion process to be utilized relies on the quantity of the algal biomass, type of the biomass feedstock, economic aspects, and desired purity of the end products (McKendry 2002a; Brennan and Owende 2010).

8.5.1 Thermochemical Conversion

Microalgae serve as a promising feedstock for producing various carbon-neutral fuels such as biodiesel, bioethanol, and biohydrogen, due to the presence of

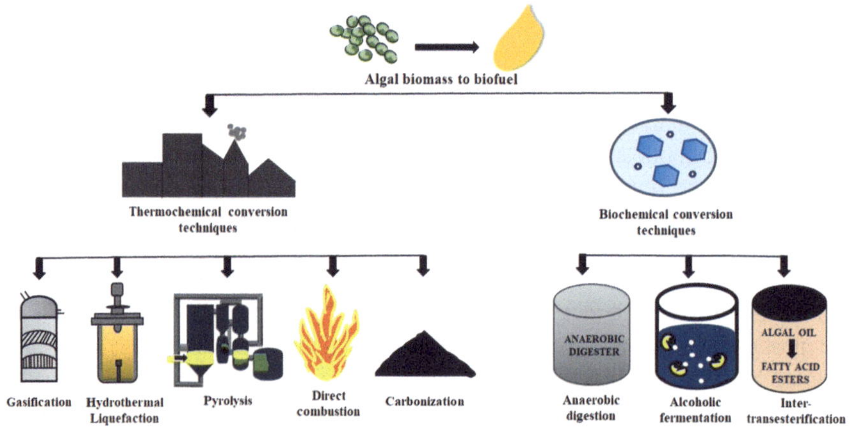

Fig. 8.6 Representation of various methods used for the conversion of biomass to biofuel

carbohydrates, proteins, and lipids in the microalgal cells (Mathimani et al. 2018; Li et al. 2008). Various advantages are offered by the thermochemical conversion as it is environment friendly, achieves better recovery of nutrition, and involves shorter processing time. The emission of fugitive gases is eliminated and is applicable for both wet and dry biomass. Various feedstocks and blends can be handled and owing to the higher temperatures employed and a small amount of residue are obtained after the conversion process, and efficient elimination of pathogens and pharmaceutically active compounds is observed (Razzak et al. 2013; Milano et al. 2016).

8.5.1.1 Gasification

A large array of feedstock can be processed by the method of gasification and is specifically suitable for feedstocks with low moisture content such as lignocellulosic biomass. In the microalgal biofuel production process, two types of gasification methods are used: conventional and supercritical. In conventional method, the biomass undergoes partial oxidation in the air, oxygen, or steam that acts as a gasification medium. Temperatures between 700 and 1000°C are employed, and the following steps occur in a gasifier (Basu 2010; Mathimani et al. 2018): the residual moisture is first removed through proper drying, and the solid structure of the biomass is broken down to yield less complex molecules through the process of pyrolysis. This step is followed by the process of oxidation which involves burning some of the incoming biomass and production of heat in order to sustain further endothermic processes. Lastly, the pyrolysis products are then converted to high-energy-containing smaller molecules through the process of gasification. In case of supercritical method, high water content of the microalgal feedstock demands extensive drying when processed by the conventional method. Hence, the supercritical method of gasification provides a more suitable alternative for the conversion of algal biomass into biofuels. The temperature and pressure conditions employed are in the supercritical range for water as 400–500 °C and 24–36 MPa. Similar to the hydrothermal liquefaction (HTL), this process is also performed in an aqueous media, but employs supercritical conditions, whereas HTL involves subcritical conditions. The microalgae structures are decomposed into smaller molecules by the supercritical water and involve C–C bond breakage (Mathimani et al. 2018).

8.5.1.2 HTL

In case of HTL, temperatures from 250 to 380°C and 5–20 MPa of pressure are employed to obtain biofuel. These subcritical conditions are provided in an aqueous medium and result in the breakdown of algal structures to simpler and smaller molecules. As compared to pyrolysis, a more deoxygenated and vicious liquid blend (or bio-oil) is produced from HTL process. Nitrogen, hydrogen, carbon dioxide, carbon monoxide, and light hydrocarbons constitute the bio-oil or the bio-crude (Brown et al. 2010), while the secondary products are characterized as being ashes and char. The secondary products are obtained as gases or as solid particles. HTL carbonization takes place at temperature < 250 °C and results in production of hydro-char as the primary product. The carbohydrate and protein portions of the microalgal cells are often involved in hydro-char production, thereby allowing the extraction of lipids prior to the carbonization process (Heilmann et al. 2010; Mathimani et al. 2018).

8.5.1.3 Pyrolysis

The algal biomass is subjected to thermal decomposition in an atmospheric pressure inert environment. It involves temperatures as high as 400–600 °C and oxygen-free environment. In slow pyrolysis, the biomass is exposed to a lower heating rate of 5–10 °C/minute and for longer residence times of 1 h, while in fast pyrolysis, it consists of shorter residence time of a few seconds. The method of fast pyrolysis employing heating rates as high as 600°C/min is more suitable with respect to the process of slow pyrolysis and allows for operation of a continuous process. Syngas, also called pyrogas, is the principle product obtained from the process of pyrolysis and comprises of non-condensable gases, solid char, and bio-oil (Mathimani et al. 2018). Flash pyrolysis is a promising substitute for biofuel production and future replacement of fossil fuels by the produced biofuels. It employs temperature of 500.8 °C and a short residence time of about 1 s. Approximately 95.5% biomass-to-liquid conversion ratios can be achieved through this method (Brennan and Owende 2010; Demirbas 2006; Clark and Deswarte 2008).

8.5.1.4 Direct Combustion

In this method, microalgae are burned in an oxygen-rich environment in a boiler or furnace at around 850 °C. The biomass having a moisture content greater than 50% is loaded in the boilers, and 10% excess of air (relative to the feedstock) is pumped in the combustor. This scenario favors the release of heat and further allows reaction to achieve completion. During the process of combustion, the photosynthetically obtained chemical energy in microalgae is transformed to hot gases. Combustion results in the production of a large amount of heat which cannot be stored feasibly and is thus further converted into other valuable products, e.g., generation of electricity in a turbine (Mathimani et al. 2018). In addition, combustion can be employed from very small-scale utilities (as for the domestic purposes) to large-scale industrial processes (McKendry 2002b; Goyal et al. 2008; Brennan and Owende 2010). Various pretreatments of biomass, such as drying, grinding, or chopping, are needed for this conversion process. This process leads to additional energy and cost, thereby delimiting the use of this technique.

8.5.1.5 Carbonization

Carbonization involves the production of carbon or carbon-rich residues from the organics in an exothermic process. This process further results in the release of a large amount of heat out of the system. Although this technique demands lesser energy in comparison to other conversion methods (Benavente et al. 2017; Mathimani et al. 2018), its use remains restricted owing to the high costs and high nitrogen contents involved in cultivation of microalgae as a feedstock for the carbonization process. When the operating parameters are 150–250°C and the pressure is less than 100 bars, the carbonization process is designated as hydrothermal carbonization (HTC). The residence time of this mild-treatment method is usually longer (>1 h). This conversion technique is aimed at producing high-energy-density solid fuels from the process of carbon concentration (Mathimani et al. 2018).

8.5.2 Biochemical Conversion

Anaerobic digestion, alcoholic fermentation, and inter-transesterification are the three main biological processes for the production of biofuels from the microalgal biomass. Since this conversion technique involves generation of other energy sources along with various coproducts, it is technically more viable as compared to the remaining conversion methods (Adeniyi et al. 2018). A renewable feedstock for biofuel production is produced from this environmentally feasible process. These biofuels serve as efficient sources of energy that can be utilized at the industrial scale (Ehimen et al. 2013; Adeniyi et al. 2018).

8.5.2.1 Anaerobic Digestion (AD)

The process comprises of three consecutive steps (Adeniyi et al. 2018): the first step is hydrolysis where soluble sugars are formed from the breakdown of complex molecules, the second step is fermentation where these sugars are further converted by the fermentative bacteria, and the third step is methanogenesis where the fermentation products are then acted upon by the methanogens, and they metabolize them to provide methane and carbon dioxide as the principal products (Saratale et al. 2018). Various advantages offered by the AD of microalgal cells for biofuel production include great amounts of carbon that are fixed by this process owing to the efficient nutrient extraction from the harvested biomass and followed by their subsequent transfer back into the cultivation system (Hallenbeck et al. 2016; Adeniyi et al. 2018). Biogas, particularly carbon dioxide and methane, can be produced by the decomposition of organic matter present in the biomass waste. This biogas can be further utilized for domestic cooking or power generation (Lee and Lee 2016; Adeniyi et al. 2018). Energy recovery from sunlight is encouraged by this process and is achieved through photosynthesis, further leading to the integration of efficiency into the biofuel production process (Adeniyi et al. 2018).

8.5.2.2 Alcoholic Fermentation

This biochemical process relies on the production of alcohol from an organic solvent through various enzymes (Suganya et al. 2016; Adeniyi et al. 2018). This metabolic reaction basically results in the formation of bioethanol from starch/alginate/cellular sugar/laminarin/mannitol stored in the microalgal cells employing the activities of yeast (Lee and Lee 2016). Acetone and butanol can also be produced from this process by utilizing acidogenesis and solventogenesis (Trivedi et al. 2015; Adeniyi et al. 2018). However, this conversion technique is associated with a drawback that pretreatment processes, such as milling, saccharification, and liquefaction, are required to achieve an efficient fermentation of the algal biomass (Lee and Lee 2012; Adeniyi et al. 2018). *Chlorella vulgaris* is effectively used for alcoholic fermentation and thus ethanol production. The high concentration of starch found in the *Chlorella vulgaris* cells is responsible for providing conversion efficiencies as high as 65% (Adeniyi et al. 2018).

8.5.2.3 Inter-transesterification

This method involves enzymatic production of fatty acid esters from algal oil. The easy removal of the by-products and high purity of the final product so obtained encourages the use of this conversion technique. However, its use is restricted by high cost of the enzymes that further reduces the economic feasibility of the process (Razzak et al. 2013; Milano et al. 2016).

8.6 Transesterification

The diesel production from algal oil involves a method called transesterification which is a chemical conversion process of the microalgal biomass. This chemical reaction proceeds with the formation of FAME and glycerol due to the reaction of triglyceride with alcohol. The glycerol so produced could be used in cosmetic and pharmaceutical industry (Suganya et al. 2016; Kandiyoti et al. 2017; Adeniyi et al. 2018). The type of alcohol employed, the kind of catalyst used, and the molar ratios determine this chemical reaction that forms low molecular weight FAMEs from raw algal lipids (Adeniyi et al. 2018). This type of transesterification is vital in biodiesel production because it reduces the viscosity of the algal oil, thereby enhancing its fluidity (Adeniyi et al. 2018). Two main types of transesterification processes are utilized in the biodiesel production, namely, the direct transesterification and the conventional method and supercritical methanol transesterification (Bahadar and Khan 2013; Adeniyi et al. 2018) (Fig. 8.7).

8.6.1 Direct Transesterification

This method of simultaneous lipid extraction is also called the in situ method or the single-stage method (Lee and Lee 2016). The reaction system is fed directly with the wet and unwashed algae, thereby allowing the transesterification to proceed directly (Jazzar et al. 2015). Pretreatment methods like degumming and extraction are not needed in this kind of transesterification process. Further, some amount of water is

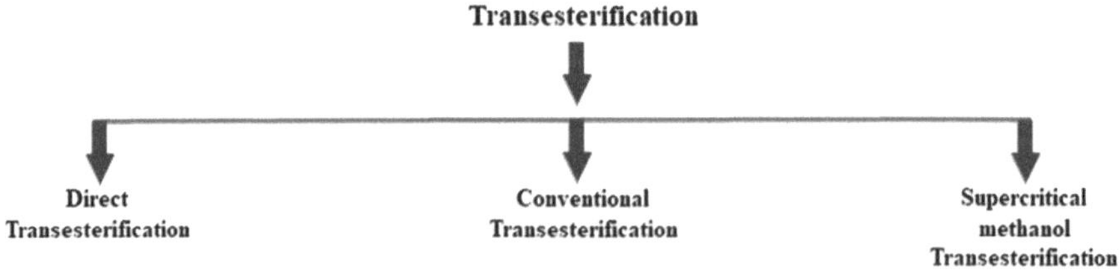

Fig. 8.7 Schematic representation of various kinds of transesterification processes

tolerable owing to the large quantity of methanol employed in the system. As compared to the conventional method, relatively higher biomass yields are obtained through this direct form of transesterification (Al-lwayzy et al. 2014; Park et al. 2015; Adeniyi et al. 2018).

8.6.2 Conventional Transesterification

This method is comprised of two stages and involves the mechanical extraction of lipids prior to the transesterification process. Degumming and extraction methods employing nonpolar, cheap, and unreactive solvents are significantly involved in this kind of transesterification process (Hossain et al. 2018). The product obtained from this method is highly refined and utilized in high-speed diesel engines (Salam et al. 2016; Martinez-Guerra and Gude 2016; Adeniyi et al. 2018).

8.6.3 Supercritical Methanol Transesterification

In this method, methanol is introduced to 100 mL algae culture containing cylinder under supercritical conditions which are defined as 350–400 °C, 10–25 MPa (Demirbas 2008). Higher biomass yields are obtained through this economically feasible method (Bahadar and Khan 2013); however, this method is not applied on a commercial scale and is still being explored by the researchers.

8.7 Applications of Algal Biomass in Various Biotechnological Sectors

The algal biomass can have immense applications in various biotechnological and industrial sectors which are as follows and has been represented in Fig. 8.8.

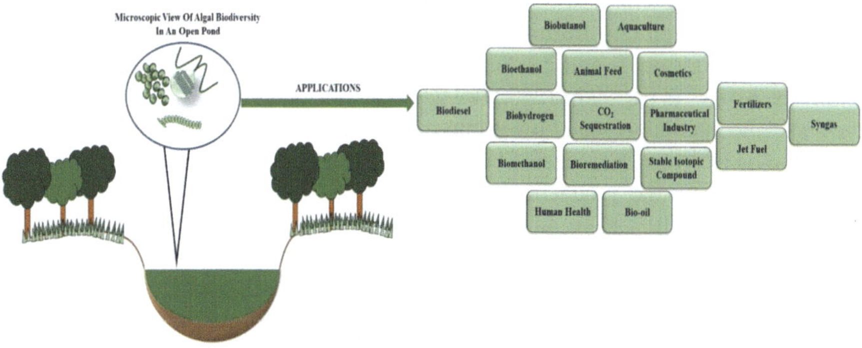

Fig. 8.8 Application of microalgae in various industries and biotechnological sectors

8.7.1 Biodiesel Production via Algal Biomass

It is a nontoxic fuel exhibiting high flash point than diesel (50 °C and 65 °C), has 10% built-in oxygen which allows complete burning, fails to emit sulfur oxides, and reduces the level of carbon monoxide, unburned hydrocarbons, and other pollutants (Hemaiswarya et al. 2012). The most commonly used algal feedstocks used for biofuel production include green algae *Chlorella* sp. and *Chlorococcum* sp. In particular strains such as *Haematococcus* and *Neochlorosis* are perfect for biodiesel and *Chlorella*, *Scenedesmus*, and *Spirulina* for biomethanol production (Maity et al. 2014). Marine microalgae *C. vulgaris* yielded 22% lipid content and 61% biodiesel (Mathimani et al. 2015). Moreover, *C. vulgaris* produced biodiesel, and freshwater *Chlorella strain* accumulated high lipid in comparison to cyanobacterial strains (Mathimani and Nair 2016). As per Maity et al. (2014) *Chlorella* and *Nannochloropsis* strains showed 100 mg/L/day lipid productivity.

8.7.2 Bioethanol Production from Microalgae

The extensive use of sugar and starch materials for the production of bioethanol has created major competition within the food market in terms of land for cultivation, making the bioethanol production from these sources economically less feasible (de Farias Silva and Bertucco 2016; Shuba and Kifle 2018). Thus, microalgae have been used as an effective alternative as the issues which are present in S^{nd}GB have been overcome in T^{rd}GB. Presently, *Chlorella vulgaris* has been a good source of ethanol production due to its high starch content, where $\geq$65% ethanol conversion efficiency has been reported (Shuba and Kifle 2018).

8.7.3 Biohydrogen Production from Microalgae

Hydrogen has been regarded as the "future energy carrier" as it excludes the use of carbon dioxide in combustion, generates huge energy per unit, and can be transformed to electricity by fuel cells. The current production of hydrogen is a fossil fuel-based process and produces large amounts of greenhouse gases (Shuba and Kifle 2018). As per Melis (2002) by depleting the quantity of sulfur available to the algae, the internal oxygen flow is interrupted, thereby allowing the production of hydrogen by hydrogenase. Later, Chochois et al. (2009) stated that direct photolysis is responsible for the production of hydrogen in *C. reinhardtii*. The cells are illuminated after they have adapted to anaerobic conditions, and the electrons originating from the splitting of water at PSII are driven by photosynthetic electron transport chain to ferredoxin and then to a reversible iron hydrogenase, thus enabling the production of hydrogen from water and solar energy (Fig. 8.3). Three methods can be used to produce hydrogen from algae (i.e., biochemical process, gasification, and steam reforming).

8.7.4 Biomethanol: Efficient Biofuel from Microalgae

The biomethanol can be blended with petrol or can be used as a feedstock for other environmentally friendly fuels. The fuels burning methane produce less carbon dioxide per unit of heat released, thereby reducing the pollution which is practically not feasible in the conventional process (Shuba and Kifle 2018).

8.7.5 Production of Biobutanol from Microalgae

The green waste left from the algae oil extraction can be used for the production of butanol. It has an energy density similar to gasoline and higher than either ethanol or methanol. It can be used in as a replacement of gasoline in gasoline engines without any modifications (Ullah et al. 2015; Maiti et al. 2016).

8.7.6 Utilization of Microalgae as Animal Feed

The microalgae have been used as animal feed, and in order to prove it is harmless and safe for human consumption, various toxicological tests were performed. Microalgae are categorized as unconventional sources of protein and for these the toxicological tests are necessary. The various investigations confirm that the algal proteins have high quality as compared to the plant proteins. A series of nutritional and toxicological tests have demonstrated that microalgae can be used as supplement of protein in the animal feed. The algal biomass is highly effective and can easily substitute the conventional sources of protein such as fish meal, soybean meal, rice bran, etc. The commercial use of these microalgae is mainly in the poultry as these can be easily incorporated into the poultry rations and provide suitably efficient results (Brennan and Owende 2010). Among various algae, *Arthrospira* is the most common strain which is used as a protein source for domestic animals (e.g., dogs, horses, cats, aquarium and ornamental fish, breeding bulls, cows) (Spolaore et al. 2006). The microalgae affect the physiology of these animals in many ways, such as increasing the immune response; providing vitamins, minerals, and essential fatty acid; and increasing their fertility. Similarly, they also affect their external appearance by providing them with lustrous coat and healthy skin (Spolaore et al. 2006).

8.7.7 Carbon Dioxide Sequestration via Microalgae

Two main strategies are available for the mitigation of emitted carbon dioxide (Wang et al. 2008; Mata et al. 2010): the first strategy relies on the chemical reaction and the second strategy on biological mitigation. The former is energy consuming and thereby a costly process. It is also not eco-friendly as the carbon dioxide captured in this process needs to be disposed in an appropriate manner (Mata et al. 2010). Alternatively, the biological mitigation produces biomass energy while mitigating carbon dioxide through carbon dioxide fixation via photosynthesis (Pulz and Gross 2004). Thus, the utilization of industrial emissions as carbon dioxide source for microalgal growth proves to be a promising method for reduction of the GHG emissions (Mata et al. 2010). Apart from carbon dioxide, sulfur oxides, and nitrogen oxides, some heavy metals are also present in the flue gases that demand further attention and proper removal. Microalgae play a crucial role for the removal of these substances, thereby reducing the overall emission of GHG in the ecosystem (Patel et al. 2017).

8.7.8 Nutrient Recycling from the Wastewater or Bioremediation

In the wastewater, nutrients are present which are used by the microalgae for their growth (Mulbry et al. 2008; Roberts et al. 2013). Microalgae release the free oxygen into wastewater during their growth, thereby enhancing the waste degradation by other microorganisms. This results in improvement of the biochemical oxygen demand (BOD) and chemical oxygen demand (COD) of the waste stream (Pittman et al. 2011). These microalgae remove dissolved nitrogen, toxic metals, and phosphorous from the water and are thus used in the tertiary phase of wastewater treatment (Munoz and Guieysse 2006; Rawat et al. 2011). Microalgae can also be used for the degradation of persisting molecules such as antibiotics, heavy metals, and hydrocarbons from the wastewater (Schwarzenbach 2006; Patel et al. 2017).

8.7.9 Effective Role of Microalgae in Improving Human Health

Microalgae synthesize various compounds which can be used as food colorants (Becker 2013) and are a good source of food supplements as they are rich in carbohydrates, lipid, and protein (e.g., *Chlorella vulgaris* composed of 51–58% of carbohydrates, 14–22% of lipid, and 12–17% of protein (Spolaore et al. 2006; Mathimani et al. 2018)). They are widely used as capsules or tablets and as components of pastas, snacks, and beverages (Liang et al. 2004). Microalgae also contain many types of sterols which are used to prevent the cardiovascular diseases, for example, *Spirulina* sp. contain clionasterol which leads to enhanced synthesis of plasminogen-activating factor in the endothelial cells of the vascular system (Barrow and Shahidi 2008; Mata et al. 2010). Antioxidants, such as astaxanthin, carotenoids, mycosporines, dimethylsulfoniopropionate, beta-carotene, etc., have also been extracted from the microalgae. These antioxidants can prevent oxidative stress, which is responsible for various diseases and contributes to the process of ageing. Degenerative diseases can be further prevented and treated by carotenoids such as lutein, which is commonly found in egg yolk, spinach, vegetables, kale, and other yellow-colored foods. A high level of these carotenoids is present in *Muriellopsis* sp. (Del et al. 2007; Mata et al. 2010). Owing to their high protein content,

Arthrospira is widely used for nutritional purposes (Soletto et al. 2005; Spolaore et al. 2006). Apart from this, the microalgae are also used for other health benefits, namely, growth promotion, suppression of hypertension, prevention of renal diseases, relief from hyperlipidemia, and reduction in the increased level of serum glucose (Liang et al. 2004; Spolaore et al. 2006).

Edible algae include the microalgal species *Chlorella* and *Spirulina*. These algae are widely studied for their biological activities and component molecules (Pulz and Gross 2004; Mata et al. 2010). *Spirulina platensis* and *Spirulina maxima* are widely used for human consumption owing to the various benefits offered by them, namely, boosting the immune system, prevention of cancer and viral infections, and rise in the number of intestinal lactates (Mata et al. 2010). The polysaccharide complexes from *Chlorella pyrenoidosa* and *Chlorella ellipsoidea* possess immunomodulatory characteristics (Barrow and Shahidi 2008; Mata et al. 2010). Various species of microalgae such as *N. oculata*, *Thalassiosira pseudonana*, and *Phaeodactylum tricornutum* have the capability to produce varying amounts of long-chain polyunsaturated fatty acids (PUFAs) that have a vital therapeutic and dieticiary role (Pulz and Gross 2004; Kumar et al. 2019). *Euglena gracilis* and *Prototheca moriformis* are widely cultivated for the production of biotin, tocopherol, and ascorbic acid, respectively (Li et al. 2008; Patel et al. 2017). It also promotes antibody production and is thus a potential immunomodulator. Moreover, it is also associated with prevention of onset of cancer in the oral cavity, liver, and bladder (Mata et al. 2010).

8.7.10 Aquaculture

Microalgae are widely used in the aquaculture processes and are cultivated by two main methods: natural phytoplankton and algal monocultures. In natural phytoplankton, various nutrients are added to enrich the population of phytoplankton that are found naturally or in the form of cultures (Mata et al. 2010). However, this process of using natural phytoplanktons has many drawbacks including variable composition of algal populations, amount of nutrients available, hindrance from undesirable predators, presence of other contaminating species, and lack of monitoring and control over the production process. In case of monoculture cultivation of microalgae feed source of high quality with known nutritional properties, the desired algal cultures free of contamination from unwanted species are utilized. The major obstruction to monoculture cultivation is that the microalgae species are subjected to a large number of predators such as the crustaceans, protozoans (zooflagellates or rhizopods), larvae of benthic organisms and infection by viruses, bacteria, or fungi when cultivated on a larger scale. Predation by various types of protozoans and the high possibility of formation of toxic algal blooms are major problems associated with marine microalgae culture. For instance, growth of *Synechocystis* in freshwater and *Phaeodactylum* in seawater is unenviable for the bivalve mollusks (Mata et al. 2010).

8.7.11 Cosmetics

The two main microalgae commercialized in the field of cosmetics are *Chlorella* and *Arthrospira* (Stolz and Obermayer 2005; Spolaore et al. 2006). Extracts from these algae contain antioxidants and other regenerant chemicals which are used in skin care products such as antiaging cream. The use of microalgae as sun protection and hair care products is also common (Ariede et al. 2017). For example, extract rich in protein from *Arthrospira* is used to repair the signs of ageing and prevent formation of stria (e.g., Protulines and Exsymol by S.A.M., Monaco). Similarly, chemicals isolated from *Chlorella vulgaris* help in the synthesis of collagen, hence supporting the regeneration of tissues and reduction of wrinkles (products Dermochlorella and Codif manufactured by St. Malo, France).

Other commercially available products include Pepha-Tight (launched by Pentapharm, Switzerland) produced from *Nannochloropsis oculata*, which exhibit high skin tightening properties. PephaCtive (launched by Pentapharm, Switzerland) promotes cell proliferation, thereby influencing the skin metabolism (Stolz and Obermayer 2005). Microalgae are also responsible for the exclusive synthesis of scytonemin, a nontoxic secondary pigment which possesses high UV-absorbing capacity. In addition, it exhibits the ability to scavenge free radicals, demonstrates an efficient activity against cyclobutane purine/pyrimidine dimer (CPD) formation, and is highly stable under conditions of abiotic stresses. These potential features are utilized to counteract the harmful UV solar radiations, and scytonemin can thus be used for the production of natural sunscreens (Singh et al. 2010; Rastogi et al. 2013, 2014). It can be further exploited as a therapeutic agent in the acute inflammations due to its dual kinase inhibitory and antiproliferative activities (Stevenson 2002; Mclnnes et al. 2005).

8.7.12 Production of Pigments via Microalgae and Its Utility in Pharmaceutical Industry

The phycobiliproteins are primarily used as natural dyes and are known to possess other health benefits and thus have widespread applications in the pharmaceutical industry (Spolaore et al. 2006). The commercial sources available for phycoerythrin and phycocyanin are *Cyanobacterium*, *Arthrospira*, and *Porphyridium* (Viskari and Colyer 2003). Phycocyanins are most widely used as a food pigment, thereby replacing the synthetic pigments (Becker 2004). Phycobiliproteins are also used in various research laboratories owing to their peculiar properties that make them suitable for immunolabeling experiments and in diagnostic processes. They also have high fluorescence yield, high photostability, and high absorbance, so they are very sensitive fluorescent reagents. They are used as labels for receptors, antibodies, and other molecules in cell sorters and are used in fluorescence microscopy as well (Bermejo et al. 2002; Spolaore et al. 2006).

8.7.13 Microalgae: An ideal Source for Stable Isotope Compounds

Microalgae by the process of photosynthesis are able to incorporate various stable isotopic compounds such as ^{2}H, ^{13}C, and ^{15}N from other inexpensive inorganic compounds such as $^{2}H_2O$, $^{13}CO_2$, and $^{15}NO_3$. These stable compounds then form other essential organic compounds such as carbohydrates, lipids, amino acids, and nucleic acids. These stable isotopic biochemicals are used in two ways: first, they are used to determine the atomic structures by incorporation into proteins, carbohydrates, and nucleic acids, and second, they are used in the metabolic studies (Spolaore et al. 2006).

8.7.14 Miscellaneous Applications of Microalgae

Microalgae can also be converted into biogas (Montingelli et al. 2015), bio-oil, syngas, jet fuel, and fertilizers. Bio-oil a synthetic liquid fuel directly used in engines or in a blend is extracted by processing biomass at high temperature in the absence of oxygen (Demirbas 2006). Syngas contains very low concentrations of hydrocarbons and higher concentrations of carbon monoxide and hydrogen generated by oxygen gasification processes (Zhu et al. 2014). It can be converted into diesel fuel by Fisher–Tropsch synthesis process, thus making it possible to integrate algal feedstock into the existing thermochemical infrastructure. Microalgae-derived jet fuel has also received attention (Ghasemi et al. 2012) and is compatible for use in selected commercial jet test flights (Zhu et al. 2014). Many agents that play the role of modifying viscosity in various foods and pharmaceutical products are also obtained from different types of seaweeds. Alginate, carrageenan, and agar are some examples of such hydrocolloid compounds (Barrow and Shahidi 2008; Mata et al. 2010). Lastly, the use of algae in the agricultural fields as biofertilizers is also a common practice. Algae are used as soil conditioners as they have the potential to fix large amount of nitrogen, thus making the soil fertile (Song et al. 2005; Mathimani et al. 2018; Renuka et al. 2018).

8.8 Limitation and Future Prospect

The microalgae have numerous applications; however, it has certain limitations: first, algal strain should be carefully selected, and the production of biomass should be enhanced. The desired strain can then be improved by lipidomics, genomics, proteomics, and metabolomics which will have higher growth rate and lipid production and broader tolerance to environmental stresses and have the ability to produce many valuable coproducts (Schenk et al. 2008; Singh et al. 2011). The specific characteristics of algal strains can be modified through genetic engineering methods with an intended alternation of the algae cells, thereby improving the production of algal feedstock for biofuel (Tabatabaei et al. 2011; Adeniyi et al. 2018) and be efficiently applied in both natural and artificial methods of cultivation. Thus, with the following improvisation, the algal strain can be the most effective biological tool for the enhanced biofuel production and can thus be an effective step toward greener and cleaner environment.

8.9 Conclusion

The microalgae can be an effective solution for the ever-increasing environmental problems, as it neither competes with the food stock nor the land area as in case of FstGB and SndGB. The TrdGB along with the combination of FthGB can be a turning point from the global prospect and can help address the issues associated with the biofuel industries as well as other biotechnological application. However, intense research is required to apply the effectiveness of the microalgae beyond the laboratories, and only then can the vision of "clean and green" environment be accomplished.

Acknowledgments PV is thankful to DBT (Grant No. BT/304/NE/TBP/2012; Grant No. BT/PR7333/PBD/26/373/2012), and KA and AB are thankful to Central University of Rajasthan, Ajmer, India, for providing the financial support.

Competing Interests All the authors declare that they have no competing interests.

References

Adam F, Abert-Vian M, Peltier G, Chemat F (2012) "Solvent-free" ultrasound-assisted extraction of lipids from fresh microalgae cells: A green, clean and scalable process. Bioresour Technol 114:457–465. https://doi.org/10.1016/j.biortech.2012.02.096

Adeniyi OM, Azimov U, Burluka A (2018) Algae biofuel: current status and future applications. Renew Sust Energ Rev 90:316–335. https://doi.org/10.1016/j.rser.2018.03.067

Al-lwayzy SH, Yusaf T, Al-Juboori RA (2014) Biofuels from the fresh water microalgae *Chlorella vulgaris* (FWM-CV) for diesel engines. Energies 7:1829–1851. https://doi.org/10.3390/en7031829

Anandarajah K, Mahendraperumal G, Sommerfeld M, Hu Q (2012) Characterization of microalga *Nannochloropsis* sp. mutants for improved production of biofuels. Appl Energy 96:371–377. https://doi.org/10.1016/j.apenergy.2012.02.057

Andrade MR, Costa JAV (2007) Mixotrophic cultivation of microalga *Spirulina platensis* using molasses as organic substrate. Aquaculture 264:130–134. https://doi.org/10.1016/j.aquaculture.2006.11.021

Ariede MB, Candido TM, Jacome ALM, Velasco MVR, de Carvalho JCM, Baby AR (2017) Cosmetic attributes of algae – a review. Algal Res 25:483–487. https://doi.org/10.1016/j.algal.2017.05.019

Bahadar A, Khan MB (2013) Progress in energy from microalgae: a review. Renew Sust Energ Rev 27:128–148. https://doi.org/10.1016/j.rser.2013.06.029

Barrow C, Shahidi F (2008) Marine nutraceuticals and functional foods. CRC Press/Taylor & Francis Group, Boca Raton

Basu P (2010) Biomass gasification and pyrolysis: practical design and theory. Academic press, Amsterdam

Becker W (2004) Microalgae in human and animal nutrition. In: Richmond A (ed) Handbook of microalgal culture. Blackwell, Oxford, pp 312–351

Becker EW (2013) Microalgae for human and animal nutrition. Handbook of microalgal culture. J Appl Phycol Biotechnol:461–503. https://doi.org/10.1002/9781118567166

Benavente V, Fullana A, Berge ND (2017) Life cycle analysis of hydrothermal carbonization of olive mill waste: comparison with current management approaches. J Clean Prod 142:2637–2648. https://doi.org/10.1016/j.jclepro.2016.11.013

Bermejo Román RB, Alvárez-Pez JM, Acién Fernández FG, Molina Grima E (2002) Recovery of pure B-phyco-erythrin from the microalga *Porphyridium cruentum*. J Biotechnol 93:73–85. https://doi.org/10.1016/s0168-1656(01)00385-6

Bhatt NC, Panwar A, Bisht TS, Tamta S (2014) Coupling of algal biofuel production with wastewater. Sci World J 2014:1–10. https://doi.org/10.1155/2014/210504

Borowitzka MA, Moheimani NR (2013) Algae for biofuels and energy, vol 5. Springer, Dordrecht

Brennan L, Owende P (2010) Biofuels from microalgae—a review of technologies for production, processing, and extractions of biofuels and co-products. Renew Sust Energ Rev 14:557–577. https://doi.org/10.1016/j.rser.2009.10.009

Brown TM, Duan P, Savage PE (2010) Hydrothermal liquefaction and gasification of *Nannochloropsis* sp. Energy Fuel 24:3639–3646. https://doi.org/10.1021/ef100203u

Chen G-Q, Chen F (2006) Growing phototrophic cells without light. Biotechnol Lett 28:607–616. https://doi.org/10.1007/s10529-006-0025-4

Chen F, Zhang Y, Guo S (1996) Growth and phycocyanin formation of *Spirulina platensis* in photoheterotrophic culture. Biotechnol Lett 18:603–608. https://doi.org/10.1007/bf00140211

Chen CY, Yeh KL, Aisyah R, Lee DJ, Chang JS (2011) Cultivation, photobioreactor design and harvesting of microalgae for biodiesel production: a critical review. Bioresour Technol 102:71–81. https://doi.org/10.1016/j.biortech.2010.06.159

Chisti Y (2007) Biodiesel from microalgae. Biotechnol Adv 25:294–306. https://doi.org/10.1016/j.biotechadv.2007.02.001

Chisti Y (2013) Raceways-based production of algal crude oil. Green 3:195–216. https://doi.org/10.1515/green-2013-0018

Chochois V, Dauvillee D, Beyly A, Tolleter D, Cuine S, Timpano H, Ball S, Cournac L, Peltier G (2009) Hydrogen production in Chlamydomonas: photosystem II-dependent and -independent pathways differ in their requirement for starch metabolism. Plant Physiol 51:631–640. https://doi.org/10.1104/pp.109.144576

Clark J, Deswarte F (2008) Introduction to chemicals from biomass. In: Stevens CV (ed) Wiley series in renewable resources. Wiley, West Sussex

Daroch M, Geng S, Wang G (2013) Recent advances in liquid biofuel production from algal feedstocks. Appl Energy 102:1371–1381. https://doi.org/10.1016/j.apenergy.2012.07.031

De Farias Silva CE, Bertucco A (2016) Bioethanol from microalgae and cyanobacteria: a review and technological outlook. Process Biochem 51:1833–1842. https://doi.org/10.1016/j.procbio.2016.02.016

Del Campo JA, Garcıia-Gonza'lez M, Guerrero MG (2007) Outdoor cultivation of microalgae for carotenoid production: current state and perspectives. Appl Microbiol Biotechnol 74:1163–1174. https://doi.org/10.1007/s00253-007-0844-9

Demirbas A (2006) Oily products from mosses and algae via pyrolysis. Energy Sources Part A 28:933–940. https://doi.org/10.1080/009083190910389

Demirbas A (2008) Production of biodiesel from algae oils. Energy Sources Part A 31:163–168. https://doi.org/10.1080/15567030701521775

Demirbas MF (2011) Biofuels from algae for sustainable development. Energy 88:3473–3480. https://doi.org/10.1016/j.apenergy.2011.01.059

Ehimen EA, Holm-Nielsen JB, Poulsen M, Boelsmand JE (2013) Influence of different pre-treatment routes on the anaerobic digestion of a filamentous algae. Renew Energy 50:476–480. https://doi.org/10.1016/j.renene.2012.06.064

Eriksen NT (2008) The technology of microalgal culturing. Biotechnol Lett 30:1525–1536. https://doi.org/10.1007/s10529-008-9740-3

Geciova J, Bury D, Jelen P (2002) Methods for disruption of microbial cells for potential use in the dairy industry—a review. Int Dairy J 12:541–553. https://doi.org/10.1016/s0958-6946(02)00038-9

Ghasemi Y, Rasoul-Amini S, Naseri AT, Montazeri-Najafabady N, Mobasher MA, Dabbagh F (2012) Microalgae biofuel potentials (review). Appl Biochem Microbiol 48:126–144. https://doi.org/10.1134/s0003683812020068

Gouveia L (2011) Microalgae as a feedstock for biofuels. In: Microalgae as a feedstock for biofuels. Springer briefs in microbiology. Springer, Berlin/Heidelberg

Gouveia L, Oliveria (oliveira) AC (2009) Microalgae as a raw material for biofuels production. J Ind Microbiol Biotechnol 36:269–274. https://doi.org/10.1007/s10295-008-0495-6

Goyal HB, Seal D, Saxena RC (2008) Bio-fuels from thermochemical conversion of renewable resources: a review. Renew Sust Energ Rev 12:504–517. https://doi.org/10.1016/j.rser.2006.07.014

Graham LE, Graham JM, Wilcox LW (2009) Algae, 2nd edn. Pearson Education, Inc., San Francisco

Grima EM, Belarbi EH, Fernández FG, Medina AR, Chisti Y (2003) Recovery of microalgal biomass and metabolites: process options and economics. Biotechnol Adv 20:491–515. https://doi.org/10.1016/s0734-9750(02)00050-2

Hallenbeck PC, Grogger M, Mraz M, Veverka D (2016) Solar biofuels production with microalgae. Appl Energy 179:136–145. https://doi.org/10.1016/j.apenergy.2016.06.024

Harun R, Singh M, Forde GM, Danquah MK (2010) Bioprocess engineering of microalgae to produce a variety of consumer products. Renew Sust Energ Rev 14:1037–1047. https://doi.org/10.1016/j.rser.2009.11.004

Heasman M, Diemar J, O'connor W, Sushames T, Foulkes L (2000) Development of extended shelf-life microalgae concentrate diets harvested by centrifugation for bivalve molluscs – a summary. Aquac Res 31:637–659. https://doi.org/10.1046/j.1365-2109.2000.318492.x

Heilmann SM, Davis HT, Jader LR, Lefebvre PA, Sadowsky MJ, Schendel FJ, Von Keitz MG, Valentas KJ (2010) Hydrothermal carbonization of microalgae. Biomass Bioenergy 34:875–882. https://doi.org/10.1016/j.biombioe.2010.01.032

Hemaiswarya S, Raja R, Carvalho IS, Ravikumar R, Zambare V, Barh D (2012) An Indian scenario on renewable and sustainable energy sources with emphasis on algae. Appl Microbiol Biotechnol 96:1125–1135. https://doi.org/10.1007/s00253-012-4487-0

Hosikian A, Lim S, Halim R, Danquah MK (2010) Chlorophyll extraction from microalgae: a review on the process engineering aspects. Int J Chem Eng 2010:391632. https://doi.org/10.1155/2010/391632

Hossain FM, Rainey TJ, Ristovski Z, Brown RJ (2018) Performance and exhaust emissions of diesel engines using microalgae FAME and the prospects for microalgae HTL biocrude. Renew Sust Energ Rev 82:4269–4278. https://doi.org/10.1016/j.rser.2017.06.026

Huang Q, Jiang F, Wang L, Yang C (2017) Design of photobioreactors for mass cultivation of photosynthetic organisms. Engineering 3:318–329. https://doi.org/10.1016/j.eng.2017.03.020

Jackson BA, Bahri PA, Moheimani NR (2017) Repetitive non-destructive milking of hydrocarbons from *Botryococcus braunii*. Renew Sust Energ Rev 79:1229–1240. https://doi.org/10.1016/j.rser.2017.05.130

Janssen M, Tramper J, Mur LR, Wijffels RH (2002) Enclosed outdoor photobioreactors: light regime, photosynthetic efficiency, scale-up, and future prospects. Biotechnol Bioeng 81:193–210. https://doi.org/10.1002/bit.10468

Jazzar S, Quesada-Medina J, Olivares-Carrillo P, Marzouki MN, Acién-Fernández FG, Fernández-Sevilla JM, Molina-Grima E, Smaali I (2015) A whole biodiesel conversion process combining isolation, cultivation and in situ supercritical methanol transesterification of native microalgae. Bioresour Technol 190:281–288. https://doi.org/10.1016/j.biortech.2015.04.097

Junying Z, Junfeng R, Baoning Z (2013) Factors in mass cultivation of microalgae for biodiesel. Chin J Catal 34:80–100. https://doi.org/10.1016/s1872-2067(11)60497-x

Kagan J (2010) Third and fourth generation biofuels: technologies, markets and economics through 2015. Greentech Media, Boston-NYC-San Francisco. https://www.greentechmedia.com/research/report/third-and-fourth-generationbiofuels

Kandiyoti R, Herod A, Bartle K, Morgan T (2017) Fossil fuels and renewables. In: Solid fuels and heavy hydrocarbon liquids, pp 1–9. https://doi.org/10.1016/b978-0-08-100784-6.00001-1

Kim J, Yoo G, Lee H, Lim J, Kim K, Kim CW, Park MS, Yang JW (2013) Methods of downstream processing for the production of biodiesel from microalgae. Biotechnol Adv 31:862–876. https://doi.org/10.1016/j.biotechadv.2013.04.006

Knuckey RM, Brown MR, Robert R, Frampton DMF (2006) Production of microalgal concentrates by flocculation and their assessment as aquaculture feeds. Aquac Eng 35:300–313. https://doi.org/10.1016/j.aquaeng.2006.04.001

Kumar RR, Hanumantha Rao P, Arumugam M (2015) Lipid extraction methods from microalgae: a comprehensive review. Front Energy Res 2:61–69. https://doi.org/10.3389/fenrg.2014.00061

Kumar BR, Deviram G, Mathimani T, Duc PA, Pugazhendhi A (2019) Microalgae as rich source of polyunsaturated fatty acids. Biocatal Agric Biotechnol. https://doi.org/10.1016/j.bcab.2019.01.017

Laamanen CA, Ross GM, Scott JA (2016) Flotation harvesting of microalgae. Renew Sust Energ Rev 58:75–86. https://doi.org/10.1016/j.rser.2015.12.293

Lee SM, Lee JH (2012) Ethanol fermentation for main sugar components of brown-algae using various yeasts. J Ind Eng Chem 18:16–18. https://doi.org/10.1016/j.jiec.2011.11.097

Lee OK, Lee EY (2016) Sustainable production of bioethanol from renewable brown algae biomass. Biomass Bioenergy 92:70–75. https://doi.org/10.1016/j.biombioe.2016.03.038

Lee KE, Morad N, Teng TT, Poh BT (2012) Development, characterization and the application of hybrid materials in coagulation/flocculation of wastewater: a review. Chem Eng J 203:370–386. https://doi.org/10.1016/j.cej.2012.06.109

Li Y, Horsman M, Wu N, Lan C, Dubois Calero N (2008) Biofuels from microalgae. Biotechnol Prog 24:815–820

Liang S, Xueming L, Chen F, Chen Z (2004) Current microalgal health food R&D activities in China. Hydrobiologia 512:45–48. https://doi.org/10.1023/b:hydr.0000020366.65760.98

Lu J, Sheahan C, Fu P (2011) Metabolic engineering of algae for fourth generation biofuels production. Energy Environ Sci 4:2451–2466. https://doi.org/10.1039/c0ee00593b

Maiti S, Maiti DC, Verma M, Brar SK (2016) Biobutanol— "a renewable green alternative of liquid fuel" from algae. In: Soccol C, Brar S, Faulds C, Ramos L (eds) Green fuels technology. Green energy and technology. Springer, Cham. https://doi.org/10.1007/978-3-319-30205-8_18

Maity JP, Bundschuh J, Chen C-Y, Bhattacharya P (2014) Microalgae for third generation biofuel production, mitigation of greenhouse gas emissions and wastewater treatment: present and future perspectives – a mini review. Energy 78:104–113. https://doi.org/10.1016/j.energy.2014.04.003

Manninen K, Huttunen S, Seppälä J, Laitinen J, Spilling K (2016) Resource recycling with algal cultivation: environmental and social perspectives. J Clean Prod 134:495–505. https://doi.org/10.1016/j.jclepro.2015.10.097

Martinez-Guerra E, Gude VG (2016) Energy analysis of extractive-transesterification of algal lipids for biocrude production. Biofuels 9:139–146. https://doi.org/10.1080/17597269.2016.1195972

Mata TM, Martins AA, Caetano NS (2010) Microalgae for biodiesel production and other applications: a review. Renew Sust Energ Rev 14:217–232. https://doi.org/10.1016/j.rser.2009.07.020

Mathimani T, Mallick N (2018) A comprehensive review on harvesting of microalgae for biodiesel–key challenges and future directions. Renew Sust Energ Rev 91:1103–1120. https://doi.org/10.1016/j.rser.2018.04.083

Mathimani T, Nair BB (2016) Evaluation of microalga for biodiesel using lipid and fatty acid as a marker–a central composite design approach. J Energy Inst 89:436–446. https://doi.org/10.1016/j.joei.2015.02.010

Mathimani T, Uma L, Prabaharan D (2015) Homogeneous acid catalysed transesterification of marine microalga *Chlorella* sp. BDUG 91771 lipid–an efficient biodiesel yield and its characterization. Renew Energy 81:523–533. https://doi.org/10.1016/j.renene.2015.03.059

Mathimani T, Baldinelli A, Rajendran K, Prabakar D, Matheswaran M, Pieter van Leeuwen R, Pugazhendhi A (2018) Review on cultivation and thermochemical conversion of microalgae to fuels and chemicals: process evaluation and knowledge gaps. J Clean Prod. https://doi.org/10.1016/j.jclepro.2018.10.096

McInnes C, Mezna M, Fischer P (2005) Progress in the discovery of polo-like kinase inhibitors. Curr Top Med Chem 5:181–197. https://doi.org/10.2174/1568026053507660

McKendry P (2002a) Energy production from biomass (part 1): overview of biomass. Bioresour Technol 83:37–46. https://doi.org/10.1016/s0960-8524(01)00118-3

McKendry P (2002b) Energy production from biomass (part 2): conversion technologies. Bioresour Technol 83:47–54. https://doi.org/10.1016/s0960-8524(01)00119-5

Melis A (2002) Green alga hydrogen production: progress, challenges and prospects. Int J Hydrog Energy 27:1217–1228. https://doi.org/10.1016/s0360-3199(02)00110-6

Miao X, Wu Q (2006) Biodiesel production from heterotrophic microalgal oil. Bioresour Technol 97:841–846. https://doi.org/10.1016/j.biortech.2005.04.008

Milano J, Ong HC, Masjuki HH, Chong WT, Lam MK, Loh PK, Vellayan V (2016) Microalgae biofuels as an alternative to fossil fuel for power generation. Renew Sust Energ Rev 58:180–197. https://doi.org/10.1016/j.rser.2015.12.150

Milledge JJ, Heaven S (2012) A review of the harvesting of micro-algae for biofuel production. Rev Environ Sci Biotechnol 12:165–178. https://doi.org/10.1007/s11157-012-9301-z

Montingelli ME, Tedesco S, Olabi AG (2015) Biogas production from algal biomass: a review. Renew Sust Energ Rev 43:961–972. https://doi.org/10.1016/j.rser.2014.11.052

Morales J, de la Noüe J, Picard G (1985) Harvesting marine microalgae species by chitosan flocculation. Aquac Eng 4:257–270. https://doi.org/10.1016/0144-8609(85)90018-4

Morita M, Watanabe Y, Okawa T, Saiki H (2001) Photosynthetic productivity of conical helical tubular photobioreactors incorporating *Chlorella* sp. under various culture medium flow conditions. Biotechnol Bioeng 74:137–144. https://doi.org/10.1002/bit.10119

Mulbry W, Kondrad S, Pizarro C, Kebede-Westhead E (2008) Treatment of dairy manure effluent using freshwater algae: algal productivity and recovery of manure nutrients using pilot-scale algal turf scrubbers. Bioresour Technol 99:8137–8142. https://doi.org/10.1016/j.biortech.2008.03.073

Munoz R, Guieysse B (2006) Algal–bacterial processes for the treatment of hazardous contaminants: a review. Water Res 40:2799–2815. https://doi.org/10.1016/j.watres.2006.06.011

Park JY, Park MS, Lee YC, Yang JW (2015) Advances in direct transesterification of algal oils from wet biomass. Bioresour Technol 184:267–275. https://doi.org/10.1016/j.biortech.2014.10.089

Patel A, Gami B, Patel P, Patel B (2017) Microalgae: antiquity to era of integrated technology. Renew Sust Energ Rev 71:535–547. https://doi.org/10.1016/j.rser.2016.12.081

Pittman JK, Dean AP, Osundeko O (2011) The potential of sustainable algal biofuel production using wastewater resources. Bioresour Technol 102:17–25. https://doi.org/10.1016/j.biortech.2010.06.035

Posten C, Schaub G (2009) Microalgae and terrestrial biomass as source for fuels—a process view. J Biotechnol 142:64–69. https://doi.org/10.1007/s10295-008-0495-6

Pragya N, Pandey KK, Sahoo PK (2013) A review on harvesting, oil extraction and biofuels production technologies from microalgae. Renew Sust Energ Rev 24:159–171. https://doi.org/10.1016/j.rser.2013.03.034

Pulz O, Gross W (2004) Valuable products from biotechnology of microalgae. Appl Microbiol Biotechnol 65:635–648. https://doi.org/10.1007/s00253-004-1647-x

Rashid N, Rehman MSU, Han J-I (2013) Use of chitosan acid solutions to improve separation efficiency for harvesting of the microalga *Chlorella vulgaris*. Chem Eng J 226:238–242. https://doi.org/10.1016/j.cej.2013.04.062

Rashid N, Ur Rehman MS, Sadiq M, Mahmood T, Han J-I (2014) Current status, issues and developments in microalgae derived biodiesel production. Renew Sust Energ Rev 40:760–778. https://doi.org/10.1016/j.rser.2014.07.104

Rastogi RP, Sinha RP, Incharoensakdi A (2013) Partial characterization, UV-induction and photoprotective function of sunscreen pigment, scytonemin from *Rivularia* sp. HKAR-4. Chemosphere 93:1874–1878. https://doi.org/10.1016/j.chemosphere.2013.06.057

Rastogi RP, Sonani RR, Madamwar D (2014) The high-energy radiation protectant extracellular sheath pigment scytonemin and its reduced counterpart in the cyanobacterium *Scytonema* sp. R77DM. Bioresour Technol 171:396–400. https://doi.org/10.1016/j.biortech.2014.08.106

Rastogi RP, Pandey A, Larroche C, Madamwar D (2018) Algal Green Energy – R&D and technological perspectives for biodiesel production. Renew Sust Energ Rev 82:2946–2969. https://doi.org/10.1016/j.rser.2017.10.038

Rawat I, Kumar RR, Mutanda T, Bux F (2011) Dual role of microalgae: phycoremediation of domestic wastewater and biomass production for sustainable biofuels production. Appl Energy 88:3411–3424. https://doi.org/10.1016/j.apenergy.2010.11.025

Razzak SA, Hossain MM, Lucky RA, Bassi AS, de Lasa H (2013) Integrated CO2 capture, wastewater treatment and biofuel production by microalgae culturing—a review. Renew Sust Energ Rev 27:622–653. https://doi.org/10.1016/j.rser.2013.05.063

Renuka N, Guldhe A, Prasanna R, Singh P, Bux F (2018) Microalgae as multi-functional options in modern agriculture: current trends, prospects and challenges. Biotechnol Adv 36:1255–1273. https://doi.org/10.1016/j.biotechadv.2018.04.004

Richmond A (2000) Microalgal biotechnology at the turn of the millennium: a personal view. J Appl Phycol 12:441–451. https://doi.org/10.1023/a:1008123131307

Richmond A (2004) Handbook of microalgae culture: biotechnology and applied phycology. Blackwell Science Ltd, Oxford

Richmond A, Cheng-Wu Z, Zarmi Y (2003) Efficient use of strong light for high photosynthetic productivity: interrelationships between the optical path, the optimal population density and cell-growth inhibition. Biomol Eng 20:229–236. https://doi.org/10.1016/s1389-0344(03)00060-1

Rinaudo M (2006) Chitin and chitosan: properties and applications. Prog Polym Sci 31:603–632. https://doi.org/10.1016/j.progpolymsci.2006.06.001

Roberts DA, de Nys R, Paul NA (2013) The effect of CO_2 on algal growth in industrial waste water for bioenergy and bioremediation applications. PLoS One 8:e81631. https://doi.org/10.1371/journal.pone.0081631

Rodolfi L, Zittelli GC, Bassi N, Padovani G, Biondi N, Bonini G, Tredici MR (2009) Microalgae for oil: strain selection, induction of lipid synthesis and outdoor mass cultivation in a low-cost photobioreactor. Biotechnol Bioeng 102:100–112. https://doi.org/10.1002/bit.22033

Sahena F, Zaidul ISM, Jinap S, Karim AA, Abbas KA, Norulaini NAN, Omar AKM (2009) Application of supercritical CO2 in lipid extraction – a review. J Food Eng 95:240–253. https://doi.org/10.1016/j.jfoodeng.2009.06.026

Salam KA, Velasquez-Orta SB, Harvey AP (2016) A sustainable integrated in situ transesterification of microalgae for biodiesel production and associated co-product-a review. Renew Sust Energ Rev 65:1179–1198. https://doi.org/10.1016/j.rser.2016.07.068

Salim S, Bosma R, Vermuë MH, Wijffels RH (2010) Harvesting of microalgae by bio-flocculation. J Appl Phycol 23:849–855. https://doi.org/10.1007/s10811-010-9591-x

Saratale RG, Kumar G, Banu R, Xia A, Periyasamy S, Saratale GD (2018) A critical review on anaerobic digestion of microalgae and macroalgae and co-digestion of biomass for enhanced methane generation. Bioresour Technol. https://doi.org/10.1016/j.biortech.2018.03.030

Schenk PM, Thomas-Hall SR, Stephens E, Marx UC, Mussgnug JH, Posten C, Kruse O, Hankamer B (2008) Second generation biofuels: high-efficiency microalgae for biodiesel production. Bioenergy Res 1:20–43. https://doi.org/10.1007/s12155-008-9008-8

Schwarzenbach RP (2006) The challenge of micropollutants in aquatic systems. Science 313:1072–1077. https://doi.org/10.1126/science.1127291

Shelef G, Sukenik A, Green M (1984) Microalgae harvesting and processing: a literature review. United States. https://doi.org/10.2172/6204677

Shuba ES, Kifle D (2018) Microalgae to biofuels: "Promising" alternative and renewable energy, review. Renew Sust Energ Rev 81:743–755. https://doi.org/10.1016/j.rser.2017.08.042

Singh G, Patidar SK (2018) Microalgae harvesting techniques: a review. J Environ Manag 217:499–508. https://doi.org/10.1016/j.jenvman.2018.04.010

Singh RN, Sharma S (2012) Development of suitable photobioreactor for algae production – a review. Renew Sust Energ Rev 16:2347–2353. https://doi.org/10.1016/j.rser.2012.01.026

Singh SP, Kumari S, Rastogi RP, Singh KL, Richa SRP (2010) Photoprotective and biotechnological potentials of cyanobacterial sheath pigment, scytonemin. Afr J Biotechnol 9:580–588. https://doi.org/10.5897/AJB09.019

Singh A, Nigam PS, Murphy JD (2011) Renewable fuels from algae: an answer to debatable land-based fuels. Bioresour Technol 102:10–16. https://doi.org/10.1016/j.biortech.2010.06.032

Soletto D, Binaghi L, Lodi A, Carvalho JCM, Converti A (2005) Batch and fed-batch cultivations of Spirulina platensis using ammonium sulphate and urea as nitrogen sources. Aquaculture 243:217–224. https://doi.org/10.1016/j.aquaculture.2004.10.005

Song T, Mårtensson L, Eriksson T, Zheng W, Rasmussen U (2005) Biodiversity and seasonal variation of the cyanobacterial assemblage in a rice paddy field in Fujian. China FEMS Microbiol Ecol 54:131–140. https://doi.org/10.1016/j.femsec.2005.03.008

Sostaric M, Klinar D, Bricelj M, Golob J, Berovic M, Likozar B (2012) Growth, lipid extraction and thermal degradation of the microalga *Chlorellavulgaris*. New Biotechnol 29:325–331. https://doi.org/10.1016/j.nbt.2011.12.002

Spolaore P, Joannis-Cassan C, Duran E, Isambert A (2006) Commercial applications of microalgae. J Biosci Bioeng 101:87–96. https://doi.org/10.1263/jbb.101.87

Stevenson CS (2002) The identification and characterization of the marine natural product scytonemin as a novel antiproliferative pharmacophore. J Pharmacol Exp Ther 303:858–866. https://doi.org/10.1124/jpet.102.036350

Stolz P, Obermayer B (2005) Manufacturing microalgae for skin care. Cosmet Toilet 120:99–106

Suali E, Sarbatly R (2012) Conversion of microalgae to biofuel. Renew Sust Energ Rev 16:4316–4342. https://doi.org/10.1016/j.rser.2012.03.047

Suganya T, Varman M, Masjuki HH, Renganathan S (2016) Macroalgae and microalgae as a potential source for commercial applications along with biofuels production: a biorefinery approach. Renew Sust Energ Rev 55:909–941. https://doi.org/10.1016/j.rser.2015.11.026

Tabatabaei M, Tohidfar M, Jouzani GS, Safarnejad M, Pazouki M (2011) Biodiesel production from genetically engineered microalgae: future of bioenergy in Iran. Renew Sust Energ Rev 15:1918–1927. https://doi.org/10.1016/j.rser.2010.12.004

Taher H, Al-Zuhair S, Al-Marzouqi AH, Haik Y, Farid M (2014) Effective extraction of microalgae lipids from wet biomass for biodiesel production. Biomass Bioenergy 66:159–167. https://doi.org/10.1016/j.biombioe.2014.02.034

Trivedi J, Aila M, Bangwal DP, Kaul S, Garg MO (2015) Algae based biorefinery—how to make sense? Renew Sust Energ Rev 47:295–307. https://doi.org/10.1016/j.rser.2015.03.052

Uduman N, Qi Y, Danquah MK, Forde GM, Hoadley A (2010) Dewatering of microalgal cultures: a major bottleneck to algae-based fuels. J Renewable and Sustainable Energy 2:012701. https://doi.org/10.1063/1.3294480

Ugwu C, Ogbonna J, Tanaka H (2002) Improvement of mass transfer characteristics and productivities of inclined tubular photobioreactors by installation of internal static mixers. Appl Microbiol Biotechnol 58:600–607. https://doi.org/10.1007/s00253-002-0940-9

Ugwu CU, Aoyagi H, Uchiyama H (2008) Photobioreactors for mass cultivation of algae. Bioresour Technol 99:4021–4028. https://doi.org/10.1016/j.biortech.2007.01.046

Ullah K, Ahmad M, Sofia SVK, Lu P, Harvey A, Zafar M, Sultana S (2015) Assessing the potential of algal biomass opportunities for bioenergy industry: a review. Fuel 143:414–423. https://doi.org/10.1016/j.fuel.2014.10.064

Vandamme D, Foubert I, Meesschaert B, Muylaert K (2009) Flocculation of microalgae using cationic starch. J Appl Phycol 22:525–530. https://doi.org/10.1007/s10811-009-9488-8

Viskari PJ, Colyer CL (2003) Rapid extraction of phycobiliproteins from cultured cyanobacteria samples. Anal Biochem 319:263–271. https://doi.org/10.1016/s0003-2697(03)00294-x

Wang B, Li Y, Wu N, Lan CQ (2008) CO2bio-mitigation using microalgae. Appl Microbiol Biotechnol 7:707–718. https://doi.org/10.1007/s00253-008-1518-y

Wang B, Lan CQ, Horsman M (2012) Closed photobioreactors for production of microalgal biomasses. Biotechnol Adv 30:904–912. https://doi.org/10.1016/j.biotechadv.2012.01.019

Zhu LD, Hiltunen E, Antila E, Zhong JJ, Yuan ZH, Wang ZM (2014) Microalgal biofuels: flexible bioenergies for sustainable development. Renew Sust Energ Rev 30:1035–1046. https://doi.org/10.1016/j.rser.2013.11.003

Zittelli GC, Rodolfi L, Biondi N, Tredici MR (2006) Productivity and photosynthetic efficiency of outdoor cultures of *Tetraselmissuecica* in annular columns. Aquaculture 261:932–943. https://doi.org/10.1016/j.aquaculture.2006.08.011

Recent Trends in Biogas Upgrading Technologies for Biomethane Production

9

B. S. Dhanya, Dhruv Singh, Asim Kumar Jana, Anjani Kumar Dwiwedi, Ashok Kumar Sharma, Munusamy Chamundeeswari, and Madan Lal Verma

Abstract

Biogas, an ultimate renewable energy, is of enormous demand currently, due to increased fuel price and its fluctuations with expansive pollution emission. Biogas is environmentally feasible and viable. Biomethane production is of high impact, and hence the present chapter is concentrated on various biogas upgradation technologies conjugated with carbon dioxide and hydrogen sulphide removal strategies. The upgrading methods such as absorption, adsorption, membrane separation, biological methods, cryogenic technology, hybrid methods, supersonic separation, industrial lung, in situ methane enrichment and chemical dehydrogenation are discussed. High methane purity with minimized methane loss is the key for an effective upgradation method. A comprehensive study of comparison between various biogas upgradation technologies is analysed,

Authors "B. S. Dhanya" and "Dhruv Singh"contributed equally to this work.

B. S. Dhanya
Department of Biotechnology, Udaya School of Engineering, Nagercoil, Tamil Nadu, India

D. Singh · A. K. Dwiwedi · A. K. Sharma
Department of Chemical Engineering, Ujjain Engineering College, Ujjain, Madhya Pradesh, India

A. K. Jana
Department of Biotechnology, National Institute of Technology, Jalandhar, Punjab, India

M. Chamundeeswari
Department of Biotechnology, Udaya School of Engineering, Nagercoil, Tamil Nadu, India

Department of Biotechnology, St. Joseph's College of Engineering, Chennai, Tamil Nadu, India

M. L. Verma (✉)
Centre for Chemistry and Biotechnology, Deakin University, Geelong, VIC, Australia

Department of Biotechnology, Dr Y. S. Parmar University of Horticulture and Forestry, Hamirpur, Himachal Pradesh, India

N. Srivastava et al. (eds.), *Biofuel Production Technologies: Critical Analysis for Sustainability*, Clean Energy Production Technologies,
https://doi.org/10.1007/978-981-13-8637-4_9

showcasing the advantages and disadvantages too. It is concluded that the recently innovated technologies have wide potential advantages than the conventional biogas upgrading technologies. Although innovated technologies are so far better, detailed analysis, research and development is required for acquiring a technology which is economically, environmentally, technologically, operationally and socially feasible and acceptable.

Keywords
Biomethane · Upgrading technologies · Desulphurization · Cryogenic · Biological method · Scrubbing

9.1 Introduction

With increasing urbanisation and population growth, there has been a tremendous increase in waste generation and there has been a renewed interest in using waste as a resource for producing energy. Biogas is a valuable renewable energy produced by the anaerobic digestion of organic materials with the major product of methane and carbon dioxide along with traces of impurities like H_2S, siloxanes, water vapour, amines, ammonia etc. It is also a profitable solution for organic waste management, fertiliser production, and reduction of greenhouse gas emissions. In general, cattle dung, agricultural residues, food waste, organic fraction of municipal waste, sewage sludge and energy crops are majorly used as a substrate for anaerobic digestion process. However, the methane yield depends upon the type of feedstocks as well as operational behaviour of the digester (Bauer et al. 2013a, b; Al Mum and Torii 2015). Thus, many types of anaerobic digester designs have been implemented such as anaerobic sequencing batch reactor, anaerobic plug-flow reactor, continuous stirred tank reactor and anaerobic contact reactor. The components of biogas, CH_4, CO_2, H_2S, NH_3 and water vapour have different impacts on the basis of its utilisation. For the reduction of these impacts, biogas components should be removed (Petersson and Wellinger 2009).

The raw biogas contains approximately 55–70% of CH_4, 35–45% CO_2 and 200–30,000% ppm H_2S along with <5% traces of NH_3, siloxanes and water vapour (Sahota et al. 2018). The energy content of CH_4 described by lower calorific value (LCV) is approximately 36 MJ/m^3-CH4 (at STP conditions). Due to the fact that the presence of components other than methane has no calorific value, this leads to lowering down the LCV to 22–25 MJ/m^3- biogas. Apart from energy prospective, these components also lead to environmental pollution upon combustion. Thus, biogas upgradation (Fig. 9.1) is needed for biomethane production as a replacement of CNG.

Before going for biogas upgradation, first treatment is biogas cleaning. The biogas cleaning step involves removal of H_2S which is a hazardous and extremely corrosive acid gas. H_2S leads to damaging of engines and metal parts of the system

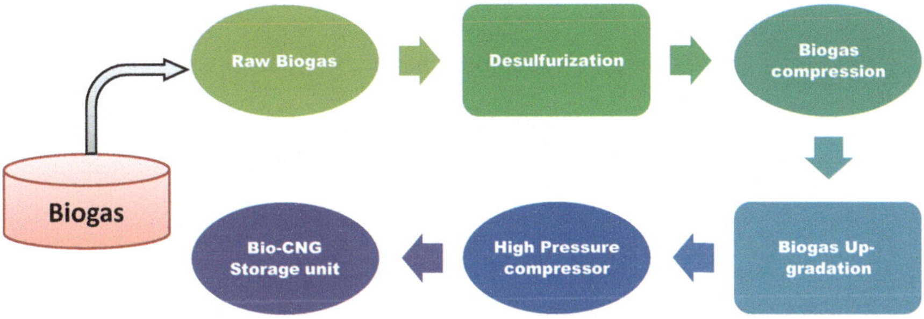

Fig. 9.1 Schematic diagram of biogas upgradation process

due to the emission of SO_2 upon combustion. The second step for biomethane production is removal of CO_2 and other trace elements by the various methods of upgradation such as absorption, adsorption, membrane separation, cryogenic technologies, in situ methane enrichment and hybrid technologies (Singhal et al. 2017). Since biogas is considered as a promising renewable fuel having high energy value and being environment-friendly, wide attention has been given to produce biomethane via cost-effective routes. This chapter emphasises on biogas cleaning techniques (H_2S removal), followed by biogas upgradation methods (CO_2 removal) (Kadam and Pawar 2017; Awe et al. 2017).

9.2 Biogas Desulphurisation

H_2S is a harmful and odorous gas which is not only hazardous to the environment but also fatal to human health and corrodes metal parts and engines. The concentration of H_2S varies from 200 to 30,000 ppm, majorly depending on the feedstock used for biogas production. The maximum allowable H_2S concentration after treatment depends upon the mode of biogas utilisation (Fig. 9.2). The removal of H_2S is a must-to-do step before biogas upgradation (Awe et al. 2017). The traditional desulphurisation technologies for biogas included majorly absorption, adsorption, membrane, biological and in situ desulphurisation methods (Zhao et al. 2010).

9.2.1 In Situ Desulphurisation

9.2.1.1 Air/Oxygen Dosing

This is one of the oldest methods used for removal of H_2S from biogas in which concentration of H_2S is controlled within the biogas digester. In this process, oxygen/air dosing is given within a digestor in which conversion of H_2S to elemental sulphur by a group of sulphur-oxidising bacteria takes place (Awe et al. 2017). The amount of oxygen/air dosing to the digester depends on the concentration of H_2S concentration. The following reaction takes place by injecting oxygen/air to the digester by Eq. (9.1):

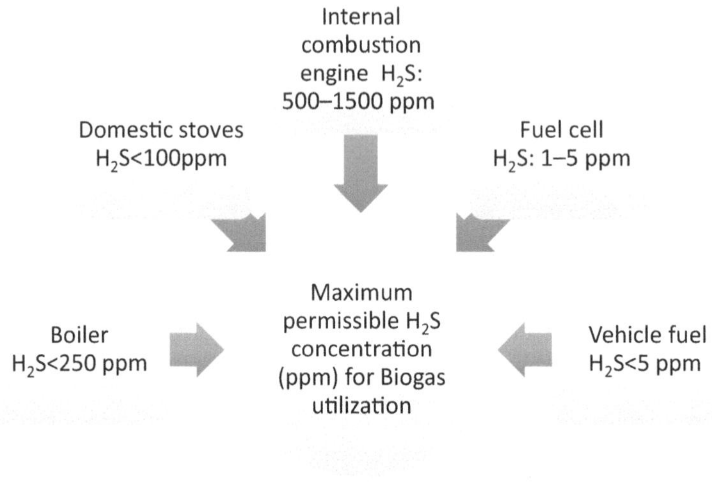

Fig. 9.2 Maximum permissible H_2S concentration (ppm) for technological recommendations and mode of biogas utilisation

$$2H_2S + O_2 \rightarrow 2S + 2H_2O \tag{9.1}$$

9.2.1.2 Iron Chloride Dosing

Similar to air/oxygen dosing, iron chloride dosing is done to the biogas digester in which H_2S can be reduced by adding $FeCl_2$ and $FeCl_3$ or which can be explained by Reaction 9.2:

$$Fe^{2+} + S^{2-} \rightarrow FeS \tag{9.2}$$

The resultant FeS can be easily removed from the system with discharge solids and can be utilised as a sulphur-rich fertiliser in the form of elemental sulphur.

9.2.2 Absorption Method

Absorption process is generally used for effective removal of H_2S from biogas. Absorption is the process by which one material (a solid or liquid) is absorbed by another substance, such as a liquid or gas, through minute pores or voids to its bulk surface. The absorption process can be classified into chemical or physical. Both the processes are discussed below. The basic absorption of H_2S can be explained by the following Eqs. (9.3, 9.4 and 9.5):

$$H_2S(g) + H_2O \rightleftharpoons H_2S\ (aq) \tag{9.3}$$

$$H_2S\ (aq) \rightleftharpoons H^+ + HS^- \tag{9.4}$$

$$HS^- \rightleftharpoons H^+ + S^{-2} \tag{9.5}$$

9.2.2.1 Physical Absorption

Physical absorption process is the most common and older process used for removal of H_2S from biogas stream. In this process, no chemical solution is used.

Physical absorption is mainly done by water and organic solvents such as cold formaldehyde propylene carbonate and methylpyrrolidone. The process is very environment friendly and could meet the demand of H_2S removal from high concentration H_2S containing biogas. For this reason, we have to move forward to chemical absorption process (Lin et al. 2013; Awe et al. 2017).

9.2.2.2 Chemical Absorption

Chemical absorption is used effectively for the removal of high-concentration H_2S at larger scale biogas plants. In chemical-based absorption system, H_2S is absorbed by various aqueous solution such as amines solution, chelated iron solutions, nitrite solutions, alkaline salt solutions and caustic solutions (Zicari 2003; Horikawa et al. 2004).

9.2.2.2.1 Amine Solution

Different amine solutions such as monoethanolamine (MEA), diethanolamine (DEA) and methyldiethanolamine (MDEA) are used for chemical absorption of H_2S. The amines are water soluble which lead to the efficient absorption of acid gas as it has one hydroxyl group and one amino group and has the capability to remove H_2S by absorption. After absorption, amine dissolves H2S in an aqueous amine (Siefers et al. 2010; Belmankhout et al. 2009).

$$RNH_2 + H_2S \rightleftharpoons RNH_3^+ + SH^- \tag{9.6}$$

9.2.2.2.2 Chelated Iron Solutions

In this process, chemical absorption of H_2S into iron-chelated solutions takes place with the formation of elemental sulphur as a by-product. This method is advantageous over others due to high H_2S removal efficiency and low use of chemicals for the reason that the iron-chelated solutions function as a pseudo-catalyst that can be regenerated (Siefers et al. 2010; Awe et al. 2017). The major advantage of using iron-chelated solution is that it can remove H_2S from the biogas stream with the formation of H_2S into a more stable or valuable product, as do processes that transform H_2S into S. Iron-chelating-based process comprises of physical absorption of H_2S onto the water undergoing dissociation which is explained in Eq. (9.5).

The formation of elemental sulphur occurs by the oxidation of sulphide by chelated iron as described by Eq. (9.7):

$$S^{-2} + 2Fe^{+3} \rightleftharpoons 2\ Fe^{+3} + 2OH \quad (9.7)$$

9.2.2.2.3 Nitrite Solutions

Nitrite solutions are used for H_2S absorption whenever a simple process configuration is required. The overall process requires only one bubble column reactor associated with mist eliminator. The nitrites of sodium and potassium are commonly used for H_2S absorption.

The overall reaction of sodium nitrite is described in Eq. (9.8):

$$3\ H_2S + NaNO_2 \rightarrow NH_3 + 3S + NaOH + \text{some } NO_x \quad (9.8)$$

Though the reaction products are ammonia and NOx which may lead to environmental pollution, the spent slurry is non-hazardous to the environment with low-cost maintenance of the system and easy change out of adsorbent.

9.2.2.2.4 Alkaline Salt Solution

Alkaline salt solutions readily react with acid gases such as H_2S. Generally, it is a regenerative process, and common alkaline salts are sodium and potassium carbonate, phosphate, borate, phenolate as well as salts of weak organic acids.

9.2.2.2.5 Caustic Scrubbing

Hydroxide solution is used for effective removal of H_2S from biogas but limits its regenerative property. Various oxido-alkaline solutions such as sodium hydroxide (NaOH) and potassium hydroxide are commonly used for H_2S absorption (Awe et al. 2017). Furthermore, selective absorption performance of hydroxide solutions is higher than that of amine solutions.

$$NaOH + H_2S\ (aq) \rightarrow NaHS\ (aq) + H_2O \quad (9.9)$$

$$NaHS\ (aq) + NaOH\ (aq) \rightarrow Na_2S(aq) + H_2O \quad (9.10)$$

9.2.3 Adsorption

9.2.3.1 Adsorption Using Metal Oxides

9.2.3.1.1 Iron Oxide

Iron oxide adsorption is the oldest known method and remains in practice to date for the removal of H_2S by the formation of insoluble iron sulphides. However, if the bed is exposed to air, it further forms elemental sulphur and regenerates the iron oxides.

Eventually, media becomes clocked with the elemental sulphur, so replacement of adsorbent is required. The most common iron oxide sorbent is iron sponge (Siefers et al. 2010). There is some commercially produced iron oxide-based system for H_2S removal that produces non-hazardous waste. There are various iron oxide alternatives which are commercially available such as Sulfa-Treat®, Sulphur-Rite® and Media-G2®

$$Fe_2O_3 + 3H_2S \rightarrow 3H_2O + Fe_2S_3 \tag{9.11}$$

$$Fe_2S_3 + \frac{3}{2}O_3 \rightarrow Fe_2O_3 + 3S \tag{9.12}$$

9.2.3.1.2 Zinc Oxide

Zinc oxides are also preferred metal oxide sorbent for the removal of H_2S at elevated temperature around 200–400 °C because zinc oxides have increased selectivity for H_2S at higher temperature. H_2S removal on the surface of zinc oxide forms insoluble layer of zinc sulphides, so that H_2S is removed from the gaseous stream (Zicari 2003). Zinc oxide has high surface area for H_2S capturing, but according to thermodynamic analysis, it loses its surface area at high temperature regeneration. The reaction mechanism is explained by the equation

$$ZnO + H_2S \rightarrow ZnS + H_2O \tag{9.13}$$

9.2.3.2 Adsorption on Zeolite

Zeolites have high adsorption capacity than metal oxides. Zeolites are used to capture H_2S molecules in its highly porous surface known as molecular sieves. ZSM, 4A, 5A and 13X zeolite are some common zeolites used for H_2S removal. Nowadays, some natural zeolites are also used for acid gas separation such as mordenite, clinoptilolite, erionite, phillipsite and ferrierite though natural zeolites need activation (Ozekmekci et al. 2015; Micoli et al. 2014)

9.2.3.3 Adsorption on Activated Carbon

Activated carbon is another effective sorbent used for the removal of H_2S which have been introduced in the recent years. Activated carbons are more preferred over other mesoporous materials, such as zeolite and metal-organic oxides due to its high surface area, microporosity, thermal stability, high removal capacity and low cost per unit volume. Presently, activated carbon comes primarily in two forms: non-impregnated and impregnated. Impregnation comprises of addition of cations to the activated surface which act as catalyst for higher adsorption of H_2S from biogas (Ozekmekci et al. 2015; Zulkefli et al. 2019; Pipatmanomni et al. 2009). The impregnation of activated carbon is done by some alkaline solvents like sodium carbonate, potassium iodide, copper sulphate, zinc acetate, sodium and potassium hydroxide which leads to higher dissociation of H_2S due to its alkalinity

9.2.3.4 Adsorption on Biomass-Based Sorbents

The requirement of high temperature for the preparation of activated carbon is the major concern. In this regard, biochar can be a promising adsorbent that can be prepared from readily available waste biomasses. Moreover, biochar waste adsorbents are relatively cheaper and environment friendly than commercially available activated carbon (Sahota et al. 2018; Zicari 2003). The most common biomass waste adsorbents are biochar (leaf waste, peat and sludge waste, slurry based, manure based)

9.2.4 Biological Desulphurisation

Microbes such as *Thiobacillus* and *Sulfolobus* are used to remove hydrogen sulphide by undergoing the process of oxidation. Biological desulphurisation occurs only in the presence of oxygen, and so a small amount of air should be added to the digester. The microorganisms are immobilised in this process. Another solution to remove the hydrogen sulphide is using the trickling filter method, where the biogas is allowed to pass through the trickling filter after leaving the digester. The microbes are packed in the trickling filter and the sulphur containing the compounds can be removed. The main disadvantage in both the biological methods is that it cannot be applied if the biogas is used for grid injection or as a vehicular fuel and it is due to the presence of trace amount of oxygen gas. This problem is rectified by developing an alternative method where the biogas is free of oxygen after the removal of hydrogen sulphide gas (Petersson and Wellinger 2009).

9.2.5 Biofiltration of H_2S

From available biological methods for hydrogen sulphide removal, biofiltration is a potential cost-effective technique which utilises living material to capture acid gas. Nowadays, some common sulphur-oxidising microorganisms mainly from the family of *Thiobacillus, Thiomonas, Paracoccus, Acidithiobacillus, Sulfurimonas* or *Halothiobacillus* are used for biofilters. However, this technique is limited for utilisation at high concentration of H_2S containing biogas streams. Mainly two commercially available methods are used for removal of high concentration: one is Thiopaq process and second is Biopuric process (Kadam and Panwar 2017; Tomas et al. 2009; Soreanu et al. 2008). The advantages of biological methods are low energy requirement at mild conditions, and production of elemental sulphur as by-product is the major advantage of the process. Better microbial attachment onto biofilter bed needs higher specific surface area, less pressure drop and better water retaining capacity.

9.3 Biogas Upgradation: CO_2 Removal

Biogas upgrading is a necessary process to be executed in all countries since the importance of upgraded biogas is wide. It requires cost-effective investment and environmentally pleasing solutions, and hence optimisation must be done in terms of providing high methane content, less energy consumption and minimised methane emissions (Petersson and Wellinger 2009).

More biogas upgrading technologies are commercially available. The fuel efficiency is important and can be predicted using the parameter calorific value which is 21.5 MJ/m^3 for biogas and 35.8 MJ/m^3 for natural gas. The deviation in the calorific values for biogas and natural gas is due to the presence of CO_2, the incombustible section of biogas, which may lead to the minimisation of heating value with high compression value. In addition to CO_2, H_2S, nitrogen and methane gases are also present that are strongly not factorable to the environment (Sahota et al. 2018).

Many techniques are available for biogas upgradation, and novel techniques rectifying all the disadvantages (GHG emissions, hydrocarbon emission, nitrogen oxide emission, carbon monoxide emission) (Zhao et al. 2010) are under development with environmental and economical perspective (Petersson and Wellinger 2009). The use of biogas as transportation fuel is a dream for developed and developing countries. Various methods of biogas upgrading technologies (Fig. 9.3) are described below:

9.3.1 Absorption Methods

In absorption method, the gaseous components have the capacity to undergo diffusion process (Report 2012). Gaseous impurity solubility is one of the crucial factors for effective absorption. The solubility of carbon dioxide is more than the methane, and so the liquid that is emitted from the column contains increased amount of carbon dioxide and the gas emitted from the column contains more amount of methane, which forms the base for the absorption principle (Petersson and Wellinger 2009).

The untreated biogas is allowed to pass through the column containing a plastic pack for increasing the area of contact between the two phases in a countercurrent manner. The solvent used in the absorption process should be selected based on the various factors (Battino and Clever 1966). For obtaining maximum absorption of components, factors such as volatility, non-hazardous nature and cost-effectiveness should be considered (Sahota et al. 2018). The efficiency of absorption process can also be enhanced by adding fresh liquid to the already used scrubbing liquid. Since, absorption process requires lower flow rates, the method is economically feasible (Singhal et al. 2017).

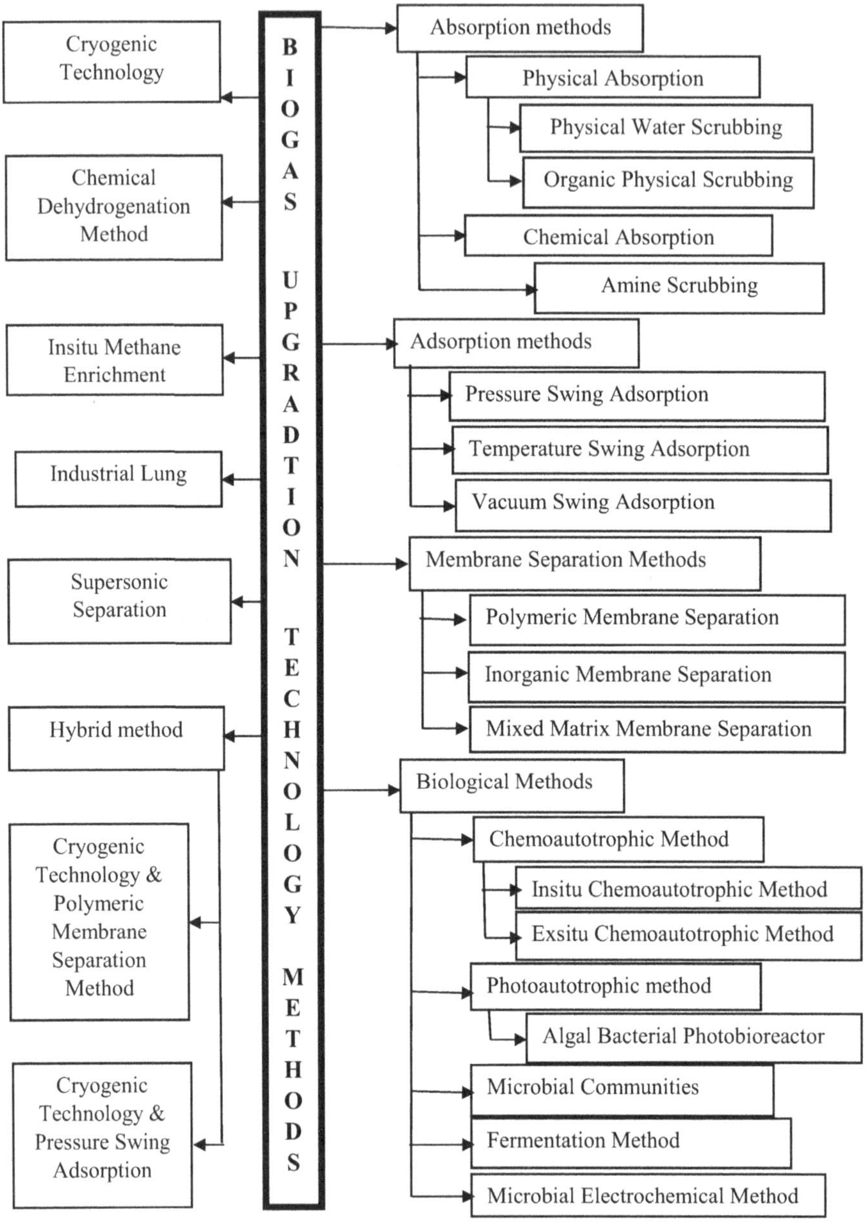

Fig. 9.3 Various methods of biogas upgradation technologies

Various types of absorbents are used, and based on the types of absorbents, the absorption methods can be categorised into the following types:

(a) Physical absorption method.
(b) Chemical absorption method.
(c) Physical absorption method.

9.3.1.1 Physical Water Scrubbing

Physical water scrubbing is one of the most commonly used biogas upgradation technology for biogas cleaning and upgradation, which is commercially available and manufactured with wide range of capacities (Thran et al. 2014). In this method, the principle of solubility is applied between two gaseous components (Kapoor et al. 2017). The two gaseous components carbon dioxide and methane vary in the property of solubility, specifically at low temperature. Carbon dioxide is more soluble when compared to methane, and while leaving the column, the carbon dioxide quantity gets increased in the liquid part and the methane quantity gets increased in the gaseous phase (Cozma et al. 2013). The water that is left over the column is passed to a flash tank with release in some amount of carbon dioxide and again retransferred to the raw biogas inlet provided. If the water is allowed to recycling process in desorption column which is filled with plastic packing, the carbon dioxide gas will be released. The high difference in the solubility of the two gases can be attained by cooling the water, before the next recycling process in the column.

Genosorb 1753 is one of the most widely applied absorbent, and by using the depressurising technique, the spent absorbents can be regenerated (Patterson et al. 2011). The hydrogen sulphide gas separation should be done since the absorption of carbon dioxide tends to reduce in the presence of hydrogen sulphide gas (Sahota et al. 2018). The operating parameters for the water scrubbing method include initial pressure of 6–10 Bar and temperature of 40 °C. From the bottom side of the column, the biogas is passed, and from the top of the column, water is passed and injected into the absorption column via the bottom side of the tank, while the water is passed that expresses the countercurrent flow between the gas and the water (Bauer et al. 2013b).

The major disadvantage of this method is that during the process of regeneration, gas components such as nitrogen and oxygen tend to get dissolved, and then it passes along with the upgraded methane gas. This technique is more valuable if:

- The plant capacity is increased.
- The decompression process is discarded.
- The heating value is reduced (Report 2012).

9.3.1.2 Organic Physical Scrubbing

In organic physical scrubbing, the absorption of carbon dioxide occurs in the organic solvents such as polyethylene glycol/methanol, and this absorption tendency is more possible because the carbon dioxide gas is more soluble in solvent than in water.

Moreover, the solvents can be regenerated using the depressurising technique. In addition to this, water, nitrogen, hydrogen sulphide and oxygen can be separated easily along with carbon dioxide. Some of the brand names for the liquids applied in this technique include Rektisol®, Genosorb®, Sepasolv®, Purisol® and Selexol® (Petersson and Wellinger 2009; Report 2012). Selexol® absorbs more carbon dioxide than water and so requires small upgrading unit. Regeneration of the organic solvent is hard because the solubility of carbon dioxide is high and also the hydrogen sulphide gas solubility is also high, requiring an additional amount of heat (Persson 2003).

The operational parameters include 7–8 bars for biogas compression, 20 °C temperature while injecting the biogas. Also, the solvent used must be cooled down before adding to the unit since temperature affects the Henry's constant (Bauer et al. 2013a). Increasing the temperature to 80 °C regenerates the organic solvent (Bauer et al. 2013b; Sun et al. 2015). About 98% methane purity is obtained (Bauer et al. 2013a; Angelidaki et al. 2018).

9.3.2 Chemical Scrubbing

Chemical scrubbing involves the absorption of gases in the scrubbing liquid followed by chemical treatment. The reaction is highly specific in not absorbing the methane gas followed by high methane recovery and purity. Amine scrubbing is an example of chemical scrubbing. Amine solutions, dimethylethanolamine, triethanolamine, mixture of piperazine and monoethanolamine, methyldiethanolamine, diglycolamines (Patterson et al. 2011) and monoethanolamine, are used widely to absorb the carbon dioxide. The carbon dioxide loading capacity is high for tertiary amines (Kadam and Panwar 2017). The mixture of piperazine and monoethanolamine has greater absorbing capacity, and hence less energy is required during the regeneration step (Bauer et al. 2013a, b; Sahota et al. 2018).

Amine scrubbing gains importance due to the reaction of amine solution with the carbon dioxide gas in addition to its absorption, leading to the reduction of methane emission to <0.1%. The liquid that is lost due to evaporation should be replaced, and regeneration of carbon dioxide is possible by increasing the temperature. The hydrogen sulphide gas should be removed before the process of absorption since it gets absorbed with the amine scrubber solutions (Petersson and Wellinger 2009).

Amine scrubbing system is made up of a stripper and an absorbing unit. The operating pressure required is very low when compared to the water scrubbing process. The pressure required for the injection of biogas into the bottom of the absorbing column is about 1–2 bars, and the amine scrubbing solutions are injected from the top of the unit in countercurrent flow method. The regeneration of the carbon dioxide and the hydrogen sulphide takes place in the stripping unit with a pressure of 1.5–3 bars and a temperature of 120–160 °C. Addition of heat to the unit breaks the chemical bonds in the absorber phase (Kapdi et al. 2005), and at last, the carbon dioxide is condensed for recirculation and the entrapped carbon dioxide is left out. In addition to the amine solutions, alkaline salts like calcium hydroxide, sodium and potassium were also added (Kougias et al. 2010; Zhao et al. 2010). The absorption capacity of carbon dioxide is greater in sodium hydroxide than in the amine solutions. For absorbing 1 ton of carbon dioxide, 0.9 tons of sodium hydroxide is required when compared to 1.39 tons of monoethanolamine (Yoo et al. 2013). About 99% methane purity can be achieved through this method with less than 0.1% methane loss (Angelidaki et al. 2018).

The advantages of chemical scrubbing include complete absorption of hydrogen sulphide by the scrubber and high selectivity of the amine solution, and the disadvantages include high investment cost initially, high energy requirement during regeneration and the chemical toxicity to the environment (Yoo et al. 2013).

The amine scrubbing process is advantageous if:

- No off-gas treatment is needed.
- Plant capacity is high.
- No compression is required (Report 2012).

9.3.3 Adsorption Method

Adsorption is defined as a process in which the components of a mixture gets attached or binds to a solid matrix which is microporous in nature but with a large surface area. Based on the types of force, the adsorption process is categorised into physisorption and chemisorption. In physisorption process, between the adsorbent and the adsorbate, weak Vander Waal's forces is involved, and in chemisorption process, strong molecular force exists. The operational parameters for adsorption methods include 3–8 bar pressure, 50–60 °C temperature and 100–200 mbar regeneration pressure. It requires 3–5 min to complete one cycle. Cycle time, purging pressure, adsorbent, feeding pressure and column interconnectivity are the characteristics of adsorption techniques.

Adsorption methods are categorised into three types as follows:

(a) Pressure swing adsorption
(b) Temperature swing adsorption
(c) Vacuum swing adsorption methods

9.3.3.1 Pressure Swing Adsorption

Adsorption phenomenon of the various gas components and pressure difference are the two key components to be considered in pressure swing adsorption method. The adsorbing materials used are zeolites or activated carbon that can effectively adsorb carbon dioxide. In pressure swing adsorption method, a significant amount of methane is generated from the column, and recycling process can be undertaken through the inlet in the pressure swing adsorption column (Awe et al. 2017). The hydrogen sulphide which gets adsorbed cannot be subjected to recycling, and so, it must be removed before the addition of biogas into the column (Zhao et al. 2010). Low energy requirement, ease of operation, equipment compactness and low investment cost are the major advantages of this method (Augelletti et al. 2017). About 96–98% methane is generated, and about 4% is lost in the off-gas (Bauer et al. 2013a; Ryckebosch et al. 2011; Angelidaki et al. 2018).

Pressure swing adsorption technology is advantageous only if:

- The plant capacity is small/medium.
- No compression is required.
- Recovered methane is utilised for other purposes (Report 2012).

9.3.4 Membrane Separation Methods

Membrane separation methods are developed for landfill gas upgrading. It is based on the property of permeability of the membranes that are mostly made of polymeric materials such as cellulose acetate (Baker 2012)/polyimide/polysulfone/polydimethylsiloxane (Report 2012). Membrane selection is very important for separation efficiency. The membranes are stacked like hollow fibres. Through the membranes, water, carbon dioxide and ammonia are highly permeable, oxygen and hydrogen sulphide are moderately permeable, and methane and nitrogen are slightly permeable. The penetration of the gas is based on the concentration gradient of permeate, its chemical affinity and the size of the molecules (Petersson and Wellinger 2009). The transport of the gases takes place during the generation of pressure differences on both sides of the membranes (Scholz et al. 2013).

Membrane separation method process can be done in two ways: dry (gas/gas separation) or wet (gas/liquid separation). In dry process, mainly polymeric membranes are used for high specificity, and the rate of permeation depends upon the membrane type and sorption coefficient (Baker 2012). The sorption coefficient is based on the condensability of the molecules. Large molecules are highly condensable. Moreover, the methane gets attached to the membrane side. There are four types of configurations in the dry method. They are two-stage with a recirculation loop, single stage, three-stage with sweep biogas stream and two-stage with sweep biogas stream (Makaruk et al. 2010). In wet process, the membrane separates the liquid, and the gas feed and the gas molecules undergo diffusion process (Angelidaki et al. 2018).

The advantages of membrane technology include requirement of low skilled labour and less maintenance, and the disadvantages include high membrane cost, membrane damage and degradation (Scholz et al. 2013).

Membrane technology is advantageous only if:

- It adapts the flexible partial load behaviour.
- The methane can be reutilised.
- It adapts to small or medium plant capacity.

- No further compression is required.
- It avoids addition of chemicals/consumables (Report 2012).

9.3.4.1 Membrane Technology

It is categorised into three types based on the membranes used. They are as follows:

(a) Inorganic membrane separation method
(b) Polymeric membrane separation method
(c) Mixed matrix membrane
(a) Inorganic membrane separation method

Inorganic membrane separation method uses inorganic membranes in porous/dense phase having high selectivity, high chemical stability, high thermal stability and high permeability. The dense membrane is made up of zirconia/palladium/silver/nickel/calcium titanate, and porous membranes are fabricated with silica/carbon/zeolite/alumina (Mallada and Menendez 2008). Inorganic membranes are capable of withstanding hard environmental problems, and hence, it is widely used for separation of methane (Zhang et al. 2013). The operational life is long for the membranes, even though the cost of the method is high. In addition, the hydrogen sulphide and water must be pretreated for efficient removal of carbon dioxide (Chen et al. 2015; Sahota et al. 2018).

9.3.4.2 Polymeric Membrane Separation Method

Polymeric membranes are dense and porous in nature made up of cellulose acetate/polyimide/polysulfone (Ahmed et al. 2010). The permeability gets altered with the size of the pore. The basic principle lies in the convective flow, molecular sieving and the Knudsen flow (Chen et al. 2015; Zhang et al. 2013). Solubility and diffusivity are important for a better transport mechanism. Firstly, the gas molecules are trapped in the membrane, and then the diffusion process is executed. The diffusion process works on the basis of difference in pressure gradient and the concentration (Ahmed et al. 2010).

9.3.4.3 Mixed Matrix Membrane

In mixed matrix membranes, the inorganic filler is mixed with the polymer matrix, and so it is heterogeneous in nature with the property of high permeability, easy scalability and economically feasible. The integration of inorganic to the polymer membrane is done to bind the advantages of both the methods (Chen et al. 2015). Metal-organic framework and zeolite are the most commonly used inorganic fillers for separating carbon dioxide and methane.

The inorganic filler zeolite is crystalline in nature with tetrahedral shape and micro-porosity. It is made up of aluminium, silicon, sodium, potassium, calcium and magnesium. The different gas molecules can be easily separated based on the pore size. Due to uniform pore size, it becomes easy for the zeolite membranes to discriminate between different gas molecules. The zeolite membrane is fabricated using flat membranes, but in large manufacturing companies, hollow fibre membranes are applied due to its low fabrication cost, increased packing density, easy handling and flexible fabrication steps (Zhang et al. 2013).

The metal organic framework is synthesised by bonding the metal ion and organic linkers, and it acts as the coordination centre that enhances the flexibility and resilience properties (Li et al. 2012). The advantages of this metal organic framework include specific surface area, unique pore volume, high adsorption capacity and wide compatibility (Sahota et al. 2018).

9.3.5 Cryogenic Technology

The difference in condensation temperature of the gases to be separated forms the basis of the cryogenic technology. For separating methane and carbon dioxide, cryogenic technology uses the principle of sublimation/boiling points of the two different gases. The carbon dioxide gas gets accumulated in the liquid phase by the process of condensation/sublimation, and the methane gets collected in the gaseous phase. During the condensation process, in addition to the carbon dioxide, the siloxanes (Munoz et al. 2015) and water also get separated and can be removed out of the unit. Carbon dioxide has the sublimation point of 194.65 K. Increased amount of methane inhibits the normal characteristics of gas. It is mandatory to undergo the cooling process in several steps, and so, the retrieval process depends upon the gases to be separated.

For example, in GPP® system, the gas is cooled to −25 °C at a pressure of 17–26 bar to remove halogens, water, sulphur dioxide, siloxanes and hydrogen sulphide from the biogas. Additional contaminants can be removed by passing it first to a coalescence filter and then to a SOXSIA® catalyst. After removing the contaminants, the carbon dioxide gas is removed, and it occurs in two steps: firstly, cooling the gas to −50 °C and −59 °C, where about 30–40% of carbon dioxide is filtered in the liquid phase, and, secondly, carbon dioxide is separated in solid form. The GPP® plus system is upgrading its technology to generate methane in the form of liquid. About 96% methane purity is achieved in a pilot plant in Canada (Petersson and Wellinger 2009).

To liquefy carbon dioxide, compression and cooling must be done. The parameters required for successful implementation of this technology includes a pressure of 80 bar and a temperature of −170 °C (Porpatham et al. 2018; Sun et al. 2015; Ryckebosch et al. 2011). Hydrogen sulphide and water must be subjected to pretreatment to avoid freezing problem, and while methane gets separated, condensation process must be done for gases like oxygen and nitrogen (Chen et al. 2015).

The major advantages include 99% methane recovery and separation of carbon dioxide. The limitations of this technique include high investment cost and operating cost, need for large number of equipment, clogging (Angelidaki et al. 2018) and high-energy requirement (Deublein and Steinhauser 2010). Green Public Procurement developed a cryogenic technology that can minimise the energy requirement and generate pure carbon dioxide and methane in liquid form (Tuinier and van SintAnnal and 2012; Sahota et al. 2018).

9.3.6 Hybrid Method

Due to the persistence of various pros and cons in the biogas upgrading technologies, hybrid methods have been developed to rectify the problems and to meet maximum efficiency. Hybrid methods are one or two methods integrated together to mitigate the challenges, to develop novel methods and to innovate a profitable technology. Currently, two hybridised methods have been developed:

(a) Hybrid cryogenic technology and polymeric membrane separation method.
(b) Hybrid polymeric membrane separation method and pressurised water scrubbing method.

9.3.6.1 Hybrid Cryogenic Technology and Polymeric Membrane Separation Method

Integrating polymeric membrane separation method with the cryogenic technology enhances the cost reduction and is energy intensive when compared to the cryogenic technology. Various simulation studies were performed in hybrid methods for increasing the energy requirements by comparing it with the chemical scrubbing method (Belaissaoui et al. 2012). Temperature-membrane-cryogenic technology method is hybridised to incorporate the techno-economic feasibility, and the novel technology consumed less energy of 1.7 MJ/kg CO_2 when compared to 2.5 to 3.5 MJ/kg CO_2 (Song et al. 2017; Sahota et al. 2018).

9.3.6.2 Hybrid Polymeric Membrane Separation and Pressurised Water Scrubbing Method

Integrating polymeric membrane separation method with the pressurised water scrubbing can overcome the burden of upgrading costs when compared to the conventional pressurised water scrubbing, and about seven different types of hybrid membranes have been processed as apart of biogas enrichment (Scholz et al. 2013).

9.3.7 Chemical Hydrogenation Method

Sabatier reaction is the basis for chemical dehydrogenation method. Carbon dioxide is reduced with hydrogen chemically by adding catalysts such as ruthenium and nickel at high temperature of 300 °C and pressure of 5–20 MPa (Xia et al. 2016). The advantage of this method lies in its high selectivity option (Jurgensen et al. 2014), and the disadvantages include regular replacement of catalysts which is degenerated by the presence of trace amount of gases (Guebitz et al. 2015), the requirement for pure gases and a possible increase of energy-related costs (Angelidaki et al. 2018).

9.3.8 In Situ Methane Enrichment

In situ methane-enrichment method is based on the counterflow of gaseous components such as nitrogen and oxygen, favouring desorption of the dissolved carbon dioxide in the sludge by the recirculation concept. Additional carbon dioxide present in the column is absorbed by sending back the circulated sludge to the digester (Kadam and Panwar 2017). This compound has been tested efficiently at pilot level (Sun et al. 2015). Increased recirculation rate causes increased methane loss in the environment. The buffering capacity of the sludge tends to be altered in this technique (Petersson and Wellinger 2009).

A small-scale plant of digester volume of 15 m^3 and with a bubble column of 140 dm^3 has been developed (Nordberg et al. 2005). Some amount of carbon dioxide gets dissolved in the fluid phase of the digester tank. Continuous withdrawal of sludge carbon dioxide generates high amount of methane (Lindberg 2003). Further simulations can be worked out to reach high methane purity of up to 95%.

The advantages include economically suitable lower upgrading costs when compared to the conventional techniques and less requirement for ancillary equipment. The disadvantages include limitation to pilot plants and mostly suitable for sludge.

9.3.9 Industrial Lung

Industrial lung is one of the hybrid processes that involves the use of enzymes such as carbonic anhydrase to dissolve the carbon dioxide. The carbon dioxide is forced to pass through the aqueous phase, and it is absorbed by the absorbent in the absorber column. By applying heat, the absorbent can be regenerated (Petersson and Wellinger 2009; Scholwin et al. 2013). Carbonic anhydrase enzyme is prepared using six molecules of histidines for effective attachment of enzyme to the matrix by a leading research organisation in Lund, Sweden, and this enzyme can enhance 99% of methane recovery (Mattiasson 2005). CO_2 Solution Inc., a Canadian company, used this technique and was patented for focussing on biogas upgradation, and their current research is on bioreactor mechanics, enzyme immobilisation, production and technology and enzyme cloning (Petersson and Wellinger 2009).

The advantages include 95–99% methane purity, withstanding higher temperatures up to 85 °C, and the disadvantages include increased enzyme production costs and constrained life time of enzymes (Sahota et al. 2018).

9.3.10 Supersonic Separation

Supersonic separation is one of the novel methods invented in the field of biogas upgradation technology. The capacity of this method is wide with the facilities of recompression, expansion and gas-liquid separation. In this method, the expansion of the raw biogas to the supersonic velocity is facilitated by the convergent-divergent nozzle that leads to the decrease in pressure and temperature, causing the condensation and separation of hydrocarbons and water from the biogas (Zhang et al. 2006). The main disadvantage of this method is the expensive cost (Sahota et al. 2018).

(A) *Biological Methods*

The biological biogas upgrading technologies are categorised into four types:

(a) Chemoautotrophic method
(b) Photoautotrophic method
(c) Fermentation method
(d) Microbial electrochemical method

(a) Chemoautotrophic Method

In chemoautotrophic method, the hydrogenotrophic methanogens utilise hydrogen and convert carbon dioxide to methane gas. Since it is a biological method, the hydrogen to be utilised is generated from the renewable resources, and so, the water is hydrolysed using renewable electricity to produce hydrogen. Converting electricity to chemical energy is highly promising. For methane, the energy required is about 36 MJ/m^3 and for hydrogen it is about 10.88 MJ/m^3. Also, the investment cost is less when compared to other upgrading techniques. The major advantage of this method is that carbon dioxide is not separated; instead, it is directly converted to methane (Kougias et al. 2017).

It can be again classified into two types:

1. In situ chemoautotrophic method
2. Ex situ chemoautotrophic method
3. Microbial communities

1. *In Situ Chemoautotrophic Method*

In in situ chemoautotrophic method, the hydrogen is allowed to couple with the carbon dioxide that is generated in the anaerobic digester, and it is directly converted into methane by the involvement of autochthonous methanogenic archaea (Kougias et al. 2017). The methanogenesis process tends to be inhibited, when the pH is about 8.5. Increase in pH leads to the bicarbonate ion removal. Inhibition of methanogenesis is a challenge to be tackled, and to overcome this, codigestion process was performed, thereby stopping increase in pH (Luo and Angelidaki 2013) or the pH level is controlled completely (Luo et al. 2014). Alcohols and fatty acids undergo oxidation which leads to decreased concentration of hydrogen (Batstone et al. 2002).

2. *Ex Situ Chemoautotrophic Method*

Using external sources, the carbon dioxide is provided, and the hydrogen in the anaerobic reactor containing hydrogenotrophic culture converts hydrogen to methane (Kougias et al. 2017). Simpler biochemical process, biomass-dependent process, maintenance of the biogas process stability, usage of external sources, feasibility of generating current to remote areas and increased stirring speed (Luo and Angelidaki 2012) are some of the advantages of ex situ chemoautotrophic method (Bassani et al. 2017; Angelidaki et al. 2018).

3. *Microbial Communities in Biological Biogas Upgrading Systems*

In this method, the biogas upgrading is done in two different ways. One is by using hydrogenotrophic methanogenic archaea that converts the carbon dioxide to methane with the hydrogen donated from external sources, and this process is known as hydrogenotrophic methanogenesis (Stams and Plugge 2009). The other way is by using homoacetogenic bacteria for converting carbon dioxide to acetate. Widely used hydrogenotrophic methanogenic genera includes *Methanobacterium, Methanothermobacter, Methanoculleus and Methanomicrobium* (Agneessens et al. 2017; Bassani et al. 2017; Luo and Angelidaki 2013; Mulat et al. 2017) and rarely used genera include Methanosarcina (Agneessens et al. 2017; Mulat et al. 2017).

(b) *Photoautotrophic Methods*

To obtain a gas rich in methane, the photoautotrophic method is the most suitable method with maximum carbon dioxide sequestration. In addition, the impurity-hydrogen sulphide can be removed by using this method. About 97% methane is recovered in this method, and the recovery depends on the selected species of algae and the type of the reactor. Algae is widely used in the conversion process, and it is mass cultured in a closed- or open-type ponds. The photosynthetic efficiency is high when the process is undertaken in a closed-type pond. During the process of upgradation, the biogas is allowed to pass through the photobioreactor for efficient conversion of the gas to methane. The major drawback lies in the high investment cost (Angelidaki et al. 2018).

(c) *Fermentation Method*

In fermentation method, the carbon dioxide is converted to valuable products such as ethanol, acetate and butyrate (Agler et al. 2011; Kennes et al. 2016). The synthesised fatty acids like butyrate and acetate can be used for the production of biofuels (Agler et al. 2011; Martin et al. 2016; Zhang et al. 2013). Microbes such as *Butyribacterium methylotrophicumn*, *Acetobacterium woodii* and *Clostridium scatologenes* can convert carbon dioxide and hydrogen to liquid products (Schiel-Bengelsdorf and Durre 2012) by undergoing Wood-Ljungdahl pathway/reductive acetyl-CoA pathway (Latif et al. 2014; Angelidaki et al. 2018).

(d) *Microbial Electrochemical Method*

Microbial electrochemical method is considered as one of the most environmentally sustainable and cost-effective biogas upgradation methods to produce methane and to remove carbon dioxide (Lovley and Nevin 2013; Van Eerten-Jansen et al. 2012). An example for microbial electrochemical method is microbial electrolysis cell. In microbial electrolysis cell, the oxidation of organic compounds by bacteria releases electrons in the anode chamber and is combined with the protons in the cathode chamber to synthesise hydrogen that can be used for upgradation of biogas (Lu and Ren 2016; Zhang and Angelidaki 2014). Utilising the biocathode in microbial electrolysis cell, methane can be produced by undergoing reduction of carbon dioxide, attaining 80% energy efficiency (Cheng et al. 2009). The reduction of carbon dioxide to methane was based on the electron transfer and the hydrogen produced. Depending upon the cathode potential, the reduction process occurs (Villano et al. 2010). In situ (the bioreactor is microbial electrolysis cell's cathode) and ex situ methods (introduction of biogas to the cathode) of biogas upgradation technology using microbial electrolysis cell are experimentally tested to prove the efficiency. The end result proves that the efficiency of in situ biogas upgrading method is better with greater carbon dioxide removing capacity. Moreover, it was found that removal of carbon dioxide is associated with both generation of methane and ionisation of carbon dioxide. The ionisation is because of the generation of alkalinity in the cathode (Xu et al. 2014).

Carbon dioxide can also be removed to another chamber for separation. Interestingly, it was found that the CO_2 removal was attributed to not only the production of methane but also the CO_2 ionisation due to alkalinity generated in the cathode. Recently, another method was presented, in which the CO_2 was removed from the gas to a separate chamber. Microbial electrolysis cell consists of two membranes: anion exchange membrane and proton exchange membrane. Comparing both membranes for the removal of carbon dioxide, the proton exchange membrane shows better removal of carbon dioxide of about $78 \pm 7\%$ and with 83 ± 24 meq/Ld methane production. But this attributes to high energy requirement (Zeppilli et al. 2016). The removed carbon dioxide is converted to bicarbonate in the presence of an alkaline environment, and the generation of this bicarbonate leads to less production of methane content from the biogas. The electrons are provided by the water during current generation accompanied with the reduction of carbon dioxide (Van Eerten Jansen et al. 2012). The efficiency of biocathode is questionable if the oxygen that is formed during oxidation gets diffused into the cathode, but it does not affect the rate of methane production (Sadhukhan et al. 2016; Wang and Ren 2013). The usage of cobalt tetra-amino phthalocyanine and carbon nanotubes as cathode enhances the conversion of carbon dioxide to formic acid (Zhao et al. 2012). In the same way, the formic acid can be synthesised by immobilising the Methylobacterium extorquens AM1 (Hwang et al. 2015). Also, from carbon dioxide, 2-oxybutyrate and acetate can be synthesised in microbial electrolysis cell containing *Sporomusa ovata* (Nevin et al. 2010). As a whole, microbial electrochemical method is an environmentally pleasing method for biogas upgrading (Angelidaki et al. 2018).

9.4 Removal of Methane from the Off-Gas

During biogas production, the off-gas is generated which contains methane, and the concentration of methane depends upon the amount of recovered methane. High methane recovery is not acceptable because it requires high investment, maintenance and operational cost. To avoid the cost problem, a certain amount of methane is left out in most of the biogas plants (Report 2012).

Reducing the emission of methane content in to the environment is logically important for an innovative biogas upgradation technology. In addition to the methane emission reduction, methane slip too should be considered, since methane is an effective greenhouse gas.

Hence, the emitted methane content that leaves a PSA column should be reduced by off-gas treatment. There are few solutions that are mentioned below:

1. One of the solution to reduce the methane slip is to combine the air which is utilised for combustion with off-gas (Petersson and Wellinger 2009). The basic way to remove the methane in the off-gas is oxidation/combustion with production of excess heat that can be utilised in anaerobic digestion plants since this plant requires heat (Report 2012).
2. The methane gas mission can be minimised by using the process of catalytic combustion/thermal oxidation (Petersson and Wellinger 2009). Commercially, many technologies have been developed by the manufacturers with combustion of methane even at low content (wp3). Megtec developed VOCSIDIZER, a device with ceramic media containing heat transfer bed developed for regenerative thermal oxidation. The off-gas containing methane is allowed to pass through the ceramic media, and heat is applied. On the way of heating process, the methane in the off-gas is oxidised with oxygen with the generation of carbon dioxide and water vapour. The VOCSIDIZER can be maintained by the heat that is generated during the process of oxidation, and also the off-gas flow can be reversed periodically.
3. Flameless oxidation is another device based on the thermal oxidation process, in which the biogas is passed through the oxidation chamber and heated at 650 °C using raw gas, and preheat is done at 450 °C using the exhaust gas. Excess heat from the exhaust gas is regenerated, and it is used for any heating purpose. Catalyst can also be used to oxidise the methane content in such a way that the energy and temperature required for the oxidation process are less. The oxidation happens at the catalyst's surface and palladium, and cobalt or platinum acts as the active component (Petersson and Wellinger 2009; Report 2012).

A countable number of companies provide high methane recovery, leaving off-gas into the atmosphere directly.

9.5 Removal of Trace Components

9.5.1 Removal of Ammonia

Protein degradation process results in the formation of ammonia, and the ammonia content quantity depends upon the pH settings in the digester and the substrate composition. Hence, the biogas upgradation must be done to rectify the problem of ammonia generation. The separation of ammonia can be done by drying the biogas by cooling technique without the involvement of separate cleaning process (Petersson and Wellinger 2009; Report 2012).

9.5.2 Removal of Siloxanes

Siloxanes, the compound with silicon-oxygen bond, are widely used in shampoos and deodorants, and its presence can be easily detected in biogas produced from a sewage sludge treatment plants. The problem with siloxanes exists when it is burned to form silicon oxide in gas engines, and hence it must be removed by adsorption process or by cooling the gas or by absorption process or during hydrogen sulphide separation process. The adsorption process can be performed using activated aluminium or activated silica gel or activated carbon. For absorption process, liquid mixture of hydrocarbons is used (Petersson and Wellinger 2009; Report 2012).

9.5.3 Removal of Water

Corrosion is one of the major problems arising due to the reaction of water condensate (formed during the saturation of biogas with the water vapour while leaving the digester) in the gas pipeline with that of the sulphur oxides. So, it is necessary to remove the water vapour from the biogas itself before leaving the digester by reducing the temperature and maximising the pressure. Reduction in temperature leads to cooling, and it is achieved using an electric cooler (refrigeration) or condensate trap. Activated charcoal, silicon dioxide or molecular sieves (zeolites) (Report 2012) can also be applied to remove the water vapour by the process of adsorption. Hygroscopic salts or glycol solutions can also be used to remove water vapour by uploading the absorption process (Petersson and Wellinger 2009).

9.5.4 Removal of Particulates

In gas turbines and gas engines, the presence of particulates in biogas leads to mechanical wear, and it can be eradicated by using special mechanical filters sized 0.01–1 μm (Report 2012; Petersson and Wellinger 2009).

9.6 Comparison of Different Biogas Upgrading Technologies

Elevated fuel price and the demand for alternative unlimited renewable fuel source pave the way for many biogas upgrading technologies. Viable biogas upgrading technologies are available that overcome the disadvantages in the traditional techniques. The biogas upgrading technologies (i.e. physical/chemical methods (pressurised water scrubbing, organic physical scrubbing, chemical scrubbing, pressure swing adsorption method, membrane technology, cryogenic technology, in situ methane enrichment method and hybrid technology (Sahota et al. 2018)) and biological methods (chemoautotrophic and photoautotrophic) (Angelidaki et al. 2018)) are compared based on the factors such as economic, technological, environmental (Sahota et al. 2018), operational and social indicators (Toledo-Cervantes et al. 2017) (Table 9.1a and 9.1b). Based on these factors, the pros and cons of those biogas upgrading technologies are analysed (Table 9.2).

(a) *Economic Indicators*

The economic indicators include investment cost, operation cost, maintenance cost and sales cost (Toledo-Cervantes et al. 2017). Cheap electricity price is one of the cost factors to be considered while selecting an upgrading technique. Also, the cost of the upgrading technique depends upon the size of the plant (Petersson and Wellinger 2009).

Comparing the absorption technologies pressurised water scrubbing, organic physical scrubbing and amine scrubbing, the investment is about 1,000,000 €/year, 1,000,000 €/year and 2,000,000 €/year, respectively. The operation cost and maintenance cost are 14 ct/100 m^3h and 15,000 €/year, respectively; for pressurised water scrubbing, 13.8 ct/100 m^3h and 39,000 €/year, respectively; and for organic physical scrubbing and amine scrubbing, 14.4 ct/100 m^3h and 59,000 €/year, respectively. While comparing the cost analysis of absorption methods with adsorption methods, the adsorption method requires high investment cost (1,750,000 €/year), less maintenance cost (12.8 ct/100 m^3h) and comparatively high maintenance cost (56,000 €/ year). Various technologies such as pressurised water scrubbing and in situ methane enrichment are economically suitable when compared to the amine scrubbing and cryogenic method (Bauer et al. 2013a, b).

The fabrication cost is minimum, investment cost is maximum (20,000 €/year), operation cost is comparatively high (10.8–15.8 ct/100 m^3h) and maintenance cost is comparatively less (25,000 €/year) for polymeric membrane technology, and fabrication cost is maximum for inorganic membrane technology among the membrane separation technologies. The membrane cost is too high for polymeric membrane technology (Vrboba and Karel 2017).

The plant capacity is high for polymeric membrane technology and pressure swing, temperature swing and vacuum swing adsorption method. But, low operational cost is pronounced more with the hybrid technologies (Sahota et al. 2018). The investment cost is 1.6 times more for the biological method than the conventional physical/chemical method, and so the biological method is economically not

Table 9.1a Comparison of different biogas upgrading technologies: physical/chemical methods

	Economic indicators			Technological indicators			Environmental indicators			Operational indicators				
Methods	IC €/year	OC ct/100 m^3h	MC€/ year	CH_4 purity %	H_2S removal	N_2/O_2 removal	H_2O used m^3/ year	Energy utility	CH_4 loss %	P Bar g	T °C	Consum-ables	PLR %	References
PWS	1,000,000	14	15,000	95–98	Yes	No	600	0.2–0.5 kWh/Nm^3	<2	4–8	25–40	Antifouling and drying agent	50–100	Sahota et al. (2018) and Angelidaki et al. (2018)
OPS	1,000,000	13.8	39,000	93–98	Yes	No	Nil	0.10–0.33 kWh/Nm^3	<4	4–8	70–80	Organic solvent	50–100	Sahota et al. (2018) and Angelidaki et al. (2018)
AS	2,000,000	14.4	59,000	>98	Contam-inant	No	90–180	0.05–0.18 kWh/Nm^3	<0.	0	120–160	Amine solution	50–100	Sahota et al. (2018) and Angelidaki et al. (2018)
PSA	1,750,000	12.8	56,000	>96–98	Yes	Yes	–	0.16–0.43 kWh/Nm^3	<3%	4–7	Low	Activated carbon	85–115	Sahota et al. (2018), Zhao et al. (2010) and Angelidaki et al. (2018)
TSA											High			
VSA										Atmospheric pressure	Low			
PMT	2,000,000	10.8–15.8	25,000	90–99	Yes	Partial	Less	0.18–0.35 kWh/Nm^3	<5	4–7	–	Oil and activated carbon	50–105	Vrboba and Karel (2017), Sahota et al. (2018) and Angelidaki et al. (2018)
CT	High			99	Yes	Yes	300	0.18–0.25 kWh/Nm^3	<0.1	80 bar	−196	–	Yes	Porpatham et al. (2018), Bauer et al. (2013a, b), Deublen and Steinhauser (2010), Sahota et al. (2018), Angelidaki et al. (2018), Ryckebosch et al. (2011), Munoz et al. (2015)

(continued)

Table 9.1a (continued)

	Economic indicators			Technological indicators			Environmental indicators			Operational indicators				
Methods	IC €/year	OC ct/100 m^3h	MC€/year	CH_4 purity %	H_2S removal	N_2/O_2 removal	H_2O used m^3/year	Energy utility	CH_4 loss %	P Bar g	T °C	Consum-ables	PLR %	References
IME	High			Up to 95	–	–	–	–	Up to 8	–	–	–	–	Petersson and Wellinger (2009) and Sahota et al. (2018)
Hybrid CT and PMT	Less			High	–	–	Less	Less	Low	–	–	–	-	Sahota et al. (2018)
Hybrid PMT and PWS	Less			High	–	–	Less	Less	Low	–	–	–	–	(Sahota et al. (2018)
CDM	Medium			97–99	No	No	–	–	–	8–10	270	–	–	Angelidaki et al. (2018)
IL	High			99	–	–	–	High	–	–	85	–	-	Sahota et al. (2018)
SS	High			–	–	–	Less	–	–	Low	Less	–	-	Sahota et al. (2018)

Table 9.1b Comparison of different biogas upgrading technologies: biological methods

	Economic indicators			Technological indicators			Environmental indicators			Operational indicators				
Methods	IC	OC	MC	CH_4 purity %	H_2S removal	N_2/O_2 removal	H_2O use	Energy utility	CH_4 loss	P	T °C	Consum-ables	PLR	References
In situ chemoautotrophic method	Low			65–98.9	No	No	Less	Less	Low	–	37–55	Nil	No	Angelidaki et al. (2018)
Ex situ chemoautotrophic method	Low			88–98	–	–	–	–	–	–	35–55	–	–	Angelidaki et al. (2018)
Photoautotrophic method	>1.6 times PWS	<7 times PWS	Less	80–97	No	No	Less	208,837 kW-hy⁻	Less	–		–	Nil	Toledo-Cervantes et al. (2017)
Fermentation method	–			95	–	–	–	High	–	–	–	–	–	Angelidaki et al. (2018)
Microbial electrochemical method	Minimum			80	–	–	–	–	–	–	–	–	–	Angelidaki et al. (2018)

Table 9.2 Advantages and disadvantages of various biogas upgrading technologies

Methods			Advantages	Disadvantages	References
Absorption method	Physical scrubbing	Pressurised water scrubbing	Simple, high solubility of carbon dioxide, and H_2S, pre-cleaning is not required, reduced H_2S and NH_3, works at lower flow rates	Oxygen and nitrogen easily dissolves in water, possibility of biological contamination, more amount of water is needed	Kapoor et al. (2017) and Sahota et al. (2018)
		Organic physical scrubbing	Higher solubility of carbon dioxide, high CH_4 purity	Generation of H_2S, complex, investment cost is high, chemicals are used	Patterson et al. (2011) and Sahota et al. (2018)
	Chemical scrubbing	Amine scrubbing	High methane recovery, low methane slip, low operating pressure, energy efficient, less CH_4 loss	High energy and temperature required during regeneration	Kapdi et al. (2005) and Sahota et al. (2018)
Adsorption method	Pressure swing Temperature swing Vacuum swing		Low energy requirements, safety and flexibility, dry process with seldom microbial contamination, chemical usage is nil	Plant capacity is high, compression is required, 15–18% methane loss, investment cost is high	Sun et al. (2015) Augelletti et al. (2017) and Sahota et al. (2018)
Membrane separation method	Inorganic		Longer operational life	Fabrication cost is high	Chen et al. (2013) and Sahota et al. (2018)
	Polymeric		High methane separation selectivity, lower fabrication cost, ease of scaling up, no	Plant capacity is high, plasticisation causes deterioration of membranes,	Scholz et al. (2013), Zhang et al. (2013), Sahota et al. (2018), Bauer et al. (2013a, b)

		chemical usage, dry process	investment cost and energy demand is high, high cost of membranes	and Vrboba and Karel (2017)
	Mixed matrix	Higher values of adsorption capacities, specific surface area	Fabrication is done for ease in handling	Zhang et al. (2013) and Sahota et al. (2018)
Cryogenic technology		High purity methane is obtained, chemical requirement is nil, no compression is required	High capital and operating cost, high energy requirement	Deublen and Steinhauser (2010), Sahota et al. (2018) and Ryckebosch et al. (2011)
In situ methane enrichment		Cost-effective technology, easy to operate	High methane loss, suitable for small-scale plants	Petersson and Wellinger (2009) and Sahota et al. (2018)
Hybrid technologies	Cryogenic and Membrane technology	Less energy intensive, enhanced energy performance, high CO_2 and S-capturing tendency, low operating cost	–	Sahota et al. (2018), Belaissaoui et al. (2012) and Song et al. (2017)
	Membrane technology and pressurised water scrubbing	Cost minimisation, high CO_2 and S-capturing tendency, less energy intensive, enhanced energy performance	–	Scholzet al. (2013) and Sahota et al. (2018)

(continued)

Table 9.2 (continued)

Methods				Advantages	Disadvantages	References
Chemical dehydrogenation method				Complete conversion of CO_2 and H_2, high process efficiency	Catalyst degenerated due to the presence of trace gases	Angelidaki et al. (2018) and Guebitz et al. (2015)
Industrial lung				95–99% methane purified, can withstand higher temperature	Limited enzyme lifetime	Sahota et al. (2018)
Supersonic separation				Simple and reliable	Expensive investments	Sahota et al. (2018)
Biological method		Chemoautotrophic method	In situ	Initial investment cost is minimum, final energy output is high, 99% methane recovery, energy content of CH4 is high	Biomass-dependent process	Angelidaki et al. (2018) and Mulat et al. (2017)
			Ex situ	Biomass-dependent process, biochemical process is simple	Low gas–liquid mass transfer rate	
	Photoautotrophic method	Algal-bacterial photo bioreactor		97% methane recovery, low energy consumption	High investment cost Large area requirement	Angelidaki et al. (2018) and Toledo-Cervantes et al. (2017)
	Fermentation method		Increased CO_2 solubility, more CO_2 consumption rate, high methane purity	–	Angelidaki et al. (2018)	
	Microbial electrochemical method		Sustainable and cost-effective method	–	Angelidaki et al. (2018) and Lovley and Nevin (2013)	

feasible. Also, the operation cost is seven times higher for the conventional type (Toledo-Cervantes et al. 2017). Further, the investment cost is high for adsorption method, membrane separation method and water scrubber and low for amine scrubber method (Bauer et al. 2013a, b).

In addition, the supersonic method is simple and reliable, but the investment cost is not adaptable (Sahota et al. 2018). The cryogenic method faces an issue of high operation and maintenance cost. While comparing all the techniques, it can be predicted that the conventional upgrading technologies are economically feasible. The hybrid technologies cryogenic/membrane technology and membrane technology/pressurised water scrubbing method need reduced operating cost (Song et al. 2017). Also, the in situ methane-enrichment technique and chemical dehydrogenation methods are too cost-effective (Fig. 9.4).

In biological methods, the microbial electrochemical method is cost-effective. Also, while comparing the chemoautotrophic method with that of the photoautotrophic method, the investment cost is high for the photoautotrophic method with requirement of mass area for algal culturing (algal-bacterial photobioreactor). In the cost comparison of biological method with conventional method (pressurised water scrubbing), it was found that the investment cost is >1.6 times pressurised water scrubbing, the operational cost is <7 times pressurised water scrubbing and the maintenance cost is less than the pressurised water scrubbing. The in situ and ex situ chemoautotrophic method needs a minimal initial investment cost (Angelidaki et al. 2018).

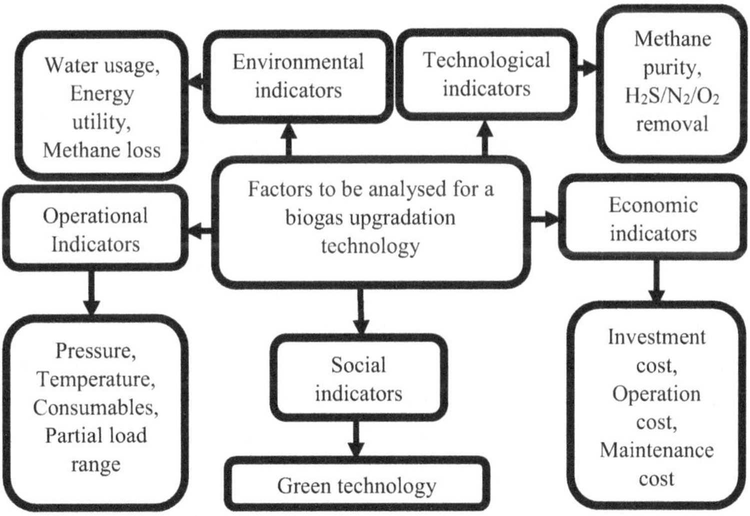

Fig. 9.4 Factors to be analysed for a biogas upgradation technology

(b) *Technological Indicators*

Technological indicators include the methane purity, removal of O_2 and N_2 along with CO_2 and power consumption (Sahota et al. 2018). High amount of methane is required for utilising biogas as a fuel to be used in the transport systems. High methane purity is ensured in amine scrubbing method (>98%), organic physical scrubbing method (93–98%), pressurised water scrubbing method (95–98%), polymeric membrane technology (90–99%), adsorption methods (>96–98%), in situ methane enrichment technique (up to 95%), chemical dehydrogenation method (97–99%), industrial lung method (99%) and cryogenic technology (99%). But the cryogenic technology and amine scrubbing method require high energy which is a major disadvantage of this method. Industrial lung method provides methane purity of about 95–99%.

The upgrading technologies, membrane technology and pressure swing adsorption techniques, have the ability to remove nitrogen and oxygen along with carbon dioxide (Bauer et al. 2013a, b), and hence the purity of the gas is pronounced more. The disadvantage of polymeric membrane separation technology is high energy demand. 15–18% methane loss is observed, and compression is required for adsorption method. Removal of hydrogen sulphide gas is possible in pressurised water scrubbing, organic physical scrubbing, polymeric membrane technology and cryogenic technology. Nitrogen/oxygen removal is possible and partially possible in cryogenic and polymeric membrane technology, respectively. $H_2S/N_2/O_2$ removal is impossible in chemical dehydrogenation method (Sahota et al. 2018).

The chemical dehydrogenation method involves in CO_2 and H_2 conversion completely but with a drawback of catalyst degeneration by the trace gases present (Guebitz et al. 2015). The advantages for the hybrid technologies cryogenic/membrane technology and membrane technology/pressurised water scrubbing method include enhanced energy performance, low energy intensity, high methane purity and increased CO_2 and S-capturing effect (Belaissaoui et al. 2012).

In addition to this, pretreatment is required for the removal of H_2S in all the upgrading technologies with the exception of pressurised water scrubbing method (Bauer et al. 2013a, b). The nitrogen and H_2S removal are a necessary step to be executed in the conventional physical/chemical method, which is an additional barrier during the operation, but in contrast to this, the nitrogen generated is utilised as a nutrient in the biological method (Toledo-Cervantes et al. 2017).

The technological indicators point out that the suitable and highly efficient method is the biological method which can overcome the disadvantages of many techniques by providing high methane purity (95%) and CO_2 consumption rate (in fermentation method), 80–97% CH_4 recovery (no need for $H_2S/N_2/O_2$ removal) in algal-bacterial photobioreactor, 65–98.9% CH_4 recovery (no need for $H_2S/N_2/O_2$ removal) in in situ chemoautotrophic method, 88–89% CH_4 recovery in ex situ chemoautotrophic method, and 80% CH_4 recovery in microbial electrochemical method (Angelidaki et al. 2018).

(c) *Environmental Indicators*

The environmental indicators include emissions, effluents and wastes in the form of water eutrophication/solid waste generation/global warming and resource usage in the form of land occupation/material depletion/energy demand/water depletion (Toledo-Cervantes et al. 2017). Processes like chemical absorption method and cryogenic separation reduce the methane slip, but the electricity demand is high in cryogenic process.

Pressurised swing adsorption method is environmentally suitable since the energy consumption is very low. Hybrid technologies encompassing two technologies combined together enhance more environmentally sustainable features such as high carbon dioxide and sulphur capturing efficiency and low energy consumption (Sahota et al. 2018).

Methane loss is up to 8% (high), $<2\%$, $<4\%$, $<0.1\%$, $<3\%$, $<5\%$ and $< 0.1\%$ for in situ methane enrichment technique, pressurised water scrubbing method, organic physical scrubbing method, amine scrubbing method, adsorption method, membrane technology and cryogenic technology, respectively. But, the methane loss is very less for hybrid technologies, cryogenic/membrane technology and membrane technology/pressurised water scrubbing method.

About 0.2–0.5 kWh/Nm^3, 0.1–0.33 kWh/Nm^3, 0.05–0.18 kWh/Nm^3, 0.16–0.43 kWh/Nm^3, 0.18–0.35 kWh/Nm^3 and 0.18–0.25 kWh/Nm^3 energy are utilised for pressurised water scrubbing method, organic physical scrubbing method, amine scrubbing method, adsorption method, membrane technology and cryogenic technology, respectively. For industrial lung method, energy utilised is high and for hybrid technologies, cryogenic/membrane technology and membrane technology/ pressurised water scrubbing method, the energy utilised is very less. The amount of water used is about 600m^3/year, 90–180 m^3/year and 300m^3/year for pressurised water scrubbing method, amine scrubbing method and cryogenic technology, respectively. However, water usage is very minimum for supersonic method, polymeric membrane technology and hybrid technologies, cryogenic/membrane technology and membrane technology/pressurised water scrubbing method (Sahota et al. 2018).

While comparing the biological method (algal-bacterial photobioreactor and physical/chemical method), 1860 times more land is utilised for the biological method with less energy consumption (208,837 kW-hy^{-1}), less water consumption, no solid waste generation, less GHG emissions and minimum environmental impacts. For physical and chemical upgrading methods, 100 m^2 (3 m^2 (Nm^3/h)$^{-1}$ treated biogas) land is required, 99.8% energy is consumed, high effluent discharge is manifested, solid waste generation is observed, GHG emissions occurs and more water consumption is required.

The sustainability of the biological method is favoured by the replenished nutrient, N and P, and the harmful occurrence of the harmful H_2S gas (Toledo-Cervantes et al. 2017). During the process of upgrading, the methane gas is lost, and it can be minimised using the latest technologies (Petersson and Wellinger 2009). Methane slip is high for pressure swing adsorption method and low for amine scrubbing method. Also, energy demand is low for amine scrubbing method (Bauer et al. 2013a, b). But the energy requirement is low for adsorption method with low microbial contamination, safety and flexibility—a guaranteed measure. Meanwhile, the cryogenic technology ensures the power consumption to be nil (Collet et al. 2017), thus making it a viable efficient fuel (Sahota et al. 2018). On the whole, the biological methods are environmentally sustainable with minimised greenhouse gas emissions.

The environmental indicators prove that the highly efficient method is the biological method which can yield negative outcomes such as low CH_4 loss, less water use and 208,837 kW-hy$^-$ energy consumption (in algal-bacterial photobioreactor), low CH_4 loss, low energy consumption and less water use (in in situ chemoautotrophic method) (Angelidaki et al. 2018).

(d) *Operational Indicators*

The operational indicators include the pressure, temperature, consumables and partial load range. Pressurised water scrubbing technique works at lower flow rates, and pre-cleaning is not necessary, but biological contamination adds to its disadvantage with high amount of water requirement. The chemical dehydrogenation method encompasses the high process efficiency. For the process of pressure swing adsorption technology, purging pressure, adsorbents used, feeding pressure and cycle time (Grande 2011) contribute to the operational issues, and those issues must be refactored to achieve the desired end product.

In contrast to this, foaming, amine loss and corrosion persist as the operational issues for amine scrubbing method. Hence, amine solution must be added as consumables. Addition of antifouling agent + drying agent, organic solvent, activated carbon, oil + activated carbon must be added as consumables for pressurised water scrubbing method, organic physical scrubbing method, adsorption method and membrane technology, respectively (Sahota et al. 2018; Bauer et al. 2013a, b).

Operating pressure (4–8 bar g, 4–7 bar g, 4–8 bar g, 4–8 bar g, 4–7 bar g, 80 bar and 8–10 bar g) is required for pressurised water scrubbing method, pressure swing adsorption method, organic physical scrubbing method, membrane technology, cryogenic technology and chemical dehydrogenation method, respectively. The pressure requirement is very less for supersonic method. Also, for vacuum swing adsorption method, atmospheric pressure is required. Among the membrane separation methods, the operational life is longer for inorganic membrane separation technique. In polymeric membrane separation method, the major problem lies in the formation of plasticisation that degrades the membranes. The adsorption capacity and specific surface area are high in mixed matrix membrane separation technology. Also, the polymeric membrane separation technology seems to be a dry process with less chemical usage (Vrboba and Karel 2017). The compression and the additional chemical requirement is nil for cryogenic technology (Sahota et al. 2018).

The temperature requirement is about 25–40 °C, 70–80 °C, 120–160 °C, −196 °C, 270 °C, 85 °C and 40 °C for pressurised water scrubbing method, organic physical scrubbing method, amine scrubbing method, membrane technology, cryogenic technology, chemical dehydrogenation method and industrial lung method, respectively. For supersonic method and pressure swing adsorption method, low temperature is demanded. In contrast to this, high temperature is a demand for temperature swing adsorption method. The partial load range is about 50–100%, 85–115%, 50–100%, 50–105% and 50–100% for pressurised water scrubbing method, amine scrubbing method, organic physical scrubbing method, pressure swing adsorption method and membrane technology (Angelidaki et al. 2018).

The in situ methane enrichment technique is operatively easy and with a major disadvantage of its use only in small-scale plants (Petersson and Wellinger 2009). In industrial lung upgrading technology, the major drawback lies in the enzyme used. As the enzyme used has limited life time, the validity of this technique is challenged, but this method can overcome high temperature (Sahota et al. 2018).

Consequently, the problem of operational issue exists for all upgraded technology, and it can be moved out of the context only if the technologies are combined giving rise to the hybrid type of technologies. Comparing the operational indicators of conventional methods and biological method, the biological methods are a biomass-dependent process and require consumables but has a disadvantage of low gas–liquid mass transfer rate (in ex situ chemoautotrophic method). The temperature requirements for ex situ chemoautotrophic and in situ chemoautotrophic method are 35–55 °C and 37–55 °C, respectively (Angelidaki et al. 2018).

(e) *Social Indicators*

The acceptance of a technology is based on the environmental impacts generated. Green technologies are forthcoming, and such upgrading technologies are acceptable since it overcomes the problem of greenhouse gas emissions, indirectly pleasing the environment. Based on this, the biological methods are accepted more by the society than the conventional physical/chemical methods such as organic physical scrubbing that requires a large amount of chemicals (Toledo-Cervantes et al. 2017).

The microbial electrochemical method is found to be one of the sustainable methods that can be upgraded for generating environmentally pleasing solutions (Lovely and Nevin 2013). In fermentation method, the CO_2 solubility is high that favours a sustainable environment.

9.7 Future Perspective in Biogas Upgradation

Currently, biogas finds novel application in wide areas besides power and heat production. Most of the countries rely upon biomethane production from the generated biogas. Among the biogas upgradation technologies, physical and chemical methods are in demand, while biological methods are not commercialised even though it provides technological easiness and high feasibility and that seems to be the major challenge to be unwinded (Fig. 9.5). Biological upgradation paves new

Fig. 9.5 Future perspectives in biogas upgradation

way for combining varied forms of renewable energy and, beyond its upgradation, may provide decoupling bioenergy production from the available biomass (Angelidaki et al. 2018).

The absorption technology can be enhanced by increasing its absorbency, removing the contaminant gas mixtures and undergoing the water recycling process which can be further used for the water scrubbing process. Also, optimisation process can be undertaken in water scrubbing-depended biogas upgrading method for generating high-quality biomethane with low operational costs and energy consumption of the upgrading plant. Novel absorption columns that can induce increased mass transfer performance with low pressure drop can be developed for undergoing energy-efficient and cost-effective absorption process (Andriani et al. 2014).

The PSA method can be technologically advanced by decreasing the PSA units, making the technology applicable for small-scale use, minimising the energy use, providing additional performance indicators for integrating CO_2 and H_2S removal in the same column with adsorption efficiency evaluation system (Singhal et al. 2017). The PSA unit size can be altered (two PSA units: first unit is fed with biogas and the second unit is being fed with the off-gas obtained) for obtaining the maximum technological feasibility that includes minimised amount of methane gas in the off-gas stream, further avoiding the emission treatment step, thereby reducing the cost associated with the technology (Augelletti et al. 2017). The PSA technology can also be improved by implementing the use of promising adsorbents with high selectivity and working capacity such as zeolitic imidazolate frameworks and metal organic framework which are efficient in removing H_2S from CO_2 (Liu et al. 2017).

In membrane separation technology, membranes, epoxyamine-based ion gel membranes that favour adverse reactions (less compressed gas, humid feed), are in need for an efficient membrane separation technology (Friess et al. 2017). A polypropylene hollow fibre membrane contractor biogas upgrading unit is used in membrane separation technology to yield high methane content and increased purity (Park et al. 2017). The cost involved in membrane separation technology can be greatly reduced by gaining knowledge about the effect of resonance radiation on mass transfer (Levdansky and Izak 2017). The operational parameters such as the pressure, retentate flow and temperature must be regulated to achieve high biomethane concentration using polysulfone and polyimide fibre membranes (Vrbova and Karel 2017).

Also, inclining towards hybridised form of upgrading technologies may show an optimistic sign with high success rate, minimised operational cost, increased CO_2, more S-capture efficiency and low energy consumption (Sahota et al. 2018). It is also to be mentioned that the hybrid technologies, PWS/PMT, PMT/CT, are techno-economically viable (Scholz et al. 2013); the hybrid technology, CT/PMT/TSA, is less energy consumable (Song et al. 2017); and the hybrid technology, TSA/PMT, is with high methane purity and reduced loss of CH_4 and CO_2 (Pinghai et al. 2012). Further, exploration of more hybrid technologies is an urgent need by integrating the good features of two or more technologies with improvised techno-economic dimension for biogas upgradation.

The most predominant challenges that all the biogas upgrading technologies face are to utilise the methane present in off-gas, to make small-scale upgrading plants economical, the support policies (Sahota et al. 2018), the innovation requirements in research and development, implementing novel biogas reforming technologies and producing liquefied biogas. Those challenges are briefly described:

(a) *Proper Utilisation of Methane Content Present in Off-Gas*

The methane present in the off-gas may be released into the surrounding and pose a serious threat to global warming, and hence the off-gas must be treated before leaving the plant. The conventional techniques, PSW and PSA, and membrane technologies suffer from the formation of a huge amount of methane (depending upon the method, the methane content varies) in the off-gas, and oxidation process must be performed to avoid the methane loss into the atmosphere. The oxidation process generates heat which can be utilised at the anaerobic digestion plant or can be wasted by cooling process. Alternatively, the liberated off-gas collated with raw biogas can also be fed to an existing combined heat and power gas engine. The future holds promise for capturing methane in off-gas and its utilisation in biogas upgradation plants (Sahota et al. 2018).

(b) *Provision for Economically Feasible Small-Scale Upgrading Plants*

Biogas upgrading costs are inversely proportional to the plant size/capacity, and a small-scale plant upgradation is too expensive to be afforded because of the upgrading equipment's high investment costs. All the instrumental parts are the same for small-scale and large-scale plants. The future challenge lies in upgrading a biogas production technology at a small scale that is economically feasible and sustainable with minimum cost, low methane content and reduced complexity of all controllable operational parameters. Also, the generated methane can be converted to a vehicular fuel as an inevitable remedy to make the small-scale plants economically sustainable (Sahota et al. 2018).

(c) *Supportive Government Policies*

Mass biogas production is available only at countries that favour the supportive governmental policies. The future endeavours include the use of green transportation fuels such as compressed and liquefied biogas for enhancing the sustainable development. Additionally, biogas purification can be performed to be applicable in the field of combustion as a household fuel for heating, cooking and generating small-scale electricity. The field of converting the livestock waste in to biogas is still uncovered (Lima et al. 2018; Owusu and Banadda 2017; Moraes et al. 2017); thus, the government should provide policy interventions to enable research and development in the field of biogas upgradation. Large-scale biogas can be generated from renewable sources with favourable supportive policies from the government in the form of subsidies to manufacturers and implementing regulatory environment for enhancing biogas-driven engine technology at the market level, providing attractive feed-in-tariff in generating their own electricity using biogas technologies, providing seed capital to start-up companies that work on renewable energy source conversion to biogas, encouraging the communities to be involved in the collection and transport of biowaste residues to biogas plants, establishing training centres for generating skilled workforce and providing resources for producing family-type biogas plants in the future emerging markets (Huttunen et al. 2014; Chen et al. 2017).

(d) *Implementing Biogas Reforming Technologies*

Biogas reforming technologies are authenticated processes for producing the green hydrogen that favours the reduction of natural gas production. CFD modelling studies, catalyst characteristic and throughput studies are required for effective implementation of biogas reforming technologies (Verma and Samanta 2016).

(e) *Commercial Distribution of Liquefied Biogas*

Liquefied biogas can be one of the future perspectives in biogas upgradation. The energy density of biogas can be increased significantly by liquefied biogas, facilitating a wide range of applications such as long-distance transport. Mass production of liquefied biogas is a challenge to be solved (Bauer et al. 2013a, b).

The future perspective is wide for biogas upgradation technologies which can be updated and executed only if all the open challenges can be met by humans with the available resources.

9.8 Conclusions

Upgraded biogas is a sustainable renewable energy option with better replacement of CNG utilisation. Due to increasing rigorous environmental rules and regulation, there is a need to develop efficient and low energy input and environment-friendly and low-cost technology for biogas up-gradation. Although the cost minimisation of the system is not the main criteria for the selection of upgradation technology, it is more important to choose the technology which delivers high methane purity and less methane losses as a final product. Apart from some conventional upgradation technologies, some recently invented technologies cryogenic separation, in situ methane enrichment, industrial lung and polymeric membrane separation method come under the recent developments in biogas upgrading technologies. However, these technologies remain under development stage developed only at laboratory scale. Hence, more efforts are required to form linkages between laboratory and commercial scale technologies. Similarly, desulphurisation of biogas is a must-to-do step prior to upgradation. Among all the available desulphurisation technologies, physical adsorption technologies using carbon-based adsorbents paid more attention to the researchers due to its easy operation and lesser cost of the system. Though every technology has its own advantage and disadvantages, there is need of further development in R&D sector for the betterment of commercialisation of nascent upgradation technologies.

References

Agler MT, Wrenn BA, Zinder SH et al (2011) Waste to bioproduct conversion with undefined mixed cultures: the carboxylate platform. Trends Biotechnol 29:70–78

Agneessens LM, Ottosen LDM, Voigt NV et al (2017) In-situ biogas upgrading with pulse H2 additions: the relevance of methanogen adaption and inorganic carbon level. Bioresour Technol 233:256–263

Ahmed I, Yusof ZAM, Beg MDH (2010) Fabrication of polymer based mix matrix membrane-A short review. Int J Basic Appl Sci 10:14–19

Al Mamun MR, Torii S (2015) Enhancement of production and upgradation of biogas using different techniques-a review. Int J Earth Sci Eng 8(2):877–892

Andriani D, Arini W, Tinton DA et al (2014) A review on optimization production and upgrading biogas through CO_2 removal using various techniques. Appl Biochem Biotechnol 172:1909–1928
Angelidaki I, Treua L, Tsapekosa P et al (2018) Biogas upgrading and utilization: Current status and perspectives. Biotechnol Adv 36(2):452–466
Augelletti R, Conti M, Annesini MC (2017) Pressure swing adsorption for biogas upgrading. A new process configuration for the separation of biomethane and carbon dioxide. J Clean Prod 140:1390–1398
Awe OW, Zhao Y, Nzihou A et al (2017) A review of biogas utilisation, purification and upgrading technologies. Waste Biomass Valoriz 8:267–283
Baker RW (2012) Membrane technology and applications, 2nd edn. Wiley, Hoboken
Bassani I, Kougias PG, Treu L et al (2017) Optimization of hydrogen dispersion in thermophilic up-flow reactors for ex situ biogas upgrading. Bioresour Technol 234:310–319
Batstone DJ, Keller J, Angelidaki I et al (2002) The IWA anaerobic digestion model no 1 (ADM1). Water Sci Technol 45:65–73
Battino R, Clever HL (1966) The solubility of gases in liquids. Chem Rev 66:395–463
Bauer F, Hulteberg C, Persson T et al (2013a) Biogas upgrading – review of commercial technologies. SGC Rapp 270
Bauer F, Persson T, Hulteberg C et al (2013b) Biogas upgrading– technology overview, comparison and perspectives for the future. Biofuels Bioprod Biorefin 7:499–511
Belaissaoui B, Le Moullec Y, Willson D et al (2012) Hybrid membrane cryogenic process for post-combustion CO_2 capture. J Membr Sci 415:424–434
Belmabkhout Y, De Weireld G, Sayari A (2009) Amine-bearing mesoporous silica for CO_2 and H_2S removal from natural gas and biogas. Langmuir 25(23):13275–13278
Chen XY, Vinh-Thang H, Ramirez AA (2015) Membrane gas separation technologies for biogas upgrading. RSC Adv 5:24399–24448
Chen B, Hayat T, Alsaedi A (2017) History of biogas production in China. In: Biogas systems in China. Springer, Berlin/Heidelberg, pp 1–15
Cheng S, Xing D, Call DF et al (2009) Direct biological conversion of electrical current into methane by electromethanogenesis. Environ Sci Technol 43:3953–3958
Collet P, Flottes E, Favre A et al (2017) Technoeconomic and life cycle assessment of methane production via biogas upgrading and power to gas technology. Appl Energy 192:282–295
Cozma P, Ghinea C, Mamaliga I et al (2013) Environmental impact assessment of high pressure water scrubbing biogas upgradation technology. Clean 41:917–927
Deublein D, Steinhauser A (2010) Biogas from waste and renewable resources: an introduction, 2nd edn. Wiley, Weinheim
Friess K, Lanc M, Pilnacek K et al (2017) CO_2/CH_4 separation performance of ionic-liquid-based epoxy-amine ion gel membranes under mixed feed conditions relevant to biogas processing. J Membr Sci 528:64–71
Grande CA (2011) Biogas upgrading by pressure swing adsorption. In: dos Santos Bernardes MA (ed) Biofuel's engineering process technology. InTech, Oslo, pp 65–84
Guebitz GM, Bauer A, Bochmann G et al (2015) Biogas science and technology. Springer, Hannover
Horikawa MS, Rossi F, Gimenes ML et al (2004) Chemical absorption of H_2S for biogas purification. Braz J Chem Eng 21(3):415–422
Huttunen S, Kivimaa P, Virkamaki V (2014) The need for policy coherence to trigger a transition to biogas production. Environ Innov Soc Transit 12:14–30
Hwang H, Yeon YJ, Lee S et al (2015) Electro-biocatalytic production of formate from carbon dioxide using an oxygenstable whole cell biocatalyst. Bioresour Technol 185:35–39
Jurgensen L, Ehimen EA, Born J et al (2014) Utilization of surplus electricity from wind power for dynamic biogas upgrading: northern Germany case study. Biomass Bioenergy 6:126–132
Kadam R, Panwar NL (2017) Recent advancement in biogas enrichment and its applications. Renew Sust Energy Rev 73:892–903

Kapdi SS, Vijay VK, Rajesh SK et al (2005) Biogas scrubbing, compression and storage: perspective and prospectus in Indian context. Renew Energy 30:1195–1202

Kapoor R, Subbarao PMV, Vijay VK et al (2017) Factors affecting methane loss from a water scrubbing based biogas upgrading system. Appl Energy 208:1379–1388

Kennes D, Abubackar HN, Diaz M et al (2016) Bioethanol production from biomass: carbohydrate vs syngas fermentation. J Chem Technol Biotechnol 91:304–317

Kougias PG, Kotsopoulos TA, Martzopoulos GG (2010) Anaerobic co-digestion of pig waste with olive mill waste water under various mixing conditions. Fresenius Environ Bull 19:1682–1686

Kougias PG, Treu L, Benavente DP et al (2017) Ex situ biogas upgrading and enhancement in different reactor systems. Bioresour Technol 225:429–437

Latif H, Zeidan AA, Nielsen AT et al (2014) Trash to treasure: production of biofuels and commodity chemicals via syngas fermenting microorganisms. Curr Opin Biotechnol 27:79–87

Levdansky V, Izak P (2017) Membrane separation of gas mixtures under the influence of resonance radiation. Sep Purif Technol 173:93–98

Li JR, Sculley J, Zhou HC (2012) Metal-organic frameworks for separations. Chem Rev 112:869–932

Lima RM, Santos AH, Pereira CR et al (2018) Spatially distributed potential of landfill biogas production and electric power generation in Brazil. Waste Manag 74:323–334

Lin WC, Chen YP, Tseng CP (2013) Pilot-scale chemical–biological system for efficient H_2S removal from biogas. Bioresour Technol 135:283–291

Lindberg A (2003) Development of in-situ methane enrichment as a method for upgrading biogas to vehicle fuel standard. Licentiate thesis, KTH, Chemical Engineering and Technology, Stockholm

Liu J, Wei Y, Li P et al (2017) Selective H_2S/CO_2 Separation by metal organic frameworks based on chemical-physical adsorption. J Phys Chem C121(24):13249–13255

Lovley DR, Nevin KP (2013) Electrobiocommodities: powering microbial production of fuels and commodity chemicals from carbon dioxide with electricity. Curr Opin Biotechnol 24:385–390

Lu L, Ren ZJ (2016) Microbial electrolysis cells for waste biorefinery: a state of the art review. Bioresour Technol 215:254–264

Luo G, Angelidaki I (2012) Integrated biogas upgrading and hydrogen utilization in an anaerobic reactor containing enriched hydrogenotrophic methanogenic culture. Biotechnol Bioeng 109:2729–2736

Luo G, Angelidaki I (2013) Co-digestion of manure and whey for in situ biogas upgrading by the addition of H2: process performance and microbial insights. Appl Microbiol Biotechnol 97:1373–1381

Luo G, Wang W, Angelidaki I (2014) A new degassing membrane coupled upflow anaerobic sludge blanket (UASB) reactor to achieve in-situ biogas upgrading and recovery of dissolved CH_4 from the anaerobic effluent. Appl Energy 132:536–542

Makaruk A, Miltner M, Harasek M (2010) Membrane biogas upgrading processes for the production of natural gas substitute. Sep Purif Technol 74:83–92

Mallada R, Menendez M (2008) Inorganic membranes: synthesis, characterization and applications: synthesis, characterization and applications. Membrane science and technology, vol 13, 2nd edn. Elsevier Science

Martin ME, Richter H, Saha S et al (2016) Traits of selected Clostridium strains for syngas fermentation to ethanol. Biotechnol Bioeng 113:531–539

Mattiasson B (2005) Ekologisklunga for biogas upgrading. Nationellt Samverkans projekt Biogas iFordon

Micoli L, Bagnasco G, Turco M (2014) H_2S removal from biogas for fuelling MCFCs: new adsorbing materials. Int J of Hydrogen Energy 39(4):1783–1787

Moraes BS, Petersen SO, Zaiat M et al (2017) Reduction in greenhouse gas emissions from vinasse through anaerobic digestion. Appl Energy 189:21–30

Mulat DG, Mosbek F, Ward AJ et al (2017) Exogenous addition of H_2 for an in situ biogas upgrading through biological reduction of carbon dioxide into methane. Waste Manag 68:146–156

Munoz R, Meier L, Diaz I et al (2015) A review on the state-of-the-art of physical/chemical and biological technologies for biogas upgrading. Rev Environ Sci Biotechnol 14:727–759

Nevin KP, Woodard TL, Franks AE et al (2010) Microbial electrosynthesis: feeding microbe's electricity to convert carbon dioxide and water to multicarbon extracellular organic compounds. MBio 1:e00103–e00110

Nordberg A, Edstrom M, Uusi-Pentilla M et al (2005) Process intern metananrikning. JTI rapport Kretslopp & Avfall 33

Owusu PA, Banadda N (2017) Livestock waste-to-bioenergy generation potential in Uganda: a review. Environ Res Eng Manag 73:45–53

Ozekmekci M, Salkic G, Fellah MF (2015) Use of zeolites for the removal of H_2S: a mini-review. Fuel Process Technol 139:49–60

Park A, Kim YM, Kim JF et al (2017) Biogas upgrading using membrane contactor process: pressure-cascaded stripping configuration. Sep Purif Technol 183:358–365

Patterson T, Esteves S, Dinsdale R et al (2011) An evaluation of the policy and techno-economic factors affecting the potential for biogas upgradation for transport fuel use in the UK. Energ Policy 39:1806–1816

Persson M (2003) Evaluation of upgrading techniques for biogas. Rep SGC 142

Petersson A, Wellinger A (2009) Biogas upgrading technologies–developments and innovations. IEA Bioenergy 20:1–19

Pinghai S, Dal-Cin M, Kumar A et al (2012) Design and economics of a hybrid membrane–temperature swing adsorption process for upgrading biogas. J Membr Sci 413:17–28

Pipatmanomai S, Kaewluan S, Vitidsant T (2009) Economic assessment of biogas-to-electricity generation system with H_2S removal by activated carbon in small pig farm. Appl Energy 86 (5):669–674

Porpatham E, Ramesh A, Nagalingam B (2018) Experimental studies on the effects of enhancing the concentration of oxygen in the inducted charge of a biogas fuelled spark ignition engine. Energy 142:303–312

Report (2012) Biogas to biomethane technology review. Vienna University of Technology Austria, pp 1–15

Ryckebosch E, Drouillon M, Vervaeren H (2011) Techniques for transformation of biogas to biomethane. Biomass Bioenergy 35:1633–1645

Sadhukhan J, Lloyd JR, Scott K et al (2016) A critical review of integration analysis of microbial electrosynthesis (MES) systems with waste biorefineries for the production of biofuel and chemical from reuse of CO_2. Renew Sust Energ Rev 56:116–132

Sahota S, Shah G, Ghosh P et al (2018) Review of trends in biogas upgradation technologies and future perspectives. Bioresour Technol Rep 1:79–88

Schiel-Bengelsdorf B, Durre P (2012) Pathway engineering and synthetic biology using acetogens. FEBS Lett 586:2191–2198

Scholwin F, Held J, Kaltschmitt M et al (2013) Biomethane from anaerobic processes. In: Renewable energy systems. Springer, New York, pp 656–664

Scholz M, Melin T, Wessling M (2013) Transforming biogas into biomethane using membrane technology. Renew Sust Energ Rev 17:199–212

Siefers A, Wang N, Sindt A et al (2010) A novel and cost-effective hydrogen sulfide removal technology using tire derived rubber particles. Proc Water Environ Fed 12:4597–4622

Singhal S, Agarwal S, Arora S et al (2017) Upgrading techniques for transformation of biogas to bio-CNG: a review. Int J Energy Res 41(12):1657–1669

Song C, Liu Q, Ji N et al (2017) Reducing the energy consumption of membrane-cryogenic hybrid CO_2 capture by process optimization. Energy 124:29–39

Soreanu G, Beland M, Falletta P et al (2008) Laboratory pilot scale study for H_2S removal from biogas in an anoxic biotrickling filter. Water Sci Technol 57(2):201–207

Stams AJM, Plugge CM (2009) Electron transfer in syntrophic communities of anaerobic bacteria and archaea. Nat Rev Microbiol 7:568–577

Sun Q, Li H, Yan J et al (2015) Selection of appropriate biogas upgrading technology-a review of biogas cleaning, upgrading and utilisation. Renew Sust Energ Rev 51:521–532

Thran D, Billig E, Persson T et al (2014) Biomethane—status and factors affecting market development and trade. IEA Task 40 and Task 37 Joint Study

Toledo-Cervantes A, Estrada JM, Lebrero R et al (2017) A comparative analysis of biogas upgrading technologies: photosynthetic vs physical/chemical processes. Algal Res 25:237–243

Tomas M, Fortuny M, Lao C et al (2009) Technical and economical study of a full-scale biotrickling filter for H_2S removal from biogas. Water Pract Tech 4(2):1–545

Tuinier MJ, Van SintAnnaland M (2012) Biogas purification using cryogenic packed bed technology. Ind Eng Chem Res 51:5552–5558

Van Eerten-Jansen MCAA, Ter Heijne A, Buisman CJN et al (2012) Microbial electrolysis cells for production of methane from CO_2: long-term performance and perspectives. Int J Energy Res 36:809–819

Verma P, Samanta SK (2016) Overview of biogas reforming technologies for hydrogen production: advantages and challenges. In: Proceedings of the first international conference on recent advances in bioenergy research. Springer, New Dehli, pp 227–243

Villano M, Aulenta F, Ciucci C et al (2010) Bioelectrochemical reduction of CO_2 to CH_4 via direct and indirect extracellular electron transfer by a hydrogenophilic methanogenic culture. Bioresour Technol 101:3085–3090

Vrbova V, Karel C (2017) Upgrading biogas to biomethane using membrane separation. Energy Fuel 31:9393–9401

Wang H, Ren ZJ (2013) A comprehensive review of microbial electrochemical systems as a platform technology. Biotechnol Adv 31:1796–1807

Xia A, Cheng J, Murphy JD (2016) Innovation in biological production and upgrading of methane and hydrogen for use as gaseous transport biofuel. Biotechnol Adv 34:451–472

Xu H, Wang K, Holmes DE (2014) Bioelectrochemical removal of carbon dioxide (CO_2): an innovative method for biogas upgrading. Bioresour Technol 173:392–398

Yoo M, Sang-Jun H, Jung-Ho W (2013) Carbon dioxide capture capacity of sodium hydroxide aqueous solution. J Environ Manag 114:512–519

Zeppilli M, Lai A, Villano M et al (2016) Anion vs cation exchange membrane strongly affect mechanisms and yield of CO_2 fixation in a microbial electrolysis cell. Chem Eng J 304:10–19

Zhang Y, Angelidaki I (2014) Microbial electrolysis cells turning to be versatile technology: recent advances and future challenges. Water Res 56:11–25

Zhang S, Yaping L, Jianfeng T et al (2006) Operation appraisal and parameter optimization of imported skid-mounted natural gas dehydration unit. Nat Gas Ind 26:128–130

Zhang Y, Sunarso J, Liu S et al (2013) Current status and development of membranes for CO_2/CH_4 separation: a review. Int J Greenh Gas Control 12:84–107

Zhao Q, Leonhardt E, MacConnell C et al (2010) Purification technologies for biogas generated by anaerobic digestion. Compressed Biomethane, CSANR Research report, pp 1–24

Zhao H, Zhang Y, Zhao B et al (2012) Electrochemical reduction of carbon dioxide in an MFC-MEC system with a layer-by-layer self-assembly carbon nanotube/ cobalt phthalocyanine modified electrode. Environ Sci Technol 46:5198–5204

Zicari SM (2003) Removal of hydrogen sulfide from biogas using cow-manure compost. Doctoral dissertation, Cornell University

Zulkefli NN, Masdar MS, Isahak WNRW et al (2019) Removal of hydrogen sulfide from a biogas mimic by using impregnated activated carbon adsorbent. PLoS One 14(2):e0211713

Efficiency Analysis of Crude Versus Pure Cellulase in Industry

10

Mohammad Shahed Hasan Khan Tushar and Animesh Dutta

Abstract

Many industries including fermentation, pulp and paper industry, brewing sector, fermentation, food and animal feed industry, and detergent and textile use cellulases due to its environment benign and sustainable process. Academic and industrial researches are being done and still ongoing on cellulases due to its enormous industrial application and make these processes green. In this article, extensive review is performed on the use of cellulases in the industrial sector in both crude and pure form. It has been observed that crude cellulases are preferred in the industrial sector due to its low cost and stimulate the process using impurities present in enzymes. Pure cellulases are mainly used in laboratories and are case specific as they are costly to use in industries.

Keywords

Biomass · Bioconversion · Cellulase enzymes · Enzyme stability · Fungi

10.1 Introduction

Lignocellulosic biomass is becoming more dominating alternative due to its environment friendly nature in terms of combustion compared to fossil fuel, rapid diminishing trend of fossil fuel reserves, and ever-increasing energy demand along with the clean environment point of view. They are readily available and

M. S. H. K. Tushar (✉)
Department of Mechanical Engineering, Rajshahi University of Engineering & Technology, Rajshahi, Bangladesh

A. Dutta
Mechanical Engineering Program, School of Engineering, University of Guelph, Guelph, ON, Canada

N. Srivastava et al. (eds.), *Biofuel Production Technologies: Critical Analysis for Sustainability*, Clean Energy Production Technologies,
https://doi.org/10.1007/978-981-13-8637-4_10

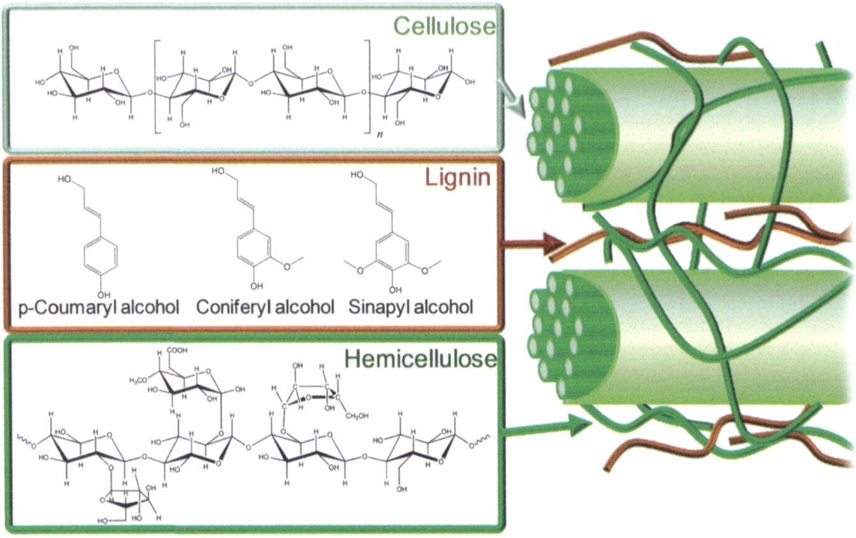

Fig. 10.1 Lignocellulosic biomass structure with major components (Brodeur et al. 2011)

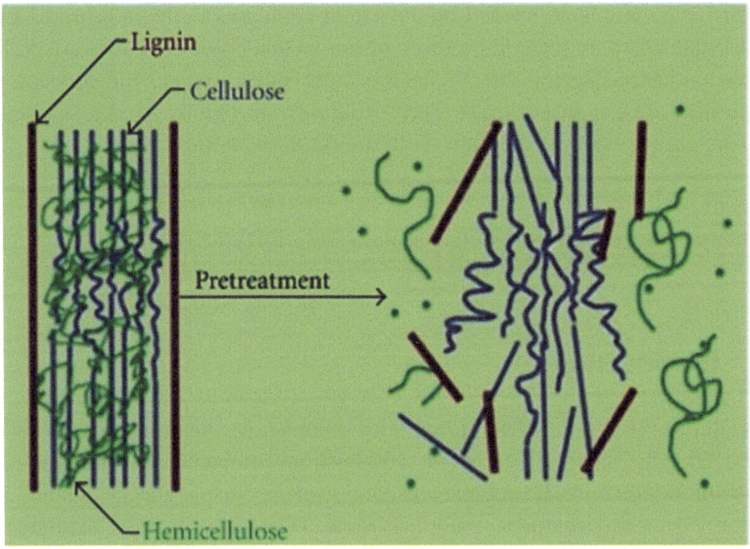

Fig. 10.2 Biomass structure after and before hydrolysis with different components (Alonso et al. 2012)

comparatively cheaper material on the Earth, such as forestry residues, livestock manures, agricultural residues, municipal solid wastes, etc. (Sánchez 2009).

Lignocellulosic biomass (plant biomass) consists of hemicellulose, cellulose, and lignin (as shown in Figs. 10.1 (Brodeur et al. 2011) and Fig. 10.2 (Alonso et al.

2012)), which means these may be used as a valuable carbon source to produce very useful value-added chemicals. Biochemical conversion, such as fermentation in the presence of microbes, is one of the conversion techniques, which needs depolymerization to attain these chemicals. Due to the use of inorganic acids during chemical hydrolysis, furfurals are produced along with fermentable sugars that inhibit successive fermentation steps which require detoxification steps to remove these inhibitors (Brodeur et al. 2011).

To improve energy security and to reduce the adverse effect of climate change, the use of bioethanol has become a major issue nowadays. Current bioethanol production is performed from sugars and processed starches obtained from different fruits and grains. Lignocellulosic biomass is a renewable and sustainable resource of producing bioethanol since they contain cellulose and hemicelluloses which are polymers of glucose. For this, they are continuously investigated of producing ethanol despite of production cost and time (Sakai et al. 2007; Sheehan and Himmel 1999).

Due to the difficulty of degrading cellulosic biomass biochemically for its complex and rigid structure, cellulose degrading enzymes have drawn the eyes of researchers. By using these complex structure as a mean of energy source, microorganisms eventually produce a complex enzyme system which is membrane bond or extracellular in nature. These enzymes are termed as cellulases and of hydrolase class. These enzymes are capable of degrading insoluble cellulose complex compounds in soluble oligosaccharides (Henrissat and Davies 1997). Cellulase converts the lignocellulosic biomass into ethanol, single cell protein, glucose, and other useful products through bioconversion by hydrolyzing (1–4) in cellulose (Chalal 1985). These enzymes are mainly generated by fungi, ruminants, insects, and plants from insects, plants, and microorganisms.

Although several species of fungi aid in producing cellulase, in terms of the quantity of production, only a few of these fungi are able to degrade the crystalline cellulose by producing cell-free enzymes. Cellulase consists of three active enzymes: endoglucanase (endo-1,4-β-glucanase), exoglucanase (exo-1,4-β-glucanase), and β-glucosidase. Endoglucanase produces shorter chains by breaking the long, crystalline glucose of cellulose at random places. Exoglucanase works on the exposed ends of the shorter chains and progress through a series of releasing cellobiose and some glucose. At the end of the degradation process, β-glucosidases wrap up the saccharification by producing soluble glucose compounds by breaking the cellobiose and cellooligosaccharides (Rodrigues et al. 2010; Lynd et al. 2002; Harrison et al. 1998). Figures 10.3 and 10.4 show the graphical explanation of enzyme activity from the beginning to end adopted from (Juturu and Wu 2014; Doi and Kosugi 2004; Desvaux 2005).

Cellulases become the most important enzymes that are used in many industries, namely, detergent, pulp and paper, textiles and laundry, food and agricultural, fruit and vegetable extraction, bioethanol production from biomass, alongside research purposes. Thus, new, more specific, and stable enzymes are in increased demand along with the screening and characterization of the novel isolates (Annamalai et al. 2013; Sohail et al. 2009; Dhillon et al. 2012). Biocatalysis is gaining the attraction of

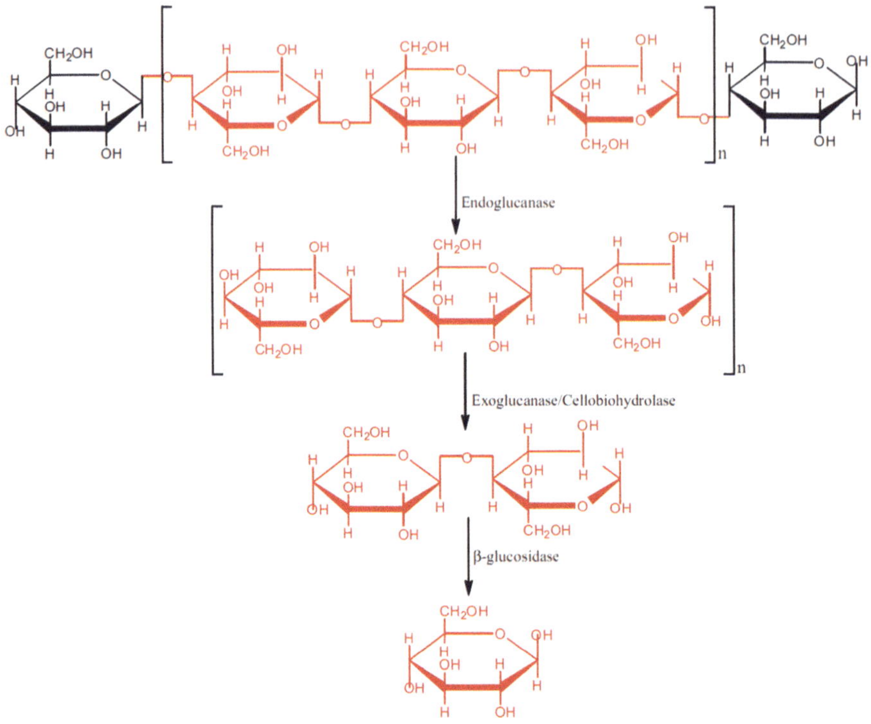

Fig. 10.3 Decomposition of cellulose structure using cellulases (crystalline cellulose shown in red color and amorphous cellulose shown in black color)

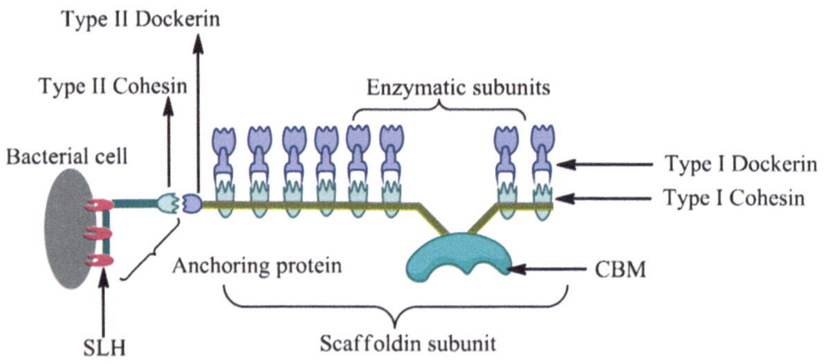

Fig. 10.4 Schematic architecture of bacterial cellulosome with various binding options CBM stands for cellulose-binding module and SLH stands for S-layer homology

the researcher due to its environment-friendly nature, and continuous advancement of biotechnology is paving the way of using these enzymes (biocatalysts) in various chemical industries such as leather, textile, fruits and vegetables processing, pulp and paper, and animal feed. Another use of these enzymes is becoming very popular to the researchers: production of biofuels through biochemical conversion processes in the presence of enzymes as catalysts in addition to their pivotal role in the food and consumer goods and products. Currently, a substantial amount of cellulase is contributing the world's enzyme market (Dogaris et al. 2009).

However, cellulases are costly enzymes and thus a pivotal hindrance to commercialize them to use in industries such as in biorefineries where a significant amount is required; e.g., for the production of a gallon bioethanol ~100 g of cellulase is required (Zhang et al. 2006; Zhu et al. 2009).

Enzyme production from microorganisms is advantageous as substantial amount of enzymes may be cultured in a relatively shorter time, and thus continuous and sufficient amount of enzymes may be supplied to the relative bioprocessing technique. Besides, they are relatively stable and may be stored in under nonideal conditions for weeks, as it does not degrade the biological activity of enzymes. As such, commercial enzymes are produced mainly through microorganisms that in turns are taking place of mechanical or cellular processes of making these enzymes. Table 10.1 shows the enzymes used for industries and processes that are replaced by them (Headon and Walsh 1994; Walsh and Headon 1994; Srivastava et al. 2015).

Cellulases are of two types: crude and pure. Cellulase in the crude form contains a complete system of enzymes, namely, endoglucanase (endo-1,4-β-glucanase), exoglucanase (exo-1,4-β-glucanase), and β-glucosidase. Pure cellulase contains only one of these enzymes at a time. Compared to pure cellulase, crude cellulases have a lower activity and are still in use mostly in mass industrial sectors since crude cellulase is cheaper (Lloyd and Wyman 2005). Pure cellulases are mainly used in laboratories for fundamental and applied research purposes (Kanmani et al. 2011). Based on the above discussion, this article is a review of crude and pure cellulases applied to various industries and relative comparison of them to check the effectiveness of these enzymes in pure and crude form. And finally, based on this, a recommendation will be made based on the cost-effectiveness and the quality of the required product.

Table 10.1 List of industrial sector use enzymes and the chemical processes replaced (Headon and Walsh 1994; Walsh and Headon 1994; Srivastava et al. 2015)

Industry segment	Enzymes	Chemical process replaced
Detergents production	Lipases, proteases, cellulases, amylases	Phosphates, silicates, high temperature
Fabrics and garment industry	Amylases, cellulases, catalases	Acids, bases, oxidizing and reducing agents, energy, new garment manufacture
Starch generation	Amylases, pullulanases	Acids, high temperatures
Baking	Amylases, proteases, xylanases	Emulsifying agents, sodium bisulfate
Pulp and paper industry	Xylanases, mannanases	Chlorine, toxic waste
Leather processing	Proteases, lipases	Sulfides, high temperature
Biocatalyst for biochemical industry	Isomerases, lipases, reductases, acylases	Acids, organic solvents, high temperature
Biofuels	Cellulase	Acids, bases, high temperature

10.2 Pulp and Paper Industry

Since early 2000, pulp and paper industry has seen an increased use of cellulase significantly to overcome the aftermath of using the mechanical pulping process, which leaves the pulps with high content of fines, bulk and stiffened (Mai et al. 2004). During the refining process, 20%–40% of energy may be saved, and the hand-sheet strength of the pulps also improved significantly (Singh et al. 2007). Crude cellulases are mainly applied for the purpose of biomodifying the fiber properties along with improving the ability of draining and beat in the paper mills after or before beating of pulp (Dienes et al. 2004).

Pure cellulase containing only endoglucanase is able to reduce the pulp viscosity during a low concentration hydrolysis (Pere et al. 1995). Additionally, pure cellulase has shown the ability to boost the bleachability of kraft pulps made from softwood with brightness compared to that obtained using xylanases (Singh et al. 2007). Pure cellulase alone or combined with xylanases is being used for deinking the paper wastes by releasing the ink from the fiber surface through the partial hydrolysis of carbohydrate molecules (Kuhad et al. 2010a). The use of enzymes for deinking the pulps reduces the use of alkali, improves the fiber brightness, increases pulp freeness and cleanliness, improves strength properties, reduced fine particles in pulp, simplifies the deinking process, changes the ink particle size distribution, and reduces environmental pollution (Kuhad et al. 2010a; Stork and Puls 1996).

10.3 Food Processing Industry

There are several steps in extracting fruit and vegetable juices from their raw materials: extraction, clarification, and stabilization. Cellulases play a very important role in these steps for quality juices. Macerating enzymes consisting of crude cellulases, xylanases, and pectinases are used to extract and clarify fruit and vegetable juices in order to enhance the yield of juices. Compared to pure cellulases, the mixture of crude cellulases, xylanases, and pectinases known as macerating enzymes enhances not only the yield but also the process performance without adding any extra cost to the process. These macerating enzymes help to enhance the texture and cloud stability of the purees and nectars of tropical fruits such as apricot, pear, papaya, plum, peach, and mango by decreasing the viscosity (De Carvalho et al. 2008). Besides some of the properties of fruits and vegetables such as aroma, flavor, and texture may be boosted using an infusion of enzymes such as pectinases and β-glucosidases (Rai et al. 2007). In addition, the flavor, texture, aroma, and volatile characteristics of vegetable and fruit may be modified by using a mixture of β-glucosidases and pectinases (Karmakar and Ray 2011).

10.4 Agricultural Sector

Enzymes also have shown ability to improve the growth of crops and to control plant diseases using various combinations of pectinases, hemicellulases, and cellulases. Hybrid species with desirable characteristics may be developed using plant or fungal protoplasts yielded through microbial hydrolases. Both crude cellulases and pure cellulase (β-glucanases produced using fungi) are able to control the plant diseases by degrading the cell walls of plant pathogens which are produced using certain fungi (Bhat 2000).

In addition to plant disease control, soil quality may also be improved by using cellulases. By burying the straw residues of the crops, the quality of soil will increase, thus reducing the dependency on mineral fertilizers (Tejada et al. 2008; Escobar and Hue 2008). Cellulose of the crop residues in the soil degrades rapidly in the presence of pure cellulase (exoglucanase (exo-1,4-β-glucanase)) (Fontaine et al. 2004) and thus may be a probable way of accelerating the straw disintegration which eventually improves the soil fertility (Han and He 2010).

10.5 Detergent Industry

The most recent innovation of the use of crude cellulases (cellulases with protease and lipase) is in detergent industry. Color brightness, feel of fabric, and dirt removal from the garments made of cotton or cotton blends may be improved by altering the cellulose fibrils as a consequence of cellulase preparation. Alkaline cellulases are already being in use in industries as potential detergent additive that removes any soil in the interfibril spaces with very precise contact with the cellulose in the presence of other detergent ingredients (Singh et al. 2007; Sukumaran et al. 2005). Currently, various measures have been taken to improve the stability of cellulases in liquid laundry detergent by using a mixture of boric acid or boric acid derivatives and propanediol, citric acid, or a salt (soluble in water), protease, cellulose, etc. To get rid of these rugged protuberances, pure cellulases are used to attain a glossier, delicate, and brighter-colored fabric (Karmakar and Ray 2011).

10.6 Textile Industry

Wet processing is an important segment in the textile industry where cellulases are applied successfully to obtain improved hand and appearance of cellulose-based textiles (Karmakar and Ray 2011; Hebeish and Ibrahim 2007). Conventional ways of stonewash treatment of denims include starch-coating removal (desizing) using amylase and abrasion using pumice stones (1–2 kg per pair of jeans) using bigger washing machines. Currently, biopolishing of cellulosic fabrics and cottons and biostoning of jeans are successfully performed using cellulases. Cellulase breaks the tiny fiber ends on the cotton fabric during the biostoning process which eventually loosens the dye, and finally they are removed during the mechanical abrasion in the wash cycle. Using cellulases is advantageous over pumice stones as it offers less damage to fabric, environment benign, and less work intensive, and more importantly, machines become more productive (Singh et al. 2007; Sukumaran et al. 2005; Uhlig 1998; Galante et al. 1998).

Biopolishing is usually a wet processing stage including desizing, scouring, bleaching, dyeing, and finishing. Crude cellulase (acidic) significantly improves the softness and water absorbance of the fabric. Besides, clean surface with less lint is obtained, and bobble formation tendency is heavily reduced. Enhanced feel, color, and look of fabrics may be achieved through biopolishing using cellulases rich in endoglucanases without any help of chemical treatment (Galante et al. 1998; Sreenath et al. 1996). As an environmentally benign process, use of cellulases offers improved hydrophilicity, moisture absorbance, and color brightness and forms a silky and glossy look of the fabrics by removing surface pills and short fibers (Bhat 2000).

The frequent washing of cotton or cotton-blended fabrics makes them furry and dull since detached microfibrils are formed on the surface. Cellulase containing endoglucanase has the ability to revive the original color and sleeky surface by removing these microfibrils, and the garment becomes soft as dirt particles trapped in the microfibrils are removed (Hebeish and Ibrahim 2007; Ibrahim et al. 2011). Several attempts were performed to enhance the surface and dyeing characteristics of cotton/polyester (50/50)-blended fabrics, mercerized cotton and bleached cotton and to promote the dimensional constancy of cellulosic garments through biological treatment using the pad-wet batch technique in the presence of cellulases and subsequently washing the garments through the mechanical process (Ibrahim et al. 2011; Cortez et al. 2002).

10.7 Olive Oil Extraction

In conventional method, slightly immature, fresh and clean olives are used through cold pressing condition, including a series of processes: (1) crushing and grinding using hammer or stone mills, (2) sending the paste through various mixture machines and horizontal carafe, and finally (3) recovering the oil using high-speed centrifuge. Nevertheless, high yield olive oil may be extracted from fully ripened olive with high acidity and poor and tainted aura (Galante et al. 1998; De Faveri et al. 2008). As such, environmentally benign and improved, but effective method using biological extraction of high-quality olive oil is required to meet the continual increased demand. Crude liquefied cellulase, named Olivex (pectinase with cellulase and hemicellulase), was first of its kind to use in improving olive oil extraction (Fantozzi et al. 1977). Liquefied crude cellulase is advantageous as it increases the olive oil yield under cold processing conditions, gives a better fractionation of the oil paste through the centrifugal process, increases vitamin E and high levels of antioxidants, curtails the induction of rancidity, improves the overall plant efficiency, and lowers the amount of oil in the waste (Galante et al. 1998).

10.8 Wine and Brewery Industry

The production of alcoholic beverages like wines and beers uses crude cellulases microbial glucanases. During the primary fermentation, glucanases are added to hydrolyze glucan, lessen the viscosity of unfermented beer or wine, and improve the quality and yields of beer and wine (Canales et al. 1988; Oksanen et al. 1985).

Beer and wine brewing is mainly dependent on enzymatic processing during malting and fermentation. Seed germination is the initial stage of malting barley through hydrolyzing the seed reserves using biosynthesis and activation of β-glucanase, carboxypeptidase and α- and β-amylases (Canales et al. 1988). Research has shown that wort viscosity and the extent of polymerization may be reduced to a maximum while the cellulase system of *Trichoderma* (endoglucanase II and exoglucanase II) is being used (Godfrey and West 1996). Experiments using crude cellulase (Cytolase 219, a mixture of xylanases, pectinase, and cellulase) in making wines showed that first wine extraction increased by 10%–35%, filtration rate increased by 70%–80%, pressing time decreased by 50–120 minutes, viscosity rate reduced by 30%–70%, energy saved in cooling of fermenter reduced by 20%–40%, and overall the stability of wine increased significantly (Galante et al. 1998).

10.9 Animal Feed Industry

Cellulases and hemicellulases have received extensive concentration due to their ability to improve animal performance and feed value (Graham and Balnave 1995). The nutritional value of grain feed and agricultural silage can be improved by pretreating with cellulases or xylanases. They have the ability to boost the nutritional value by degrading some specific feed constituents, eradicate some anti-nutritional factors, and cater with additional digestive enzymes such as glucanases, amylases, and proteases (Lewis et al. 1996). Compared to the high-quality feedstuff with low-quality feedstuffs, the latter have a high amount of cellulose and ash content with a low amount of fat and protein. Crude cellulase has the potential to increase the silage production used as cattle feed by improving the digestibility of grasses which contain a large amount of nutrients that may be digested and improve the energy values with a limited amount of carbohydrates dissolved in water. Compared to the diets of poultry and swine based on cereals, ruminants have a very complex forage diet containing pectin, hemicellulose, cellulose, and lignin. Pectinases, cellulases, and hemicellulases are used to enhance the nutritive quality of forage (Cowan 1996; Shrivastava et al. 2011). These enzymes induce partial hydrolysis of the plant cell wall of lignocellulosic materials, remove the hull from cereal grains, hydrolyze β-glucans, and make feed materials more flexible and emulsifiable, resulting improved nutritional value of the animal feed (Graham and Balnave 1995; Cowan 1996; Shrivastava et al. 2011; Pascual 2001; Fortun-Lamothe et al. 2001). Additionally, pure cellulase decreases colonization of pathogenic bacteria by boosting the generation of propionic acid which is known as bacteriostatic material and thus improving the cecal fermentation processes (Kuhad et al. 1999; Gupta et al. 2011).

10.10 Bioethanol Industry

The most exciting application of enzymes is producing biofuels from agricultural residues (e.g., sawdust and switchgrass) and forest residues (*Lantana camara* and *Prosopis juliflora*) using cellulases (Sukumaran et al. 2005; Kuhad et al. 2010b; Ghosh and Singh 1993). Lignocellulosic biomass to useful products undergoes several processes, pretreatment, hydrolysis, bioconversion, separation, and purification, to obtain the value-added products (Wyman et al. 2005; Sun and Cheng 2002; Kuhad et al. 1997). The cost associated with enzymatic hydrolysis is lower compared to that of acid or alkali hydrolysis as enzymatic hydrolysis is performed at relatively mild conditions (45–50 °C temperature and 4–6 pH). Additionally, corrosion is avoided as no alkali or acid is present during the process (Kuhad et al. 2010a; Ghosh and Singh 1993).

Currently, available bioconversion technologies of lignocellulosic biomass should be improved to give renewable biofuels and useful chemical by-products to compete with conventional methods (Kuhad and Singh 1993; Mosier et al. 2005; Baker et al. 2005; Lee et al. 1995). Two features are most widely practiced to lessen the cost associated with fuel ethanol production through enzymatic bioconversion of biomass: one is to optimize the production of cellulase and second is to develop a catalyst system based on cellulase which will be more effective. Protein engineering and aimed progression may facilitate to develop better and efficient thermophilic cellulases. Reusing and recycling of enzymes are other options for reducing the hydrolysis cost (Kuhad and Singh 1993; Lee et al. 1995; Singh et al. 1991; Bernardez et al. 1993; Yang and Wyman 2004). Enzymes become deactivated when they are adsorbed on the substrate, especially in lignin which greatly influence the recovery of these enzymes. The adsorption process is nonspecific and irreversible when cellulase is adsorbed in lignin (Yang and Wyman 2006; Kumar and Wyman 2009).

Nevertheless, compounds that attract lignin are being investigated to restrain the cellulases to get adsorbed in lignin (Tu et al. 2007; Dourado et al. 2002). Several strategies have been practiced to reuse and recycle cellulases. One is to use the ultrafiltration method to separate cellulose fraction from sugars and other tiny compounds which may suppress the enzyme activity, and the other is to recycle incapacitated enzymes which facilitate the segregation of enzymes from the system (Lee et al. 1995; Acharya and Chaudhary 2012; Lynd et al. 2005).

Figure 10.5 shows the bioethanol production flow sheet from biomass (Licht 2006). Cellulase-based ethanol production is advantageous over acid-based ethanol production as it is economically feasible and environmentally benign and has higher conversion of biomass, zero substrate loss, and neutral and noncorrosive operating conditions of the conversion process (Kotaka et al. 2008; Huang et al. 2008). It is reported that crude cellulase (combination of endoglucanase and aspergillus oryzae β-glucosidase) is used to produce ethanol from barley directly (Bhanja et al. 2009).

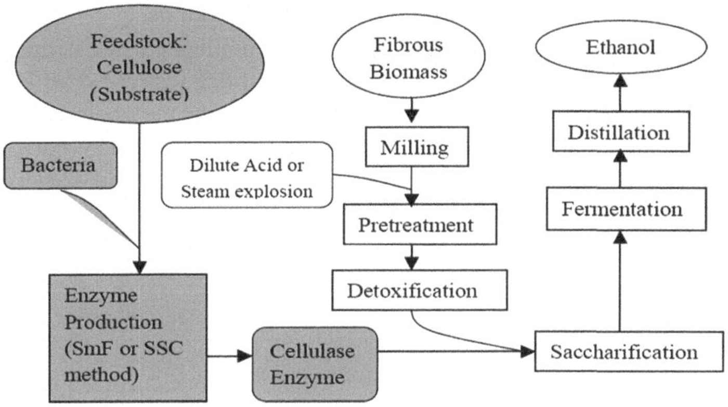

Fig. 10.5 Ethanol production showing the enzymes produced at different stages

10.11 Pharmaceutical Industry

Cellulases are becoming popular biocatalysts in pharmaceutical industry as they are able to release therapeutic compounds from plant-based biomass. Crude cellulases have shown increased ellagic acid yield through hydrolysis from ellagitannins (Do et al. 2009). Researches have shown the enhanced effect on the extraction of the biologically active polyphenols using crude cellulases from various cereals (Chen et al. 2005; Al-Ghazzewi et al. 2007). Crude cellulose increases the antioxidant properties of wheat by increasing the polyphenol extraction through free radical scavenging using enzymatic hydrolysis (Chen et al. 2005). Cellulases as supplements are gaining acceptance more and more due to the inability of digesting cellulose by human beings. As such, various digestive aids using cellulases (such as P-A-L Plus Enzymes, Digestion, etc.) are manufactured for the treatment of metabolic disorders of people (Karmakar and Ray 2011).

Prebiotics contain nondigestible fibers that help the growth of beneficial bacteria for the large intestine passing through the upper gastrointestinal tract. Glucomannans, a water-soluble polysaccharide, act as potent prebiotics when depolymerized using cellulases and mannanases which boost the prebiotic effect as well (Albrecht et al. 2009; Connolly et al. 2010).

The mixture of Konjac glucomannan (KGM) oligosaccharide, a prebiotic, was prepared using endo-β-(1,4)-glucanase and endo-β-(1,4)-mannanase from KGM polysaccharide. In vitro fermentation was observed with gut flora of human for the change in structure of the mixture of these KGM oligosaccharides for about 72 hours in the presence of matrix-assisted laser desorption/ionization-time-of-flight mass spectrometry (MALDI-TOF MS) and capillary electrophoresis-laser-induced fluorescence detection (CE-LIF). As per the observation, KGM oligosaccharide showed higher degradability produced by endo-β-(1,4)-glucanase compared to endo-β-(1,4)-mannanase. Desirable bacteria like *Lactobacilli* and *Bifidobacterium* increase in human feces by glucomannan hydrolysates (GMH) formed using cellulase (Alvaro et al. 2008; Al-Ghazzewi and Tester 2012). GMH produced using cellulases also modulate the composition of microbiota in human intestine with the short-chain fatty acids (SCFA) enriched in propionic acid which decrease the cholesterol level in humans by reducing the synthesis of cholesterol [90]. Compared to crude cellulase, pure cellulase has shown excelling growth of LAB (lactic acid bacteria) from KGM [91].

10.12 Overall Discussion

Cellulases are the important enzymes being used in various industrial applications to make the process sustainable and environmentally benign. Crude and pure forms of cellulases are used for those processes. In most industrial application, the crude cellulases are being used for low cost and ability to improve other by-products or help the process using impurities present in it. Pure forms of cellulases are case specific and mostly used in laboratory-scale research. Since they are costly and only serve some specific purpose, they are not being currently used in the industrial sector.

10.13 Conclusions

Biochemical conversion of biomass through enzymatic cellulase and microorganisms is becoming the most important part of future research. These enzymes are now commercially available in both pure and crude form and are being extensively used in fermentation, pulp and paper industry, brewing sector, fermentation, food and animal feed industry, and detergent and textile industry. As the environment and sustainability is a major concern nowadays, modern biotechnology advances in the area of novel enzymes and microbial genetics will definitely bring more industries under the enzymatic and microorganism-based green industry system. In the future, other plausible areas where cellulases may be used will become more prominent by improving the cellulase activities and enhance them by incorporating desired features through protein engineering.

References

Acharya S, Chaudhary A (2012) Bioprospecting thermophiles for cellulase production: a review. Braz J Microbiol 43(3):844–856

Albrecht S, van Muiswinkel GC, Schols HA, Voragen AG, Gruppen H (2009) Introducing capillary electrophoresis with laser-induced fluorescence detection (CE-LIF) for the characterization of konjac glucomannan oligosaccharides and their in vitro fermentation behavior. J Agric Food Chem 57(9):3867–3876

Al-Ghazzewi FH, Tester RF (2012) Efficacy of cellulase and mannanase hydrolysates of konjac glucomannan to promote the growth of lactic acid bacteria. J Sci Food Agric 92(11):2394–2396

Al-Ghazzewi FH, Khanna S, Tester RF, Piggott J (2007) The potential use of hydrolysed konjac glucomannan as a prebiotic. J Sci Food Agric 87(9):1758–1766
Alonso DM, Wettstein SG, Dumesic JA (2012) Bimetallic catalysts for upgrading of biomass to fuels and chemicals. Chem Soc Rev 41(24):8075–8098
Alvaro A, Sola R, Rosales R, Ribalta J, Anguera A, Masana L, Vallvé JC (2008) Gene expression analysis of a human enterocyte cell line reveals downregulation of cholesterol biosynthesis in response to short-chain fatty acids. IUBMB Life 60(11):757–764
Annamalai N, Rajeswari MV, Elayaraja S, Balasubramanian T (2013) Thermostable, haloalkaline cellulase from Bacillus halodurans CAS 1 by conversion of lignocellulosic wastes. Carbohydr Polym 94(1):409–415
Baker JO, McCarley JR, Lovett R, Yu CH, Adney WS, Rignall TR et al (2005) Catalytically enhanced endocellulase Cel5A from Acidothermus cellulolyticus. Appl Biochem Biotechnol 121(1-3):129–148
Bernardez TD, Lyford K, Hogsett DA, Lynd LR (1993) Adsorption of Clostridium thermocellum cellulases onto pretreated mixed hardwood, Avicel, and lignin. Biotechnol Bioeng 42 (7):899–907
Bhanja T, Kumari A, Banerjee R (2009) Enrichment of phenolics and free radical scavenging property of wheat koji prepared with two filamentous fungi. Bioresour Technol 100 (11):2861–2866
Bhat MK (2000) Cellulases and related enzymes in biotechnology. Biotechnol Adv 18(5):355–383
Brodeur G, Yau E, Badal K, Collier J, Ramachandran KB, Ramakrishnan S (2011) Chemical and physicochemical pretreatment of lignocellulosic biomass: a review. Enzyme Res:1–17
Canales AM, Garza R, Sierra JA, Arnold R (1988) The application of a beta-glucanase with additional side activities in brewing. Tech Quar Mast Brewers Assoc Am (USA) 25:27–31
Chalal D (1985) Solid state fermentation with Trichoderma reesei for cellulase production. Appl Environ Microbiol 49(1):205–210
Chen HL, Fan YH, Chen ME, Chan Y (2005) Unhydrolyzed and hydrolyzed konjac glucomannans modulated cecal and fecal microflora in Balb/c mice. Nutrition 21(10):1059–1064
Connolly ML, Lovegrove JA, Tuohy KM (2010) Konjac glucomannan hydrolysate beneficially modulates bacterial composition and activity within the faecal microbiota. J Funct Foods 2 (3):219–224
Cortez JM, Ellis J, Bishop DP (2002) Using cellulases to improve the dimensional stability of cellulosic fabrics. Text Res J 72(8):673–680
Cowan WD (1996) Animal feed. In: Godfrey T, West S (eds) Industrial enzymology, 2nd edn. Macmillan Press, London, pp 360–371
De Carvalho LMJ, De Castro IM, Da Silva CAB (2008) A study of retention of sugars in the process of clarification of pineapple juice (Ananas comosus, L. Merril) by micro-and ultra-filtration. J Food Eng 87(4):447–454
De Faveri D, Aliakbarian B, Avogadro M, Perego P, Converti A (2008) Improvement of olive oil phenolics content by means of enzyme formulations: Effect of different enzyme activities and levels. Biochem Eng J 41(2):149–156
Desvaux M (2005) Clostridium cellulolyticum: model organism of mesophilic cellulolytic clostridia. FEMS Microbiol Rev 29(4):741–764
Dhillon GS, Kaur S, Brar SK, Verma M (2012) Potential of apple pomace as a solid substrate for fungal cellulase and hemicellulase bioproduction through solid-state fermentation. Ind Crop Prod 38:6–13
Dienes D, Egyhazi A, Reczey K (2004) Treatment of recycled fiber with Trichoderma cellulases. Ind Crop Prod 20(1):11–21
Do YK, Kim JM, Chang SM, Hwang JH, Kim WS (2009) Enhancement of polyphenol bio-activities by enzyme reaction. J Mol Catal B Enzym 56(2-3):173–178
Dogaris I, Karapati S, Mamma D, Kalogeris E, Kekos D (2009) Hydrothermal processing and enzymatic hydrolysis of sorghum bagasse for fermentable carbohydrates production. Bioresour Technol 100(24):6543–6549

Doi RH, Kosugi A (2004) Cellulosomes: plant-cell-wall-degrading enzyme complexes. Nat Rev Microbiol 2(7):541–551
Dourado F, Bastos M, Mota M, Gama FM (2002) Studies on the properties of Celluclast/Eudragit L-100 conjugate. J Biotechnol 99(2):121–131
Escobar MO, Hue NV (2008) Temporal changes of selected chemical properties in three manure–Amended soils of Hawaii. Bioresour Technol 99(18):8649–8654
Fantozzi P, Petruccioli G, Montedoro G (1977) Trattamenti con additivi enzimatici alle paste di oliva sottoposte ad estrazione per pressione unica: influenze delle cultivars, dell'epoca di raccolta e della conservazione. Grasse 54:381–388
Fontaine S, Bardoux G, Benest D, Verdier B, Mariotti A, Abbadie L (2004) Mechanisms of the priming effect in a savannah soil amended with cellulose. Soil Sci Soc Am J 68(1):125–131
Fortun-Lamothe L, Gidenne T, Debray L, Chalaye F. (2001). Intake regulation, performances and health status according to feeding strategy around weaning. In Proceendings of the 2nd Meeting of COST (Vol. 848, pp. 40–41).
Galante YM, DeConti A, Monteverdi R (1998) Application of trichoderma enzymes in food and feed industries. In: Harman GF, Kubicek CP (eds) Trichoderma and gliocladium—enzymes, vol. 2 of Biological control and commercial applications. Taylor & Francis, London, pp 311–326
Ghosh P, Singh A (1993) Physicochemical and biological treatments for enzymatic/microbial conversion of lignocellulosic biomass. In: Advances in applied microbiology, vol 39. Academic Press, New York, pp 295–333
Godfrey T, West S (1996) Textiles. Indus Enzymol:360–371
Graham H, Balnave D (1995) Dietary enzymes for increasing energy availability. In: Wallace RJ, Chesson A (eds) Biotechnology in animal feeds and animal feedings. VHC, Weinheim, pp 296–309
Gupta R, Khasa YP, Kuhad RC (2011) Evaluation of pretreatment methods in improving the enzymatic saccharification of cellulosic materials. Carbohydr Polym 84(3):1103–1109
Han W, He M (2010) The application of exogenous cellulase to improve soil fertility and plant growth due to acceleration of straw decomposition. Bioresour Technol 101(10):3724–3731
Harrison MJ, Nouwens AS, Jardine DR, Zachara NE, Gooley AA, Nevalainen H, Packer NH (1998) Modified glycosylation of cellobiohydrolase I from a high cellulase-producing mutant strain of Trichoderma reesei. Eur J Biochem 256(1):119–127
Headon DR, Walsh G (1994) The industrial production of enzymes. Biotechnol Adv 12(4):635–646
Hebeish A, Ibrahim NA (2007) The impact of frontier sciences on textile industry. Colourage 54 (4):41–55
Henrissat B, Davies G (1997) Structural and sequence-based classification of glycoside hydrolases. Curr Opin Struct Biol 7(5):637–644
Huang W, Niu H, Li Z, He Y, Gong W, Gong G (2008) Optimization of ellagic acid production from ellagitannins by co-culture and correlation between its yield and activities of relevant enzymes. Bioresour Technol 99(4):769–775
Ibrahim NA, El-Badry K, Eid BM, Hassan TM (2011) A new approach for biofinishing of cellulose-containing fabrics using acid cellulases. Carbohydr Polym 83(1):116–121
Juturu V, Wu JC (2014) Microbial cellulases: engineering, production and applications. Renew Sust Energ Rev 33:188–203
Kanmani R, Vijayabaskar P, Jayalakshmi S (2011) Saccharification of banana-agro waste and clarification of apple juice by cellulase enzyme produced from Bacillus pumilis. World Appl Sci J 12(11):2120–2128
Karmakar M, Ray RR (2011) Current trends in research and application of microbial cellulases. Res J Microbiol 6(1):41
Kotaka A, Bando H, Kaya M, Kato-Murai M, Kuroda K, Sahara H et al (2008) Direct ethanol production from barley β-glucan by sake yeast displaying Aspergillus oryzae β-glucosidase and endoglucanase. J Biosci Bioeng 105(6):622–627

Kuhad RC, Singh A (1993) Lignocellulose biotechnology: current and future prospects. Crit Rev Biotechnol 13(2):151–172

Kuhad RC, Singh A, Eriksson KEL (1997) Microorganisms and enzymes involved in the degradation of plant fiber cell walls. In: Biotechnology in the pulp and paper industry. Springer, Berlin/Heidelberg, pp 45–125

Kuhad RC, Manchanda M, Singh A (1999) Hydrolytic potential of extracellular enzymes from a mutant strain of Fusarium oxysporum. Bioprocess Eng 20(2):133–135

Kuhad RC, Mehta G, Gupta R, Sharma KK (2010a) Fed batch enzymatic saccharification of newspaper cellulosics improves the sugar content in the hydrolysates and eventually the ethanol fermentation by Saccharomyces cerevisiae. Biomass Bioenergy 34(8):1189–1194

Kuhad RC, Gupta R, Khasa YP (2010b) Bioethanol production from lignocellulosic biomass: an overview. In: Lal B (ed) Wealth from waste. Teri Press, New Delhi, India

Kumar R, Wyman CE (2009) Effect of additives on the digestibility of corn stover solids following pretreatment by leading technologies. Biotechnol Bioeng 102(6):1544–1557

Lee D, Yu AH, Saddler JN (1995) Evaluation of cellulase recycling strategies for the hydrolysis of lignocellulosic substrates. Biotechnol Bioeng 45(4):328–336

Lewis GE, Hunt CW, Sanchez WK, Treacher R, Pritchard GT, Feng P (1996) Effect of direct-fed fibrolytic enzymes on the digestive characteristics of a forage-based diet fed to beef steers. J Anim Sci 74(12):3020–3028

Licht FO (2006) World ethanol markets: the outlook to 2015, Tunbridge Wells, Agra Europe Special Report, UK

Lloyd TA, Wyman CE (2005) Combined sugar yields for dilute sulfuric acid pretreatment of corn stover followed by enzymatic hydrolysis of the remaining solids. Bioresour Technol 96 (18):1967–1977

Lynd LR, Weimer PJ, Van Zyl WH, Pretorius IS (2002) Microbial cellulose utilization: fundamentals and biotechnology. Microbiol Mol Biol Rev 66(3):506–577

Lynd LR, Van Zyl WH, McBride JE, Laser M (2005) Consolidated bioprocessing of cellulosic biomass: an update. Curr Opin Biotechnol 16(5):577–583

Mai C, Kües U, Militz H (2004) Biotechnology in the wood industry. Appl Microbiol Biotechnol 63 (5):477–494

Mosier N, Wyman C, Dale B, Elander R, Lee YY, Holtzapple M, Ladisch M (2005) Features of promising technologies for pretreatment of lignocellulosic biomass. Bioresour Technol 96 (6):673–686

Oksanen J, Ahvenainen J, Home S (1985) Microbial cellulase for improving filterability of wort and beer. J Inst Brew 91(3):130–130

Pascual JJ (2001) Recent advances on early weaning and nutrition around weaning. In Proceedings of the 2nd Meeting of COST (Vol. 848, pp. 31–36).

Pere J, Siika-aho M, Buchert J, Viikari L (1995) Effects of purified T. reesei cellulases on the fiber properties of kraft pulp. TAPPI J 78(6):71–78

Rai P, Majumdar GC, Gupta SD, De S (2007) Effect of various pretreatment methods on permeate flux and quality during ultrafiltration of mosambi juice. J Food Eng 78(2):561–568

Rodrigues AL, Cavalett A, Lima AO (2010) Enhancement of Escherichia coli cellulolytic activity by co-production of beta-glucosidase and endoglucanase enzymes. Electron J Biotechnol 13 (5):5–6

Sakai S, Tsuchida Y, Okino S, Ichihashi O, Kawaguchi H, Watanabe T, Inui M, Yukawa H (2007) Effect of lignocellulose-derived inhibitors on growth of and ethanol production by growth-arrested Corynebacterium glutamicum R. Appl Environ Microbiol 73(7):2349–2353

Sánchez C (2009) Lignocellulosic residues: biodegradation and bioconversion by fungi. Biotechnol Adv 27(2):185–194

Sheehan J, Himmel M (1999) Enzymes, energy, and the environment: a strategic perspective on the US Department of Energy's research and development activities for bioethanol. Biotechnol Prog 15(5):817–827

Shrivastava B, Thakur S, Khasa YP, Gupte A, Puniya AK, Kuhad RC (2011) White-rot fungal conversion of wheat straw to energy rich cattle feed. Biodegradation 22(4):823–831

Singh A, Kumar PKR, Schügerl K (1991) Adsorption and reuse of cellulases during saccharification of cellulosic materials. J Biotechnol 18(3):205–212

Singh A, Kuhad RC, Ward OP (2007) Industrial application of microbial cellulases. In: Kuhad RC, Singh A (eds) Lignocellulose biotechnology: future prospects. I. K. International Publishing House, New Delhi, pp 345–358

Sohail M, Siddiqi R, Ahmad A, Khan SA (2009) Cellulase production from Aspergillus niger MS82: effect of temperature and pH. New Biotechnol 25(6):437–441

Sreenath HK, Shah AB, Yang VW, Gharia MM, Jeffries TW (1996) Enzymatic polishing of jute/cotton blended fabrics. J Ferment Bioeng 81(1):18–20

Srivastava N, Srivastava M, Mishra PK, Singh P, Ramteke PW (2015) Application of cellulases in biofuels industries: an overview. J Biofuels Bioenergy 1(1):55–63

Stork G, Puls J (1996) Change in properties of different recycled pulps by endoglucanase treatment. In: Proceedings of the international conference on biotechnology in the pulp and paper industry: recent advances in applied and fundamental research (Vol. 1, pp. 145–150). Facultas-Universitatsverlag

Sukumaran RK, Singhania RR, Pandey A (2005) Microbial cellulases-production, applications and challenges. J Sci Ind Res 64(11):832–844

Sun Y, Cheng J (2002) Hydrolysis of lignocellulosic materials for ethanol production: a review. Bioresour Technol 83(1):1–11

Tejada M, Gonzalez JL, García-Martínez AM, Parrado J (2008) Application of a green manure and green manure composted with beet vinasse on soil restoration: effects on soil properties. Bioresour Technol 99(11):4949–4957

Tu M, Chandra RP, Saddler JN (2007) Evaluating the distribution of cellulases and the recycling of free cellulases during the hydrolysis of lignocellulosic substrates. Biotechnol Prog 23 (2):398–406

Uhlig H (1998) Industrial enzymes and their applications. Wiley, Chicester

Walsh G, Headon D (1994) Protein biotechnology. Wiley, Chicester

Wyman CE, Dale BE, Elander RT, Holtzapple M, Ladisch MR, Lee YY (2005) Coordinated development of leading biomass pretreatment technologies. Bioresour Technol 96 (18):1959–1966

Yang B, Wyman CE (2004) Effect of xylan and lignin removal by batch and flowthrough pretreatment on the enzymatic digestibility of corn stover cellulose. Biotechnol Bioeng 86 (1):88–98

Yang B, Wyman CE (2006) BSA treatment to enhance enzymatic hydrolysis of cellulose in lignin containing substrates. Biotechnol Bioeng 94(4):611–617

Zhang YHP, Himmel M, Mielenz JR (2006) Outlook for cellulase improvement: Screening and selection strategies. Biotechnol Adv 24:452–481

Zhu Z, Sathitsuksanoh N, Zhang YHP (2009) Direct quantitative determination of adsorbed cellulase on lignocellulosic biomass with its application to study cellulase desorption for potential recycling. Analyst 134:2267–2272

Significance of Process Parameters on Fungal Cellulase Production

11

Srilakshmi Akula and Narasimha Golla

Abstract

The biotechnological production of enzymes from microorganisms was proved to generate enormous wealth that influences significant sectors of the world's economy. Among the microbial enzymes, the potentiality of cellulases in different manufactories including food, paper, biofuel, animal feed, drug, brewery, textile, agriculture and recycling of waste materials has been the compelling factor for the intense limelight on cellulases for the past several decades. Extensive studies were carried out on aerobic fungi producing cellulases and are considered as the leading workhorses in industrial processes. The enzyme production usually depends on distinct governing parameters essentially inoculum size, pH value, temperature, growth, time, aeration, inducers and medium supplements. Therefore, choosing optimum pivotal factors that throw impact on biomass of various microorganisms and build-up of the target product becomes the preliminary criteria for any profitable recovery process. Often the optimisation of multifarious criterions is a laborious and tedious chore. Hence, this chapter highlights the diverse physical and chemical parameters that immensely influence fungal cellulase production.

Keywords

Cellulase · Cellulose · Waste biomass · Renewable energy · Biofuels · Fungal microorganisms

S. Akula · N. Golla (✉)
Applied Microbiology Laboratory, Department of Virology, Sri Venkateswara University, Tirupati, Andhra Pradesh, India

N. Srivastava et al. (eds.), *Biofuel Production Technologies: Critical Analysis for Sustainability*, Clean Energy Production Technologies,
https://doi.org/10.1007/978-981-13-8637-4_11

11.1 Introduction

The tremendous wringing and discharge of fossil fuels have insistently diminished its innate resources and consequently provoked serious hazards to living world (Sajith et al. 2016). Thus, accelerated interest pertaining to environmental deterioration and the reduction of fossil fuels urges to use substitute sustainable energy reserves in order to counter the consistently growing energy consumptions (Dashtban et al. 2009). Currently, the notion of waste-to-energy became the central spotlight of numerous industries with a commercial aspect and feasible processes manifesting the utilisation of biomass accordingly (Kendry 2002). As long as plants have thrived, deposits from greenery have been substantial donors to the ecosystem of our planet. Cellulose forms the skeletal constituent of the basic structure of the green plants especially their cell coverings, that is, many forms of the oomycetes together with algae. Biofilms are produced by some species of bacteria through secretion of cellulose (Lekh Ram et al. 2014). Cellulose is the fibrous, insoluble, crystalline homopolymer constructed with glucose units cemented by glycosidic bonds of β-1,4-linkage (Jagtap and Rao 2005). From the point of energy content, cellulose is considered as the low-priced energy source in addition to the most bountiful sustainable biological reserve (Lynd et al. 2008; Zhang et al. 2009; Coral et al. 2002). Cellulose has drawn worldwide attention as it can be transformed into bioenergy and bio-based products. In recent times, due to the sky-touching price of the employment process, stupendous quantities of municipal, agricultural and industrial cellulosic leftovers are being utilised slovenly (Kim et al. 2003).

Cellulose is being preowned by the society since way back, but its capability as a storage house of energy was captured posterior to the perception of the of cellulases (Bhat and Bhat 1997). It is exploited as a nutrient stock by a variety of microbial groups inclusive of bacteria, fungi, plants and protests as well as invertebrate animals such as nematodes, crustaceans, insects, molluscs and annelids (Watanabe and Tokuda 2001; Davidson and Blaxter 2005). The aforementioned microbes employ a battery of enzymes called cellulase sequentially to support the breakdown of cellulose to simple form of energy like glucose (Beguin and Aubert 1994).

Free monosaccharide is released from cellulose being acted upon by cellulases through hydrolysis for the generation of bioethanol and various synthetic products as well, a few of which serves as future alternatives for liquid fuel by-products (Bozell and Petersen 2010). However, mechanisms of cellulose breakdown by cellulase enzymes were partially disclosed, owing to the raised degree of crystallinity and less water solubility of cellulose fibres (Yamada et al. 2005). To date, the towering challenges of the past decades remain to persist, that is, the exploration of cost-effective enzyme-based conversion of complex carbohydrate (cellulose) to monosaccharides. The three major disputes are (1) the current requisite for lengthy duration of operations to achieve an elevated cellulase release, (2) the feedback inhibition by glucose and by-products released during cellulase development, and (3) the growth associated complications at an elevated concentration of cellulose from the point of liquid kinetic impulsions.

11.2 Cellulases

Many researchers and industries chiefly focused on cellulases as they are the most unbeaten group of lignocellulolytic enzymes used in disparate economically related processes. Principally, cellulases are enzymatic proteins which operate slowly by one to two orders of amplitude compared to the rest of the carbohydrate-degrading enzymes (Himmel et al. 2007) leading to the biotransformation of lignocelluloses to efficacious requisites. The noteworthy usage of cellulose biomass is thought to be associated with economic process for the enzyme cultivation (Solomon et al. 1999; Wu and Lee 1997).

Based on their catalytic site in the cellulosic substrate, cellulases are categorised into three groups:

Endoglucanase: During the solubilisation of cellulosic material, CMCase proceeds in associated action with exoglucanases and β-D-glycosidase (Zhang et al. 2006) by breaking the internal glycosidic bonds present in the cellulosic chains producing cellooligosaccharides.

Exoglucanases: (Cellobiohydrolases (CBHs)) – It acts on cellulose by cleaving the disaccharide units initiating from the non-reducing terminal of the chain. It also acts on swollen, partially degraded amorphous substrates and cellodextrins. However, soluble derivatives of cellulose such as hydroxyl ethyl cellulose and carboxymethyl cellulose were not hydrolysed. A non-significant constituent to be noticed in few cellulase systems is glucohydrolase (Joshi and Pandey 1999).

β-Glycosidase: (Cellobiose) – *β*-Glycosidase hydrolyses cellobiose to glucose, contributing an easily metabolised source of carbon to the fungus. Numerous cellobioses were reported in wide *Aspergillus* members with different molecular weights. A great number of these enzymes except a few fail to exert action on H_3PO_4 – swollen, CM celluloses and other polymeric substrates such as cotton, Avicel and filter paper (Bhat and Hazlewood 2003).

The two foremost enzymes termed "absolute cellulases" directly act on cellulose to give off glucose. The so formed cellobiose later fragmented into glucose by the third enzyme in sequence (Dincer and Telefoncu 2006; Andersen 2007) (Fig. 11.1).

11.3 Fungal Cellulases

The most robust and widespread biomass fragmenting microbes that have evolved in nature are fungi showing assorted modes of life for the bioconversion of green cellulosic deposits on the biosphere. The secretion of multifarious set of oxidative and hydrolytic enzymes (cellulases) mould fungi to be capable of decomposing green biomass (Elzaher and Fadel 2010; Amir et al. 2011). Therefore, the scale of notable performance and potentiality of the production were presented with ease at uplifted activities on industrial scale. Industrial biorefinery applications are largely dependent on cellulase mixture of fungal origin.

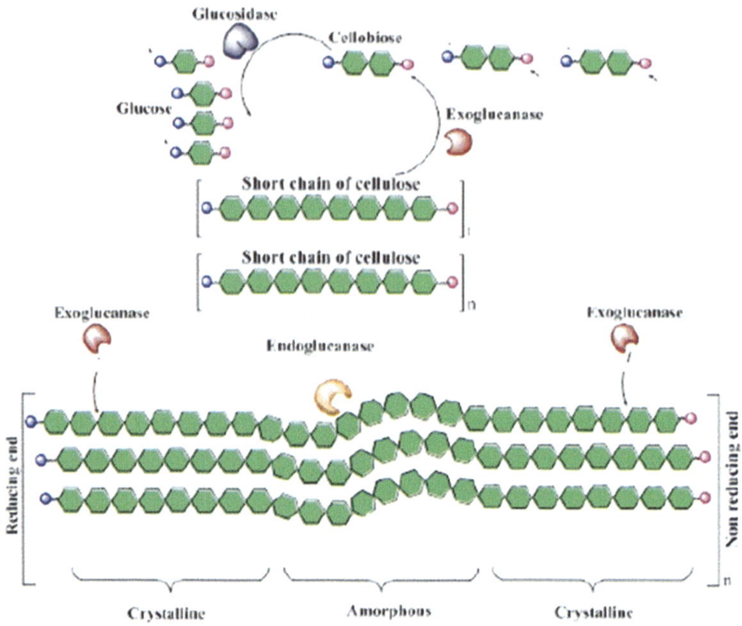

Fig. 11.1 Schematic representation of the hydrolysis of cellulose by synergistic action between (**a**) endoglucanases, (**b**) exoglucanases and (**c**) β-glycosidases

Fungal cellulases can be divided into two groups.

(a) White-rot fungi, such as *Phanerochaete chrysosporium*, and soft-rot fungi, such as *Hypocrea jecorina* (otherwise *Trichoderma reesei*) and *Penicillium pinophilum* constitute the first group. They have complete cellulolytic enzyme systems capable of converting crystalline cellulose to glucose. Moreover, they consist of many secreted enzymes acting at the terminal (exoglucanases) or in the middle (endoglucanases) of the cellulose chains. β-Glycosidases then assists the hydrolysis of so formed cellobiose to glucose.
(b) Brown-rot fungus such as *Postia placenta* constitutes the second group which is devoid of strict cellobiohydrolases (Kleman-Leyer et al. 1992) and employs oxidative components in concert with endoglucanases for the degradation of cellulose.

11.4 Structure

Solid substrates are burdensome for fungal cellulases. Most of the fungal cellulases share a common organisation at molecular level where a large catalytic domain (CD) is linked to a small carbohydrate-binding module (CBM) through a highly

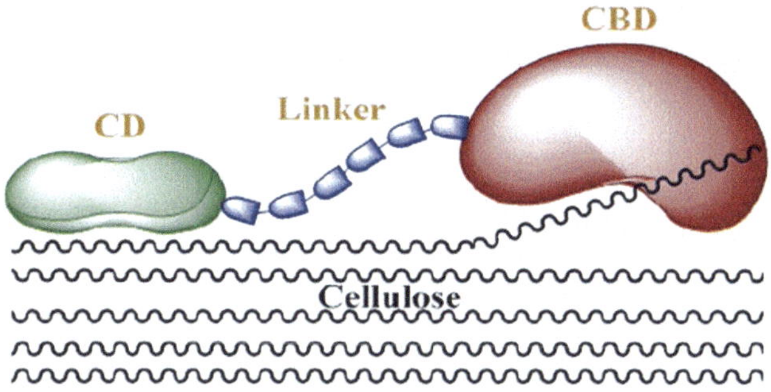

Fig. 11.2 Structure of fungal cellulases containing CD carbohydrate-binding domain attached via a linker-peptide

glycosylated linker-peptide. However, by the enzymatic treatment with papain, it can be departed easily into the two operative domains (Tomme et al. 1988) (Fig. 11.2).

The three hexagons present in the CBM pinpoint the aromatic residues accountable for interaction with the hydrophobic face of every second pyranose ring. The grey area in the figure represents the loops housing the substrate-binding sites (Hilden and Johansson 2004). The active site of a cellulase contains numerous binding sites for glucose units, which elevates the possibility for the enzyme to remain connected to the substrate after a catalytic cycle and thereby function progressively (Divne et al. 1994). These binding subsites are labelled, correspondingly to an agreement, from *–n* to *+n* with *–n* at the non-reducing whereas +n at reducing terminal. The fragmentation takes place between the −1 and + 1 subsite (Davies et al. 1997).

In general, endoglucanases have greater disclosed active site, whereas cellobiohydrolases have a tunnel-shaped active site, resulting in a cleft or groove, permitting the enzyme to bind to the centre of the substrate chain and break it down. Moreover, because some cellobiohydrolases can acquire these interior cuts, the loops covering the tunnel must be moulded to allow a cellulose chain to get into the active site. Besides CD, most fungal cellulases consist of a carbohydrate-binding module (CBM) called the cellulose-binding domain (CBD).

The CBDs are trusted to execute a pivotal role in cellulose hydrolysis. Even though these domains fail to strike the cellulase activity soluble and amorphous substrates, they remarkably elevate the enzymes' ability to hydrolyse crystalline cellulose. Currently, the CAZy classification lists 45 families of characterised CBMs depending on resemblances in amino acid sequence (Davies et al. 2005). All fungal CBDs are included in the family I containing 35–40 amino acids and show strong sequence similarity with an overall amino acid identity of 60%, few residues being totally conserved and a few exhibiting conservative substitutions (Gilkes et al. 1991).

Nuclear magnetic resonance determined the principal structure of a fungal CBD (Kraulis et al. 1989). The fungal cellulase CBDs exhibit wedge-shaped fold

Table 11.1 List of cellulolytic fungal cultures

Fungal strain	References
A. ustus	Macris (1986)
A. fumigatus	Heptinstall et al. (1986)
A. heteromorphus	Singh et al. (2009) and Singh et al. (2006)
A. terreus	Araugo and Souza (1986)
A. aculeatus	Murao et al. (1988)
T. aureoviride	Zaldivar et al. (2001)
A. flavus	Ojumu et al. (2003)
Scopulariopsis	Bharathi and Ravindra (2006)
P. chrysogenum	Chinedu et al. (2007)
T. viride	Qi et al. (2005) and Shafique et al. (2009)
Trichothecium roseum	Shanmugam et al. (2008)
A. wentii	Panda et al. (1987) and Srivastava et al. (1984)
T. harzianum	Alam (2011), Haq et al. (2006), and Shafique et al. (2009)
Alternaria alternata	Macris (1984)
A. japonicus	Sharma et al. (1985) and Sanyal et al. (1988)
A. niger	Milala et al. (2005), Narasimha et al. (2006), and Sharada et al. (2012)
A. candidus	Milala et al. (2009)
T. koningii	Wood (1988) and Wood and Mecrae (1982, 1986)
T. reesei	Latifian et al. (2007), Wang et al. (2006), and Shafique et al. (2009)
Penicillium atrovenetum	Adeleke (2013)
Purpureocillium lilacinum	Srilakshmi et al. (2017)

consisting of an elemental structure of a twisted sheet of three short antiparallel strands. One side of the wedge is placoid and contains three preserved aromatic amino acids set apart by a distance reminiscent of the length of the continuation unit in cellulose, cellobiose (Tomme et al. 1995). This interaction is recurrently supplemented by polar residues resulting in hydrogen bonds (Tormo et al. 1996). The other side is ruffled and less hydrophilic in character (Table 11.1).

11.5 Production of Cellulases

The victory of any biocommodity industry relies upon the economics of production of hydrolytic enzymes. The foremost hurdle within the exploitation of one of such enzymes is the overpriced production in addition to other factors such as the kind and cellulose reserve employed for production, the complexity of cellulose structure and dropped quantities of cellulase production by cellulose degraders due to feedback inhibition. Utilisation of fairly cheaper reserves such as lignocellulose materials in the place of high-quality cellulose is one of the effective approaches to trim the cost of enzyme production (Sridevi et al. 2008). Particularly, cellulose turnover by enzymes falls rapidly with conversion resulting in shrunken yields under extended operating periods, and inflated biocatalyst utilisation including the end product produced per amount of adsorbed enzyme evidently diminishes with hydrolysis (Nutor and Converse 1991; Wang and Converse 1992).

11.6 Submerged Fermentation

Under liquid- and solid-state fermentations, fungal cultures obtain divergent developmental patterns. Research studies on fungal growth patterns in SMF are well investigated for a numerous industrially vital fermentations. Under shaking conditions, fungi grow as free mycelia or pellets, depending on the operating strategies applied and genetic constitution of the microbial strain as well (Papagianni 1995). Environmental management is comparatively simple in the case of submerged cultures, owing to the nutrient supplements, the uniformity of the microbial cells suspension and the products in the liquid phase (Alberton et al. 2009). Seeing that ergonomics associated with enzymes are presumably hinge on constitutional enzymes, selecting appropriate carbon and nitrogen supplies and fixing their proportions are of paramount importance for upgrading enzyme performance in the biological filtrates (Gomes et al. 2000). In SMF, the production of cellulase is considerably influenced by various processes dictating physical parameters such as temperature, agitation rate and chemical parameters like crude material concentration, elicitors, production medium constituents and so on. However, the demand for a long gestation period with diminished production stands as key bottleneck in liquid-state bioprocess (Singhania et al. 2010).

11.7 Solid-State Fermentation

Recently, SSF is increasing further enthusiasm as a fitting methodology to reuse nutrient-surplus junk like cellulosic materials. In-depth knowledge about SSF demonstrated that it results in high enzyme yield requiring minimal to zero effort, utilisation of agrarian waste as a substrate and a more extensive scope of additional enzyme activities than those found in SMF (Robinson et al. 2001; Couto and Sanromán 2006). Moreover, SSF encourages not exclusively the probabilities for the bioconversion of agro-deposits to include stock, and in addition it empowers the effective reuse of lignocellulosic materials with the consumption of less vitality (Pandey 2003). In SSF, the operational criterion inclusive of temperature, pH and percentage of dampness is a fundamental component impacting the development of microbial culture and generation of cellulases. Accessibility of air in to the microbial cells and heat production are the real difficulties in SSF, which must be tended to legitimately.

11.8 Factors Affecting Cellulase Production

To establish a hit fermentation technique, it's far important to frame incubational and nutritional situations to encourage the overproduction of favoured metabolites by biological workhorses (El-Hadi et al. 2014). The cell growth and the yield of a product greatly depend on the characteristics of fermentation media. This is due to the microbial reaction to the kicks of outer environment through biochemical paths that control expression of genes and morphology of the microbes extended to their upgraded synthesis of the preferred metabolites (Bon et al. 2008). Conformational changes besides cellulose production by different isolates urge the optimisation of physicochemical parameters of the production medium (Kathiresan and Manivannan 2006; Polyanna et al. 2011; Papagianni 2004).

Majority of the fermentations rely on biomass which plays a vital role in the study of cell growth (Bon et al. 2008). The bioprocess often may be categorised into two kinds: the preliminary kind shows biomass development, while the second kind entails the synthesis of the interested products. The accumulation of microbial cells is extremely important regarding the foremost kind, because it is the vital goal of solid- or liquid-state bioprocess. The manufacturing of the target products can be in correspondence to the quantity of cell growth. For the achievement of enhanced secondary metabolite, cell growth is of prime importance (Bon et al. 2008; Alberton et al. 2009; Kang et al. 2004).

Cellulase synthesis with the aid of specific microorganisms in liquid-state methodology captures extra attention and is noted to be price restricted (Singh et al. 2009). The raw material price takes hold on economics of cellulase production, and for this reason using reasonably priced biomass sources can assist to lessen investment on cellulose (Wen et al. 2005). Lots of investigative struggles were executed to bring the price of cellulose down to earth. The paths which were considered in reducing enzyme cost are categorised into (1) sorting for organisms with contemporary enzymes, (2) microbial genetic modification and bioengineering of biocatalyst (Van Hanh et al. 2011) and (3) right choice of substrates and fermentative conditions besides enzymes reprocessing and transfiguring the production strategies (Howard et al. 2003).

11.9 Effect of pH

The buffering capacity of fermentative medium is one of the most vital environmental parameters striking the microbial cell synthesis and enzyme release. It also performs a censorious task in the transport of diverse components through the cell membrane. Both higher and lower pH result in poor consequences; however, the acidic nature of the medium near 5.0 pH is suitable for accumulation of biocatalyst. This can properly be because of the very fact that fungal organisms prefer faintly lower pH for their cellular build-up and enzyme synthesis. Different fungal species exhibit varied pH optima for cellulose production (Haltrich et al. 1996). The pH variation noticed during the microbial cell synthesis additionally impacts product stability in the medium (Gupta et al. 2003).

In an investigation study carried out by Sarkar and Aikat (2014), maximum CMCase (2.58 IU/ml) and F Passé (0.276 IU/ml) production by *Aspergillus fumigates* isolated from straw retting ground was achieved at pH 4.0. Moreover, pH 5.0 was noticed to be optimum for cell biomass and cellulose production by *Penicillium* sps. and *Aspergillus Niger* (Narasimha et al. 2006; Prasanna et al. 2016). Higher Exo and Endo-1,4-β-D-glucanases of 1.22 IU/ml and 1.82 IU/ml by *Purpureocillium lilacinum* were favoured by pH 5.5, respectively (Srilakshmi et al. 2017). The initial pH of the fermentation medium was adjusted to 6.5 inoculated with *Trichothecium roseum*, and higher CMCase and β-glycosidase activities were obtained after 7 days of incubation (Shanmugam et al. 2008), and hence initial pH of the medium exerts a sturdy influence on enzyme production. In general, the pH of the culture medium rises within 48 days of cellulose fermentation by fungi due to the utilisation of yeast extract, hemicellulose and amorphous cellulose from lignocellulosic materials for cell growth. After attaining proliferative growth, the culture pH dropped because of the release of carbonic acids and carboxylic groups from lignin (Portjanskaja et al. 2006). Production of cellulose by *Aspergillus Niger* was intensive when the starting culture pH was alerted to 6.0 or 7.0 (Sohail et al. 2009).

11.10 Effect of Temperature

Among the physical factors, temperature maintenance during the bioprocessing also occupies a major role that influences both purity level and turnout of the products of interest (Ahmed et al. 2009; Iqbal et al. 2010). In-depth investigations of many workers on fungus like *Aspergillus* spp. revealed that different temperature optima is required for similar genus of the same fungus (Akinyele et al. 2013).

Temperature optima of *A. terreus* to produce cellulase were established as 28 °C (Shahriarinour et al. 2011). According to Srilakshmi et al. (2017), growth, extracellular protein and cellulase production from *Purpureocillium lilacinum* successively inclined to 30 °C and declined at further elevated temperatures. Around 1.8 times higher CMCase (1.62 U/ml), production by *A. humicola* was achieved in the medium amended with COA as compared to the enzyme production COA lacking medium (0.91 U/ml) at 37 °C (Nipa et al. 2006). Gilna and Khaleel (2011) proved that 32 °C supported maximum cellulase activity by *Aspergillus fumigates* when inoculated on the right choice of lignocellulosic refuse when liquid state is maintained. The steady drop in the biocatalyst synthesis beyond optimum temperature may be attributed to transfiguration of the enzymes and cell growth arrest (Shazia et al. 2010). It was observed that cellulase production by *Aspergillus fumigates* gradually increased from 25 °C and catch up to maximum (CMCase 2.36 IU/ml, FPase 0.256 IU/ml) at 30 °C. Further elevated temperatures resulted in immoderate fall in cellulase production (Sarkar and Aikat 2014). Studies on thermophilic fungi by Kawamori et al. (1987) disclosed that the cultivation of

T. aurantiacus on alkali-treated bagasse at 45 °C assisted nearly 70 U/ml of CMCase following 96 h of incubation. The temperature optima for growth filter paper activity and β-glycosidases activity by *T. Aurantiacus* were recorded to be 45, 40 and 70 °C, respectively (Chin and Cole 1982).

11.11 Effect of Inoculum Size

The inoculum density is a decisive factor that commands and may either clips or extends the early lag phase of microbial growth curve. Surplus nutrient and lower inoculum ratio often leads to extravagant mycelia blooming and retarded enzyme production (Sharma et al. 1996; Haq et al. 2003). An exceeded microbial input could permit raised percentage moisture and ceasing in microbial cell synthesis and biocatalyst release, whereas diminished inoculum size may insist a longer incubation for bioprocessing to achieve the target metabolite of interest (Baysol et al. 2003).

In a study of Azzaz et al. (2012), 4% inoculum fulfilled cellulase production on modified medium using *Aspergillus niger*. Similarly, Alam et al. (2005) brought into light that the elated cellulase activity of 0.043 units was acquired when cell density of 5% (v/w) of *Trichoderma harzianum* was used on fermented oil palm biomass. As reported by Omojasola et al. (2008), the cellulase activity tends to shrink at microbial cell density exceeding 6% and 8% for pineapple peel and pulp fermentation, respectively, by *Aspergillus niger*. The decline in cellulase production with rise in inoculum might be due to aggregation of cells which could have reduced sugar and oxygen intake and enzyme release.

Aspergillus niger spores with 1×10^8 were inoculated into flasks and were incubated at 30 °C on a rotary shaker at 180 rpm for 7 days. The organism produced higher amount of cellulase with 2.478, 2.632 and 9.84 U/ml of FPase on wheat bran, rice bran and mixture of rice bran and wheat bran, respectively, in liquid-state fermentation (Praveen Kumar et al. 2015). As observed by Narasimha et al. (2006), *Aspergillus niger* gave maximum cellulase production on Czapek-Dox medium when inoculum size of 2.0×10^6 spores was used. Similarly, Sun et al. (2010) reported 2×10^8 spores of *Trichoderma* sp./flask (500 mL) were suitable for cellulase production. The optimal inoculum volume for maximum cellulase (FPase 0.344 IU/ml and CMCase 2.50 IU/ml) was 7% (v/v) containing 10^6 spores per ml beyond the limit which results in declined enzyme production attributed by the deprivation of nutrient supplements accessible for the rapid biomass and accelerated microbial growth. Optimal inoculum density of 8% v/v (12.41 mg cells/ml) for CMCase production by *Humicola insoles* was recorded by Riaz et al. (2014).

11.12 Effect of Incubation Time

The growth curve of microorganisms and the production of hydrolytic enzymes occur simultaneously; the activity increases until optimum incubation time is reached and later becomes steady or declines (Sachslehner et al. 1998). Distinct incubation period may be observed for the production of different enzymes (Smitt et al. 1996). Transient length of time affordability for cheaper enzyme biosynthesis was observed (Sonjoy et al. 1995). In an experimental study, Ojumu et al. (2003) noticed that *A. flavus* gave the highest cellulase activity when cultivated and grown on saw dust, bagasse and corn cob at 12 h of fermentation. Time duration of 72 h was set for growth of *A. Niger* to release cellulase (Azzaz et al. 2012; Gautam et al. 2010; Akinyele and Olaniyi 2013). Provision of 96 h of fermentation time was found to be favourable for enhanced cellulase activity by *Trichoderma* spp. (Khan et al. 2007). The incubation period for optimum enzyme production was correlated to substrate concentration. In the case of *P. nalgiovense*, the optimum CMCase activity in 2% pretreated wheat pollard was attained at 3 days incubation time, and then later the enzyme production was ceased due to confined nutrient concentration. When 3% and 4% pollard was used as substrates, optimal CMCase was achieved after 4 days. At further higher concentration of substrate, the culture demands longer incubation time due to less penetration of oxygen (Purwadaria et al. 2004). In the investigative reveals made by Nathan et al. (2014), *T. reesei* took 7 days of incubation to give elated FPase (0.38 U/ml) and CMCase (0.52 U/ml) activities which declined at later intervals of time. The probable speculation for weakened enzyme activity after stretched fermentation duration may be ascribed to loss of enzyme stability or self-death of the mycelia (Nipa et al. 2006).

Maximum CMCase (2.31 IU/ml) and FPase (0.261 IU/ml) were produced after 5 days of incubation beyond which induced decrease in the enzyme production (Sarkar and Aikat 2014). While in an examination made by Sun et al. (2010), enzyme activity by *Trichoderma* sp. was best at 120 h in SmF utilising apple pomace. Therefore, it is concluded that appropriate time length permits peak microbial growth and product formation to a defined level in a bioprocess approach.

11.13 Effect of Agitation Rate

Cellulase production in general was boosted with elevated shaking speed. This might be justified by the certainty that the rotational rate raised the transport of air into the microbial cells which is crucial for components of their cell membrane and synchronous of the medium components, essentially nutrients and products of catabolism (Rajagopalan and Krishnan 2008). The favourable fermentation factors for the enzymatic breakdown of brewer's spent grain were established to be 2% (w/v) substrate, 45 FPU/g maintained at 100 rpm. Under these conditions, 99.4% of cellulose bioconversion and 93.1% of glucose yield were achieved (Mussatto et al. 2008).

In liquid-state cultures, stirring speed of 180 RPM was noted to be optimal for the production of cellulase enzymes with fungal strain *T. viride* CMIT35 (Vintila et al. 2010). Cellulase production supported by 200 rpm and oil palm empty fruit bunch inoculated with *A. terreus* was four times improved as compared to the static condition (Shahriarinour et al. 2011). *Penicillium* sp. LM-HP33 and *Aspergillus* sp. LH-HP32 were the best alkaline cellulase (FPase) producers (>3 U/ml) in liquid cultures when incubated at 28 °C kept under 175 rpm agitation rate (Vega et al. 2012). The cellulolytic fungal culture designated as PSSI-3 isolated from the paper industry soil sample was noticed to show maximum CMCase activity when the cultivation was carried out at 120 rpm for 3 days (Lekh Ram et al. 2014). As regards agitation speed, 150 rpm resulted in maximum CMCase (2.40 IU/ml) and FPase (0.278 IU/ml) production. Agitation rates fewer than 150 rpm resulted in declined cellulase yields. The obstructing factor may be the insufficient dissolved oxygen level for cell growth. Higher agitation rates resulted in a negligible decline in enzyme levels, which could be due to mycelia destruction (Sarkar and Aikat 2014).

11.14 Effect of Carbon Source

The detrimental factor in any fermentation process is the carbon supplements which show its impact critically on growth and production of the desired product. Carbon sources may have either hindrance or stimulation on enzyme production. In another study on *T. reesei* C5, peak cellulase enzyme production and growth was accomplished exercising lactose as solitary carbon additive (Muthuvelayudham et al. 2004). Hartree et al. (1988) and Hanif et al. (2004) proved that growth of *Trichoderma harzianum* and *A. niger* on cellulosic residues resulted in increased cellulase production. Triggering of cellulase production by trehalose was demonstrated in *Clostridium*. (Thirumade et al. 2001). Cellulose induced cellulase synthesis by tenfold while glucose acted as inhibitor (Jahangeer et al. 2005). In the course of the growth *Trichothecium roseum*, an investigated fungus showed maximum total cellulolytic activity (1.87 FPU/ml) and extracellular protein content (234 μg/ml) on potato dextrose yeast extract broth medium amended with 1% (w/v) CMC (Shanmugam et al. 2008). Szakacs et al. (2006) and Baig (2005) opened up that fructose and glucose repress the performance of enzyme activity, whereas CMC, Avicel and lactose provoked *Trichoderma* spp. to release cellulase. Rashid et al. (2009) revealed *T. reesei* when grown on effluent from palm oil mill produced optimum cellulase at 0.5–1.5% cellulose as substrate.

The results which were drawn by Gautam et al. (2010) show that exoglucanases of 2.68 U/ml, endoglucanase of 2.17 U/ml and β-glycosidases of 2.06 U/ml were derived from filtrates comprising 1.0% sucrose succeeded by glucose, cellulose, maltose and CMC. The inclusion of glucose in medium affected pronounced repression of cellulose synthesis by *Streptomyces albaduncus and Aspergillus niger* (Jyostna et al. 2015; Narasimha et al. 2006). Xylose at 3% amended with 1% cellulose as substrate exhibited raised FPase and CMCase while it was repressed when cellulose was used as the one and only carbon supplement (Srilakshmi et al. 2017).

11.15 Effect of Nitrogen Source

Nitrogen is one of the primary constituents of proteins, and stimulatory effect of ammonium salt on cellulose activity might be related to its forthright access into synthesis of proteins (Mandels 1975). The necessity of specific nitrogen source varies from organism to organism or even among the same species for maximum enzyme production (Balaji and Sharma 2011). The fundamental requirement to be satisfied for optimal growth of any organism is the supplementation of amenable source of nitrogen. The highest cellulase activity of 87.43 U/ml was achieved in yeast extract culture. According to the assessment of Jyostna et al. (2015), the highest activities of FPase, CMCase and β-glycosidases with 4.8 IU/ml, 5.3 IU/ml and 1.6 IU/ml were accomplished with urea followed by peptone as nitrogen source. The most effective nitrogen source for FPase activity on the seventh day and CMCase on the third day of incubation by *Trichoderma viride* VKF3 was traced out to be peptone (Nathan et al. 2014). At 0.2%, tryptone concentration maximum CMCase (2.39 IU/ml) and FPase (0.446 IU/ml) activities were obtained from *A. fumigatus* (Sarkar and Aikat 2014). As per the conclusions dragged by Akinyele and Olaniyi (2013), locust beans proved efficient for the far-reached yield of cellulase activity of 0.36 μmol/min/ml, followed by soybeans, and the dropped cellulose activity was seen particularly with cotton seeds and ammonium sulphate.

Meat extract was proved to be unbeatable nitrogen supplement producing higher levels of cellulase activity by *A. niger* (0.097 U/ml), indicating that organic nitrogen should be added for better outcome (Azzaz et al. 2012). The medium amended with $NaNO_3$ as the nitrogen substituent established that mutant *T. viride* 1433 was stimulated to release higher amounts of cellulases (Khare and Upadhyay 2011). The combination of apple pomace and corn-steep solid induced towering enzyme production by *Trichoderma* sp. G/M 3.0010 (Sun et al. 2010). Peptone and yeast extract acted as the best organic nitrogen sources for *Aspergillus* sp. *and Fusarium* sp., respectively, yielding maximum endoglucanase (40–43 U/ml) when 0.06% (w/v) KNO_3 was added (Chellapandi and Abha 2009). In an explorative study carried out by Narasimha et al. (2006), *Aspergillus niger* dispersed on Czapek-Dox medium with 0.03% urea resulted in peak cellulase activity (1.682 U/ml) compared to peptone and $NaNO_3$. Ammonium sulphate appeared to be the best followed by ammonium nitrate, peptone and urea using *Aspergillus terreus* AV49 on pretreated ground nutshell exhibiting an elevated endoglucanase activity (2.833 IU/ml) and exoglucanase activity (0.282 FPU/ml) (Vyas and Vyas 2005).

11.16 Effect of Lignocellulosic Substrates

One of the foremost troubles in cellulase production by fermentation is the employment of overpriced raw materials. Trimming the cost of the substrate could be assisted by the alteration of green cellulosic deposits using microorganisms capable to generate elated cellulose production (Kotchoni and Shonukan 2002). The society prefers to adapt bio-based economy rather than fossil-depended thrift. It is under argument as to how a sustainable bio-economy can be established, where natural resources such as land, green residues and water are employed in the most productive path. Proficient utilisation of renewable lignocellulosic materials for the production of bio-based products and bioenergy would show profits to the environment, local economy and safety to national energy pools (Zhang 2008; Padmavathi et al. 2012). Agro-wastes are the most copious and inexhaustible deposits produced on earth. Forests, agricultural practices and industrial processes contribute huge quantities of agro-wastes specifically from agriculture-associated manufactories like timber, textile paper, breweries and pulp industries (Ilyas et al. 2012). The biological path seems to be very alluring and feasible for enzymes production from this lignocellulosic biomass because of various rationales, the uppermost being the pervasive and inexhaustible character of natural reserves and its competitiveness with consumable produce (Singhania et al. 2010).

Utilisation of natural and cheaper sources makes a possible route for significant kick to the rate of cellulase production (Ozioko et al. 2013). Economic analyses stipulate that fragmentation of cellulosic materials to simple sugars remains bonded with the production cost (Xu et al. 2011). A variety of unused cellulosic materials are explored to arrive at beneficial approaches associated with cellulase production by a vast array of cellulolytic fungi (Chinedu et al. 2011). The following table summarises the various lignocellulosic substrates used for fungal cellulase production (Table 11.2).

11.17 Effect of Surfactants

In an investigation study performed by Singh et al. (2007), the surfactants are reported to be hydrophilic and hydrophobic compounds whose accumulation at interface of immiscible fluids tends to minimise surface and coherence tensions, thereby increasing the mobility, solubility and bioavailability and ultimately biotransformation of insoluble organic or hydrophobic compounds. It has been put forward that the hydrophilic and hydrophobic part of the surfactant is responsible for steric hindrance and binding to lignin and consequently arresting the enzymes from unproductive binding with lignin which results inaccessibility of more enzymes for cellulose hydrolysis (Borjesson et al. 2007). Tween-80 ossifies unstable cellulase components during hydrolysis and enzyme production (Okino et al. 2013). Addition of surfactant like polyethylene glycol (PEG) has been shown to be vital to enhance the enzymatic conversion of the lignocellulosic substrate. Tween-80 at 0.02% (v/v) showed higher cellulase, protein and fungal biomass production (Srilakshmi et al. 2017). The production of cellulases was doubled in a fermentation using OPEFB carried out by the addition of Tween-80 as a surfactant compared to fermentation devoid of surfactant (Shahriarinour et al. 2011).

The use of Tween-80 is fruitful because it does not disturb the enzyme nature. Tween-80 (2 ml/L) was supportive for the production of cellulases and β-glycosidase by mixed culture of *Trichoderma reesei* and *Aspergillus phoenicis* grown on dairy manure (Wen et al. 2005). Incorporation of 1% and 0.2% (v/v) of Tween-80 induced

Table 11.2 List of various lignocellulosic substrates for fungal cellulase production

Agro-industrial residue	Microorganisms (fungal strains)	References
Coconut shell	*Aspergillus niger*	Coelho et al. (2001)
Grape marc	*Aspergillus phoenicis*	Silva (2008)
Grape marc	*Aspergillus awamori*	Botella et al. (2005)
Grape marc	*Monascus purpureus*	Daroit et al. (2007)
Grape marc	*Monascus purpureus*	Silveira et al. (2008)
Grape mark and orange peel	*Aspergillus awamori*	Diaz et al. (2012)
Wheat bran	*Trichoderma harzianum*	Haq et al. (2006)
Wheat straw	*Aspergillus heteromorphus*	Singh et al. (2009)
Banana agro-waste	*Trichoderma lignorum*	Baig (2005)
Sugar beet pulp	*Trichoderma reesei*	Nasab and Nasab (2007)
Groundnut shell	*Aspergillus terreus, Aspergillus nidulans, Trichoderma viride*	Vyas and Vyas (2005)
Sugarcane bagasse	*Trichoderma reesei QM9414, Aspergillus terreus SUK-1*	Massadeh et al. (2001)
Oat straw, wheat bran	*Thermoascus aurantiacus, Aspergillus niger*	Stoilova et al. (2005)
Sugarcane bagasse	*Humicola insolens TAS-13*	Javed et al. (2007)
Orange waste	*Saccharomyces cerevisiae*	Omojasola et al. (2008)
Palm kernel cake	*Rhizopus oryzae ME01*	Othman et al. (2013)
Solka-Floc	*Acremonium cellulolyticus* CF-2612	Fang et al. (2009)
Pea seed husk	*Purpureocillium lilacinum*	Srilakshmi et al. (2017)
Milk pack	*Acremonium cellulolyticus* C-1	Park et al. (2011)
Solka-Floc	*Acremonium cellulolyticus* C-1	Ikeda et al. (2007)
Glucose and sugarcane bagasse	*Aspergillus niger* A12	Cunha et al. (2012)
Pretreated sugarcane bagasse	*Penicillium funiculosum*	Maeda et al. (2011)
Corn cob	*Trichoderma reesei* 2U-02	Liming and Xueliang (2004)
Pretreated sugarcane bagasse and sucrose	*Trichoderma harzianum* P49P11	Delabona et al. (2012)

peak cellulase production by *Trichoderma reesei* QM-9414 and *Streptomyces flavogriseus*, respectively (Hari Krishna et al. 2000). Tween-80 also acted as a good surfactant in the case of CMCase production by *Aspergillus glaucus* and *Trichoderma viride* (Chang et al. 2006; Liu et al. 2006). Thus, it looks for various reasons; Tween-80 has the potential for cutting the price of the biocatalyst-based utilisation of cellulosic depositions. On the contrary, Micales (Micales 1991) proved that the inclusion of Tween-80 had no impact on the secretion of CMCase by *Postia placenta*.

11.18 Applications of Cellulases

Cellulase occupies the third position in the world enzyme market and is grabbing rejuvenated interests because of their boundless and diverse range of applications, thereby dominating the rest of the plant cell wall degrading enzymes (Chandel et al. 2012; Singhania et al. 2010). Cellulases of microbial origin have been exploited for a broad range of industrial applications. Moreover, these enzymes have been economically feasible for more than 30 years now and emerged as a goal for academic and industrial exploration (Singh et al. 1999, 2007). Currently, cellulases from *Trichoderma* and *Aspergillus* account for roughly 20% of the world enzyme market (Bhat 2000). The following table highlights some of the large-scale benefits of cellulases.

Industry	Application	References
Food	Filtration and clarification of fruit juices, extraction of oil seeds	Bhat and Bhat (1997), Bhat (2000) and Mussatto et al. (2007)
Animal feed	To improve feed nutritive value for animal growth	Bhat (2000)
Textile	To improve tenderness and water uptake of the fibres and to reduce the ability to form pills and providing a luminous surface with reduced piles	Sreenath et al. (1996)
Detergent	To remove rough protrusions for a fluffy, glossy and glistening fabric	Bhat (2000)
Paper and pulp	To enhance deinking of paper, pulp solubility	Bhat (2000)
Biofuel	Bio-transformation of cellulose to ethanol	Goldschmidt (2008) and Zhang et al. (2006)
Laundry	Biopolish tissue process	Bon et al. (2008)
Agriculture	To control the plant diseases by degradation of the cell wall of plant pathogens	Bhat (2000)
Wine and beer	To increase the extraction, filtration rate, reducing pressing time and must	Galante et al. (1998)
Others	generate plant protoplast for genetic engineering to control industrial slime production of cellulase-based chitosan especially with immunomodulatory, antitumor and antibacterial activities	Liu and Zhu (2000), Wiatr (1990), Qin et al. (2004) and Wu and Tsai (2004)

11.19 Future Outlook: The Challenges In Cellulase Research

Hidden aspects related to the efficient development of the biocatalyst and advancement of technical knowledge for yielding and employment of cellulase brought into light by the efforts of numerous explorative works. No solitary methodologies stood cheaper and efficient in the transformation of the native cellulosic material for the generation of worthy target outputs or biofuel. In the present trend, biotransformation of lignocellulosic waste by the use of cellulase build-ups available from the market is not economically advantageous.

The following are prime objectives for forthcoming investigations on cellulose:

1. Upgrading the activity of cellulase for greater potential of decreasing the concentration of the desired enzyme (Qin et al. 2004).
2. Trimming cellulase production cost.
3. Economic feasibility of cellulase production by guarded genetic modifications into the metabolisms of cellulose degraders for upgraded productions.
4. Negating the feedback loops by glucose and achievements in cohesive bioprocess for the production of cellulases.
5. More fundamental research is required to produce designer enzymes good enough for clear-cut applications.
6. Extended studies on enhancement of cellulase action or conferring of required attributes to enzymes via manipulation of protein structures.

   The aspects regarding the bioconversion of cellulose materials become the bottom line of forthcoming researches associated with cellulase and cellulose degraders. The problem which raises the attention is not restricted to enzyme production alone but a concerted effort to figure out the basic physiology of cellulolytic microbes, and the utilisation of this knowledge merged with the engineering principles to achieve greater heights in processing and exploitation of the most copious innate reserves. The aspects disclosed to consideration include pretreatment methods of cellulosic materials for an easy microbial access, cost-effective approaches for production of hydrolytic enzyme products, organism development of metabolic engineering and eventually protein engineering to improve the characteristics of enzymes to increase their specific activities, process tolerance and stability (Sukumaran et al. 2005).

11.20 Conclusion

Among the various commercially available cellulases, fungal cellulases gain a cutting-edge advantage owing to pronounced yield, efficiency, effortless crystallisation and a higher degree of purity. In the development of any bioprocess technology, choosing efficient, modest, and handy raw materials for the production of biocatalyst besides optimisation of media components and fermentation circumstances is prerequisite due to the fact that the production of protein machines of microbial origin has an immense striking effect on the entire worth of the operating processes.

References

Adeleke AJ (2013) Endoglucanase production by *Penicillium atrovenetum* using plantain peels as substrate. Assumpt Univ J Technol 16(3):140–146

Ahmed S, Bashir A, Saleem H, Saadia M, Jamil A (2009) Production and purification of cellulose degrading enzymes from a filamentous fungus *Trichoderma harzianum*. Pak J Bot 41 (3):1411–1419

Akinyele JB, Olaniyi OO (2013) Investigation of the cellulase production by *Aspergillus niger* nspr 002 in different cultivation conditions. Innov Rom Food Biotechnol 13:71–79

Akinyele B, Fabunmi AO, Olaniyi OO (2013) Effect of variations in growth parameters on cellulase activity of *Trichoderma viride* NSPR006 cultured on different wood – dusts. Malays J Microbiol 9(3):193–200

Alam (2011) Factors affecting endoglucanase production by *Trichoderma reesei* RUT C-30 from solid state fermentation of oil palm empty fruit bunches using Plackett-Burman design. Afr J Biotechnol 10(46):9402–9409

Alam MZ, Muhammad N, Mahmat ME (2005) Production of cellulase from oil palm biomass as substrate by solid state bioconversion. Am J Appl Sci 2(2):569–572

Alberton LR, Vandenberghe LPDS, Assmann R, Fendrich RC, Rodriguez-Leon J, Soccol CR (2009) Xylanase production by *Streptomyces viridosporus* T7a in submerged and solid state fermentation using agro-industrial residues. Braz Arch Biol Technol 52:171–180

Amir I, Zahid A, Yusuf Z, Igbal H, Aish M, Mohammed I, Sajid M (2011) Optimization of cellulase enzyme production from corn cobs using *Alternaria alternata* by solid state fermentation. J Cell Mol Biol 9(2):51–56

Andersen N (2007) Enzymatic hydrolysis of cellulose. PhD dissertation, Department of Chemical Engineering, Technical University of Denmark, Copenhagen, Denmark

Araugo A, Souza JD (1986) Characterization of cellulolytic enzyme components from *Aspergillus terreus* and its mutant derivatives. J Ferment Technol 64:463–467

Azzaz HH, Murad HA, Kholif AM, Hanfy MM, Abdel Gawad MH (2012) Optimization of culture conditions affecting fungal cellulase production. Res J Microbiol 7(1):23–31

Baig MMV (2005) Cellulolytic enzymes of *Trichoderma lignorum* produced on banana agro-waste: optimisation of the culture medium and conditions. J Sci Ind Res 64:57–60

Balaji BK, Sharma P (2011) An –alkalithermotolerant extracellular protease from newly isolated *Streptomyces* sp.DD2. New Biotechnol 28:725–732

Baysol Z, Uyar F, Aytekin C (2003) Solid state fermentation for production of alpha amylase by a thermotolerant *Bacillus subtilis* from hot spring water. Process Biochem 38:1665–1668

Beguin P, Aubert J-P (1994) The biological degradation of cellulose. FEMS Microbiol Rev 13 (1):25–58

Bharathi K, Ravindra P (2006) Pretreatment studies of rice bran for the effective production of cellulase. EJEAF Che 5(2):1253–1264

Bhat MK (2000) Cellulases and related enzymes in biotechnology. J Biotechnol Advances 18:355–383

Bhat M, Bhat S (1997) Cellulose degrading enzymes and their potential industrial applications. Biotechnol Adv 15:583–620

Bhat MK, Hazlewood GP (2003) Enzymology and other characteristics of cellulases and xylanases. In: Bedford MR, Partridge GG (eds) Enzymes in farm animal nutrition. CAB International, Cambridge, pp 11–60

Bon EPS, Ferrara MA, Corvo ML (2008) Enzymes in biotechnology: production, applications and market. Rio de Janeiro. Interciencia; UFRJ: CAPES: FAPERJ: FCT (Portugal), Rio de Janeiro

Borjesson J, Engqvist M, Sipos B, Tjerneld F (2007) Effect of poly(ethylene glycol) on enzymatic hydrolysis and adsorption of cellulase enzymes to pre-treated lignocellulose. Enzyme Microbial Technol 41(1–2):186–195

Botella I, De Ory I, Webb C, Cantero D, Blandino A (2005) Hydrolytic enzyme production by *Aspergillus awamori* on grape pomace. Biochem Eng J 26:100–106

Bozell JJ, Petersen GR (2010) Technology development for the production of bio- based products from biorefinery carbohydrates-The US Department of Energy's "Top10" revisited. Green Chem 12(4):525–728

Chandel AK, Chandrasekhar G, Silva MB, da Silva SS (2012) The realm of cellulases in biorefinery development. Crit Rev Biotechmol 32(3):187–202

Chang X, Minnan L, Xiaobing W, Huijuan X, Zhongan C, Fengzhang Z, Liangshu X (2006) Screening and characterization of the high cellulase producing strain *Aspergillus glaucus* XC9. Front Biol China 1:35–40

Chellapandi P, Abha AJ (2009) Enhanced endoglucanase production by soil isolates of *Fusarium* sp. and *Aspergillus* sp. through submerged fermentation process. Turk J Biochem 34 (4):209–214

Chin TC, Cole Anthony LJ (1982) Cellulase production by the thermophilic fungus, Thermoascus aurantiacus. Pertanika 5(2):255–262

Chinedu NS, Okachi VJ, Smith HA, Okafor UA, Onyegema Okerenta BM, Omidiji O (2007) Effect of carbon sources on cellulase production by *Penicillium chrysogenum* PCL 501.*Afr.* J Biochem Res 1(1):006–010

Chinedu SN, Okochi VI, Omidiji O (2011) Cellulase Production by wild strains of *Aspergillus niger*, *Penicillium chrysogenum* and *Trichoderma harzianum* grown on waste cellulosic materials. IFE J Sci 13(1):57–62

Coelho MAZ, Leite SGF, Rosa MF, Furtado AAL (2001) Utilization of agro-industrial residues: production of enzymes from the green coconut shell. CEPPA News lett 19:33–42

Coral G, Arikan B, Unaldi MN, Guvenmes H (2002) Some properties of crude carboxy-methyl cellulase of *Aspergillus niger* Z10 wild-type strain. Turk J Biol 26:209–213

Couto SR, Sanromán MA (2006) Application of solid-state fermentation to food industry-a review. J Food Eng 76:291–302

Cunha FM, Esperanca MN, Zangirolami TC, Badino AC, Farinas CS (2012) Sequential solid-state and submerged cultivation of *Aspergillus niger* on sugarcane bagasse for the production of cellulase. Bioresour Technol 112:270–274

Daroit DJ, Silveira ST, Hertz PF, Brandelli A (2007) Production of extracellular β-glucosidase by *Monascus purpureus* on different growth substrates. Process Biochem 42:904–908

Dashtban M, Schraft H, Qin W (2009) Fungal bioconversion of lignocellulosic residues; opportunities & perspectives. Int J Biol Sci 5(6):578–595

Davidson A, Blaxter M (2005) Ancient origin of glycosyl hydrolase family 9 cellulase genes. Mol Biol Evol 22(5):1273–1284

Davies GJ, Wilson KS, Henrissat B (1997) Nomenclature for sugar-binding sub-sites in glycosyl hydrolases. Biochem J 321:557–559

Davies GJ, Gloster TM, Henrissat B (2005) Recent structural insights into the expanding world of carbohydrate-active enzymes. Curr Opin Struct Biol 15(6):637–645

Delabona PS, Farinas CS, Silva MR, Azzoni SF, Pradella JGC (2012) Use of a new *Trichoderma harzianum* strain isolated from the Amazon rainforest with pre-treated sugar cane bagasse for on-site cellulase production. Bioresour Technol 107:517–521

Diaz AB, De Ory I, Caro I, Blandino A (2012) Enhance hydrolytic enzymes production by *Aspergillus awamori* on supplemented grape pomace. Food Bioprod Process 90(1):72–78

Dincer A, Telefoncu A (2006) Improving the stability of cellulase by immobilization on modified polyvinyl alcohol coated chitosan beads. J Mol Catal B Enzyme 45:10–14

Divne C, Stahlberg J, Reinikainen T, Ruohonen L, Pettersson G, Knowles JKC, Teeri TT, Jones TA (1994) The three-dimensional crystal structure of the catalytic core of cellobiohydrolase I from *Trichoderma reesei*. Science 265:524–528

El-Hadi A, Abu S, El-Nour B, Ali Hammad B, Zeinat Kamel C, Mai Anwar B (2014) Optimization of cultural and nutritional conditions for carboxymethylcellulase production by *Aspergillus hortai*. J Radiat Res Appl Sci 7(1):23–28

Elzaher AFH, Fadel M (2010) Production of bio-ethanol via enzymatic saccharification of rice straw by cellulase produced by *Trichoderma reesei* under solid state fermentation. Int J Appl Microbiol Biotechnol Res 43(3):72–78. New York Science

Fang X, Yano S, Inoue H, Sawayama S (2009) Strain improvement of *Acremonium cellulolyticus* for cellulase production by mutation. J Biosci Bioeng 107(3):256–261

Galante YM, DeConti A, Monteverdi R (1998) Application of *Trichoderma* enzymes in food and feed industries. In: Harman GF, Kubicek CP (eds) *Trichoderma* and *Gliocladium*—Enzymes, vol. 2 of Biological control and commercial applications. Taylor & Francis, London, pp. 311–326

Gautam SP, Bundela PS, Pandey AK, Awasthi MK, Sarsaiya S (2010) Screening of cellulolytic fungi for management of municipal solid waste. J Appl Sci Environ Sanit 5(4):391–395

Gilkes NR, Henrissat B, Kilburn DG, Miller RC, Warren RAJ (1991) Domains in microbial β-1,4-glycanases: sequence conservation, function, and enzyme families. Microbiol Rev 55 (2):303–315

Gilna VV, Khaleel KM (2011) Biochemistry of cellulase enzyme activity of *Aspergillus fumigatus* from mangrove soil on lignocellulosics substrate. Recent Res Sci Technol 3(1):132–134

Goldschmidt F (2008) From cellulose to ethanol: engineering microorganisms to produce biofuel. Institute of Biogeochemistry and Pollutant Dynamics, pp 1–17

Gomes I, Gomes J, Gomes DJ, Steiner W (2000) Simultaneous production of high activities of thermostable endoglucanase and beta-glucosidase by the wild thermophilic fungus *Thermoascus aurantiacus*. Appl Microbiol Biotechnol 53(4):461–468

Gupta RP, Gigras H, Mahopatra VK, Goswani, Chauhan B (2003) Microbial α-amylase. A biotechnological perspective. Process Biochem 38:1599–1616

Haltrich D, Nidetzky B, Kulbe KD, Steiner W, Zupaneie S (1996) Production of fungal xylanases. Bioresour Technol 58(2):137–161

Hanif A, Yasmeen A, Rajoka MI (2004) Induction, production, repression, and de-repression of exoglucanase synthesis in *Aspergillus niger*. Bioresour Technol 94(3):311–319

Haq I, Ali S, Qadeer MA, Iqbal J (2003) The kinetics basis of the Ca^{+2} ions for the yield of citric acid in a repeated batch cultivation system. World J Microbiol Biotechnol 19(8):817–823

Haq I, Shahzadi K, Hameed U, Javed MM, Qadeer MA (2006) Solid state fermentation of cellulases by locally isolated *Trichoderma harzianum* for the exploitation of agricultural byproducts. Pak J Biol Sci 9(9):1779–1782

Hari Krishna S, Sekhar Rao KC, Suresh Babu J, Srirami Reddy D (2000) Studies on the production and application of cellulase from *Trichoderma reesei* QM-9414. Bioprocess Biosyst Eng 22 (5):467–470

Hartree MM, Hogan CM, Saddler JN (1988) Influence of growth substrate on production of cellulase enzymes by *Trichoderma harzianum* E-58. Biotechnol Bioeng 31(7):725–729

Heptinstall J, John C, Stewart MS (1986) Fluorimetric estimation of exo-cellobiohydrolase and β-D- glucosidase activities in cellulase from *Aspergillus fumigates Fresenius*. Enzym Microb Technol 8(2):70–74

Hilden L, Johansson G (2004) Recent developments on cellulases and carbohydrate-binding modules with cellulose affinity. Biotechnol Lett 26(22):1683–1693

Himmel ME, Ding SY, Johnson DK, Adney WS, Nimlos MR, Brady JW, Foust TD (2007) Biomass recalcitrance: engineering plants and enzymes for bio fuels production. Science 315:804–807

Howard RL, Abotsi E, Jansen van Rensburg EL, Howard S (2003) Lignocellulose biotechnology: issues of bioconversion and enzyme production. Afr J Biotechnol 2(12):602–619

Ikeda Y, Hayashi H, Okuda N, Park EY (2007) Efficient Cellulase Production by the filamentous fungus *Acremonium cellulolyticus*. Biotechnol Prog 23(2):333–338
Ilyas U, Ahmed S, Majeed A, Nadeem M (2012) Bihydrolysis of Saccharum Spontaneum for cellulase production by *Aspergillus terreus*. Afr J Biotechnol 11:4914–4920
Iqbal HMN, Asgher M, Ahmed I, Hussain S (2010) Media optimization for hyper-production of carboxymethyl cellulase using proximally analyzed agro-industrial residue with *Trichoderma harzianum* under SSF. Int J Agro Vet Med Sci 4:47–55
Jagtap S, Rao M (2005) Purification and properties of a low molecular weight 1,4-beta-d-glucan glucohydrolase having one active site for carboxymethyl cellulose and xylan from an alkalothermophilic *Thermomonospora* sp. Biochem Biophys Res Commun 329(1):111–116
Jahangeer S, Khan Jahangeer NS, Sohail M, Shahzad S, Ahmad A, Khan SA (2005) Screening and characterization of fungal cellulases isolated from the native environmental source. Pak J Bot 37 (3):739–748
Javed MM, Khan TS, Ul-Haq I (2007) Sugar cane bagasse pretreatment: an attempt to enhance the production potential of cellulose by *Humicola insolens* TAS-13. Elec J Environ Agric Food Chem 6:2290–2296
Joshi V, Pandey A (1999) Biotechnology: food fermentation: microbiology, biochemistry, and technology. Educational Publishers & Distributors, New Delhi
Jyostna PK, Ramakrishna Rao A, Devaki K (2015) Effect of nutritional factors on cellulase production by *Streptomyces albaduncus* from the gut of earthworm, *Eisenia foetida*. Pest Manag Hortic Ecosyst 21(1):75–80
Kang SW, Park YS, Lee JS, Hong SI, Kim SW (2004) Production of cellulases and hemicellulases by *Aspergillus niger* KK2 from lignocellulosic biomass. Bioresour Technol 91(2):153–156
Kathiresan K, Manivannan S (2006) Cellulase production by *Penicillium fellutanum* isolated from coastal mangrove rhizosphere soil. Res J Microbiol 1(5):438–442
Kawamori M, Takayama K-i, Takasawa S (1987) Production of cellulases by a thermophilic fungus, *Thermoascus aurantiacus* A- 131[+]. Agric Biol Chem 51(3):647–654
Kendry Mc P (2002) Energy production from biomass (part 1): overview of biomass. Bioresour Technol 83:37–46
Khan MMH, Ali S, Fakhrul-razi A, Alam MD (2007) Use of fungi for the bioconversion of rice straw into cellulase enzyme. J Environ Sci Health Part B 42(4):381–386
Khare A, Upadhyay RS (2011) Influence of some cultural factors on production of cellulase and β-1, 3-glucanase by the mutant strains of *Trichoderma viride* 1433. J Agric Technol 7 (2):403–412
Kim KC, Seung-Soo Y, Oh Young A, Seong-Jun K (2003) Isolation and characterization of *Trichoderma harzianum* FJ1 producing cellulase and xylanase. J Microbiol Biotechnol 13 (1):1–8
Kleman-Leyer K, Agosin E, Conner AH, Kirk TK (1992) Changes in molecular size distribution of cellulose during attack by white rot and brown rot fungi. Appl Environ Microbiol 58:1266–1270
Kotchoni OS, Shonukan OO (2002) Regulatory mutations affecting the synthesis of cellulase in B. Pumilus. World J Microbiol Biotechnol 18(5):487–491
Kraulis J, Clore GM, Nilges M, Jones TA, Pettersson G, Knowles J, Gronenborn AM (1989) Determination of the three-dimensional solution structure of the C-terminal domain of cellobiohydrolase I from *Trichoderma reesei*. A study using nuclear magnetic resonance and hybrid. Biochemistry 28(18):7241–7257
Latifian M, Esfahani ZH, Barzegar M (2007) Evaluation of culture conditions for cellulase production by two *Trichoderma reesei* mutants under solid state fermentation conditions. Bioresour Technol 98:3634–3637
Lekh R, Kuar K, Sharma S (2014) Screening isolation and characterization of cellulase producing microorganisms from soil. Int J Pharm Sci Invent 3(3):12–18
Liming X, Xueliang S (2004) High-yield cellulase production by *Trichoderma reesei* ZU-02 on corn cob residue. Bioresour Technol 91:259–262

Liu W, Zhu WM (2000) Production and regeneration of *Trichosporon cutaneum* protoplasts. Process Biochem 35:659–664

Liu J, Yuan X, Zeng G, Shi J, Chen S (2006) Effect of biosurfactants on cellulase and xylanase production by *Trichoderma viride* in solid substrate fermentation. Process Biochem 41 (11):2347–2351

Lynd LR, Laser MS, Bransby D, Dale BE, Davison B, Hamilton R, Himmel M, Keller M, McMillan JD, Sheehan J, Wyman CE (2008) How biotech can transform biofuels. Nat Biotechnol 26:169–172

Macris BJ (1984) Enhanced cellulases and β-glucosidases production by a mutant of *Aspergillus alternate*. Biotechnol Bioeng 26:194–196

Macris BJ (1986) Enhanced cellobiohydrolase production from *Aspergillus ustus* and *Trichoderma harzianum*. Enzym Microb Technol 68(3):141–144

Maeda RN, Serpa VI, Rocha VA, Mesquita RAA, Santa Anna IMM, Castro AM, Driemeier DE, Pereira N, Polikarpov I (2011) Enzymatic hydrolysis of pretreated sugar cane bagasse using *Penicillium funiculosum* and *Trichoderma harzianum* cellulases. Process Biochem 46 (5):1196–1201

Mandels M (1975) Microbial source of cellulase. Biotechnol Bioeng 5:81–105

Massadeh MI, Yusoff WMW, Omar O, Kader J (2001) Synergism of cellulase enzymes in mixed culture solid substrate fermentation. Biotechnol Lett 23:1771–1774

Micales JA (1991) Increased recovery of β-D-glucosidase from *Postia placenta* in presence of tween surfactants. Special print from: material and organisms 26. Heft Verlag Duncker and Humblot, 1000 Berlin, p.41. Microbiol Rev 13(1):25–58

Milala MA, Shugaba A, Gidado A, Ene AC, Wafar JA (2005) Studies on the use of agricultural wastes for cellulase enzyme production by *Aspergillus niger*. Res J Agric Biol Sci 1(4):325–328

Milala MA, Shehu BB, Zanna H, Omosioda VO (2009) Degradation of agro-waste by cellulase from *Aspergillus candidus*. Asian J Biotechnol 1(2):51–56

Murao S, Sakamoto R, Arai M (1988) Cellulases of *Aspergillus aculeatus*. Methods Enzymol 160:274–299

Mussatto I, Fernandes M, Ines Roberto C (2007) Lignin recovery from brewer's spent grain black liquor Solange. Carbohydr Poly 70(2):218–223

Mussatto S, Fernandes M, Milagres A, Roberto I (2008) Effect of hemicellulose and lignin on enzymatic hydrolysis of cellulose from brewer's spent grain. Enzym Microb Technol 43:124–129

Muthuvelayudham R, Viruthagiri T, Deiveegan S (2004) Enhanced cellulase production using mutant strain *Trichoderma reesei* growing on lactose in batch culture. CHEMCON, Ind Chem Engr Congress, pp 15–18

Narasimha G, Sridevi A, Viswanath B, Subhosh Chandra M, Rajasekhar Reddy B (2006) Nutrient effects on production of cellulolytic enzymes by *Aspergillus niger*. Afr J Biotechnol 5 (5):472–476

Nasab MM, Nasab MM (2007) Cellulase Production by *Trichoderma reesei* using sugar Beet Pulp. Iran Agric Res 25(1–2):107–116

Nathan VK, Rani ME, Rathinasamy G, Dhiraviam KN, Jayavel S (2014) Process optimization and production kinetics for cellulase production by *Trichoderma viride* KF3. Springer Plus 3:92

Nipa MN, Sultana S, Hakim MA (2006) Induction of cellulase biosynthesis by cellobiose octaacetate in *Aspergillus humicola*. Banglad J Microbiol 23(2):174–176

Nutor JRK, Converse AO (1991) The effect of enzyme and substrate levels on the specific hydrolysis rate of pretreated poplar wood. Appl Biochem Biotechnol 28:757

Ojumu TV, Solomon BO, Betiku E, Layokun SK, Amigun B (2003) Cellulase production by *Aspergillus flavus* Linn isolate NSPR 101 fermented in saw dust, bagasse and cornconb. Afr J Biotechnol 2(6):150–152

Okino S, Ikeo M, Ueno Y, Taneda D (2013) Effects of Tween-80 on cellulase stability under agitated conditions. Bioresour Technol 142:535–539

Omojasola PF, Jilani OP, Ibiyemi SA (2008) Cellulase production by some fungi cultured on pineapple waste. Nat Sci 6(2):64–79

Othman MF, Kalil MS, Sahri MM (2013) Solid State fermentation of Palm Kernel Cake (Pkc) by newly isolated *Rhizopus oryzae* Me01. Asian J Exp Biol Sci 4(1):84–88

Ozioko PC, Ikeyi AP, Ugwu OPC (2013) Cellulases, their substrates, activity and assay methods. The Experiment 12:778–785

Padmavathi T, Nandy V, Agarwal P (2012) Optimization of the medium for the production of cellulases by *Aspergillus terreus* and *Mucor plumbeu.* Eur J Exp Biol 2(4):1161–1170

Panda T, Bisaria VS, Ghose TK (1987) Effect of culture phasing and a polysaccharide on production of xylanase by mixed culture of *Trichoderma reesei* D1-6 and *Aspergillus wentii* Pt 2804. Biotechnol Bioeng 30(7):868–874

Pandey A (2003) Solid-state fermentation. Biochem Eng J 13:81–84

Papagianni M (1995) Morphology and citric acid production of *Aspergillus niger* in submerged culture. PhD thesis, University of Strathclyde

Papagianni M (2004) Fungal morphology and metabolite production in submerged mycelial processes. Biotechnol Adv 22(3):189–259

Park EY, Naruse K, Kato T (2011) Improvement of cellulase production in cultures of *Acremonium cellulolyticus* using pretreated waste milk pack with cellulase targeting for biorefinery. Bioresour Technol 102(10):6120–6127

Polyanna NH, Porto TS, Moreira KA, Pinto GAS, Cristina MSM, Ana LFP (2011) Cellulase production by *Aspergillus japonicus* URM5620 using waste from castor bean (Ricinus communis L.) under solid state fermentation. Appl Biochem Biotechnol 165:1057–1067

Portjanskaja E, Preis S, Kallas J (2006) Aqueous photocatalytic oxidation of lignin and humic acids with supported TiO2. Int J Photoene 8:1–7

Prasanna HN, Ramanjaneyulu G, Rajasekhar Reddy B (2016) Optimization of cellulase production by *Penicillium* sp. *3*. Biotech 6(2):162

Praveen Kumar Reddy G, Narasimha G, Kanderi Dileep Kumar G, Ramanjaneyulu A, Ramya B, Shanti Kumari S, Rajasekhar Reddy B (2015) Cellulase production by *Aspergillus niger* on different natural lignocellulosic substrates. Int J Curr Microbiol Appl Sci 4(4):835–845

Purwadaria T, Agnes Kumalasari T, Tutiharyati, Pius Ketaren P, Arnold Sinurat P (2004) Optimization of cellulase production with *Penicillium nalgiovense* all grown on pretreated wheat pollard. Biotropia 23:1–12

Qi BC, Aldrich C, Lorenzen L, Wolfaardt GW (2005) Acidogenic fermentation of lignocellulosic substrate with activated sludge. Chem Eng Commun 192(9):1221–1242

Qin C, Zhou B, Zeng L, Zhang Z, Liu Y et al (2004) The physicochemical properties and antitumor activity of cellulase-treated chitosan. Food Chem 1:107–115

Rajagopalan G, Krishnan C (2008) α-Amylase production from catabolite depressed *Bacillus subtilis* KCC103 utilizing sugarcane bagasse hydrolysate. Bioresour Technol 99(8):3044–3050

Rashid SS, Alam MZ, Karim MIA, Sallah MH (2009) Optimization of the nutrient supplements for cellulose production with the basal medium palm oil mill effluent. World Acad Sci Eng Technol 60:809–815F

Riaz H, Jabbar A, Rashid MH, Riaz S, Latif F (2014) Endoglucanase production by Humicola insolens: effect of physiochemical factors on growth kinetics and thermodynamics. Int J Agric Biol 16:1141–1146

Robinson T, Singh D, Nigam P (2001) Solid-state fermentation: a promising microbial technology for secondary metabolite production. Appl Microbiol Biotechnol 55:284–289

Sachslehner A, Nidetzky B, Kulbe KD, Haltrich D (1998) Induction of Mannanase, Xylanase, and Endoglucanase activities in *Sclerotium rolfsii*. Appl Environ Microbiol 64(2):594–600

Sajith S, Priji P, Sreedevi S, Benjamin S (2016) An overview on fungal cellulases with an industrial perspective. J Nutr Food Sci 6(13):1–13

Sanyal A, Kanda RK, Sinha SN, Dube DK (1988) Extracellular cellulolytic enzyme system of *Aspergillus japonicus*. ibid 10(2):85–90

Sarkar N, Aikat K (2014) *Aspergillus fumigates* NITDGPKA3 provides for increased cellulase production. Int J Chem Eng 5:1–9

Shafique S, Bajwa R, Shafique S (2009) Cellulase biosynthesis by selected *Trichoderma* species. Pak J Bot 41(2):907–916

Shahriarinour M, Wahab MNA, Mohamad R, Mustafa S, Ariff AB (2011) Effect of medium composition and cultural condition on cellulase production by *Aspergillus terreus*. Afr J Biotechnol 10(38):7459–7467

Shanmugam P, Mani M, Narayanasamy M (2008) Biosynthesis of cellulolytic enzymes by *Trichothecium roseum* with citric acid mediated induction. Afr J Biotechnol 7(21):3917–3921

Sharada R, Venkateswarlu G, Narsi Reddy M, Venkateshwar S, Anand Rao M (2012) Production of cellulase by solid state fermentation. Int J Pharm Res Dev 4(1):224–230

Sharma A, Milstein O, Vered Y, Gressel J, Flowers HM (1985) Effects of aromatic compounds on hemicellulose-degrading enzymes in *Aspergillus japonicas*. Biotechnol Bioeng 27 (8):1095–1101

Sharma DK, Tiwari M, Behere BK (1996) Solid state fermentation of new substrates for production of cellulase and other biopolymer hydrolyzing enzymes. Appl Biochem Biotechnol 15:495–500

Shazia KM, Hamid M, Ammad AF, Ikram UH (2010) Optimization of process parameters for the biosynthesis of cellulose by *Trichoderma viride*. Pak J Bot 6(42):4243–4251

Silva LAD (2008) Production and characterization of cellulolytic enzymes by *Aspergillus pheonicis*. Master's thesis, Federal University of Rio Grande do Sul

Silveira ST, Daroit DJ, Brandelli (2008) Pigment production by *Monascus purpureus* in grape waste using factorial design. Food Sci Technol 41(1):170–174

Singh BK, Kuhad RC, Singh A, Lal R, Triapthi KK (1999) Biochemical and molecular basis of pesticide degradation by microorganisms. Crit Rev Biotechnol 19:197–225

Singh R, Singh N, Parkash A, Poonia S (2006) Saccharification studies by a thermophilic fungus *Sporotricum thermophile* isolated from agriculture waste of Bhopal. Ind J Environ Ecoplan 12 (1):97–104

Singh A, Van Hamme J, Ward O (2007) Surfactants in microbiology and biotechnology. Appl Aspects Biotechnol Adv 25(1):99–121

Singh A, Singh N, Narsi R, Bishnoi LI (2009) Production of cellulases by *Aspergillus heteromorphus* from wheat straw under submerged fermentation. Int J Civil Environ Eng 1:1

Singhania RR, Sukumaran RK, Patel AK, Larroche C, Pandey A (2010) Advancement and comparative profiles in the production technologies using solid state and submerged fermentation for microbial cellulases. Enz Microbial Technol 46(7):541–549

Smitt JP, Rinzema A, Tramper J, Van HM, Knol W (1996) Solid state fermentation of wheat bran by *Trichoderma reesei* QMQ 414. Appl Microbiol Biotechnol 46(5–6):489–496

Sohail M, Siddiqi R, Ahmad A, Khan SA (2009) Cellulase production from *Aspergillus niger* MS82: effect of temperature and pH. New Biotechnol 25(6):6437–6441

Solomon BO, Amigun B, Betiku E, Ojumu TV, Layokun SK (1999) Optimization of cellulase production by *Aspergillus flavus* Linn Isolate NSPR101 grown on bagasse. JNSChE 16:61–68

Sonjoy S, Bill BEX, Houston KH (1995) Cellulase activity of *Trichoderma reesei* (RUT-C30) on municipal solid waste. Appl Biochem Biotechol 51-52(1):145–153

Sreenath HK, Shah AB, Yang VW, Gharia MM, Jeffries TW (1996) Enzymatic polishing of jute/ cotton blended fabrics. J Ferment Bioeng 81(1):18–20

Sridevi A, Narasimha G, Rajasekhar RB (2008) Production of cellulase by *Aspergillus niger* on natural and pretreated lignocellulosic wastes. Internet J Microbiol 7(1):1–8

Srilakshmi A, Saigopal DVR, Narasimha G (2017) Impact of bioprocess parameters on cellulase production by *purpureocillium lilacinum* isolated from forest soil. Int J Pharm Bio Sci 8(1): B157–B165

Srivastava SK, Gopal KS, Ramachandran KB (1984) Kinetic characterization of crude-β-D-glucosidase from *Aspergillus wentii* Pt. 2804. Enzym Microb Technol 6:508–512

Stoilova IS, Gargova SA, Krastanov AI (2005) Production of enzymes by mixed culture from mycelial fungi in solid-state fermentation. Biotechnol Biotechnol Equip 19(1):103–108

Sukumaran KR, Singhania RR, Pandey A (2005) Microbial cellulases production, applications and challenges. J Sci Ind Res 64:832–844

Sun H, Ge X, Hao Z, Peng M (2010) Cellulase production by *Trichoderma* sp. on apple pomace under solid state fermentation. Afr J Biotechnol 9(2):163–166

Szakacs G, Tengerdy RT, Nagy V (2006) Cellulases. In: Pandey A, Colin W, Soccoi CR, Lorroche C (eds) Enzyme technology. Spring Science Business Media, Inc. and Asia Tech

Thirumade SD, Rani K, Anand K (2001) Control of cellulose formation by trehalose in *Clostridium papyrosolvens* CFR-703. Process Biochem 37(3):241–245

Tomme P, Van Tilbeurgh H, Pettersson G, Van Damme J, Vandekerckhove J, Knowles J, Teeri T, Claeyssens M (1988) Studies of the cellulolytic system of *Trichoderma reesei* QM 9414. Eur J Biochem 170(3):575–581

Tomme P, Warren RA, Gilkes NR (1995) Cellulose hydrolysis by bacteria and fungi. Adv Microb Physiol 37:1–81

Tormo J, Lamed R, Chirino AJ, Morag E, Bayer EA, Shoham Y, Steitz TA (1996) Crystal structure of a bacterial family-III cellulose-binding domain: a general mechanism for attachment to cellulose. EMBO J 15(21):5739–5751

Van Hanh V, Pham T, Kim K (2011) Improvement of fungal cellulase production by mutation and optimization of solid state fermentation. Mycobiology 39(1):20–25

Vega K, Villena GK, Sarmiento VH, Ludena Y, Vera N, Gutierrez-Correa M (2012) Production of alkaline cellulase by fungi isolated from an undisturbed rain forest of Peru. Biotechnol Res Int 2012:1–7

Vintila T, Croitoriu V, Dragomirescu M, Nica D (2010) The effects of bioprocess parameters on cellulase production with *Trichoderma viride* CMIT35. Anim Sci Biotechnol 43(1):337–340

Vyas A, Vyas D (2005) Production of fungal cellulases by solid state bioprocessing of groundnut shell Wastes. J Sci Ind Res 64:767–770

Wang SS, Converse AO (1992) On the use of enzyme adsorption and specific hydrolysis rate to characterize thermal-chemical pretreatment. Appl Biochem Biotechnol 34-35:61–74

Wang JSH, Wang J, Gulfraz M (2006) Efficient cellulase production from corn straw by *Trichoderma reesei* LW 1 through solid state fermentation process. Ethnobot Leafl 2005(1): Article 7

Watanabe H, Tokuda G (2001) Animal cellulases. Cell Mol Life Sci 58(9):1167–1178

Wen Z, Liao W, Chen S (2005) Production of cellulase/β-glucosidase by the mixed fungi culture of *Trichoderma reesei* and *Aspergillus phoenicis* on dairy manure. Appl Biochem Biotechnol 121-124:93–104

Wiatr CL (1990) Application of cellulase to control industrial slime: Google Patents

Wood TM (1988) Preparation of crystalline, amorphous and dyed cellulose substrates. Methods Enzmol 166:19–45

Wood TM, Mecrac I (1986) The cellulase of *Penicillium pinophilum*. Synergism between enzyme components in solubilizing cellulose with special reference to the involvement of two immunologically distinct cellobiohydrolases. Biochem J 234:93–99

Wood TM, Mecrae SI (1982) Purification of some properties of a 1,4-β-D-glucan glucohydrolase associated with the cellulase from the fungus *Penicillium funiculosum*. Carbohydr Res 110 (2):291–303

Wu Z, Lee YY (1997) Inhibition of the enzymatic hydrolysis of cellulose by ethanol. Biotechnol Lett 19(10):977–979

Wu GJ, Tsai GJ (2004) Cellulase degradation of shrimp chitosan for the preparation of a water-soluble hydrolysate with immunoactivity. Fish Sci 70:1113–1120

Xu F, Wang J, Chen S, Qin W, Yu Z, Zhao H, Xing X, Li H (2011) Strain Improvement for Enhanced Production of Cellulase in *Trichoderma viride*. Appl Biochem Microbiol 47(1):53–58

Yamada M, Amano Y, Horikawa E, Kanda T (2005) Mode of action of cellulases on dyed cotton with a reactive dye. Biosci Biotechnol Biochem 69(1):45–50

Zaldivar M, Velasquez JC, Contreras I, Perez LM (2001) *Trichoderma aureoviride* 7-121, a mutant with enhanced production of lytic enzymes: its potential use in waste cellulose degradation and/or biocontrol. Electron J Biotechnol 4(3):13–14
Zhang YHP (2008) Reviving the carbohydrate economy via multi-product biorefineries. J Ind Microbiol Biotech 35(5):367–375
Zhang PYH, Himmel ME, Mielenz JR (2006) Outlook for cellulase improvement: screening and selection strategies. Biotechnol Adv 24(5):452–481
Zhang Y-HP, Hong J, Ye X (2009) Cellulase assays. Methods Mol Biol 581:213–231

Modelling and Simulation of Pyrolysis of Teak (*Tectona Grandis*) Sawdust

12

Sathyanarayanan Aswin, Selvi Pandiyan Ranjithkumar, and Selvaraju Sivamani

Abstract

Pyrolysis is used to produce bio-char, bio-oil and syngas from industrial residues through thermochemical processing route. Pyrolysis decomposes biomass at an elevated temperature in an inert atmosphere. The aim of this study is to develop a model and evaluate activation energy for pyrolysis of teak sawdust. Teak sawdust was pyrolysed at four different temperatures from 300 to 600 °C. The mathematical model was developed for pyrolysis of teak sawdust. Kinetic constants were calculated by fitting the data from pyrolysis experiments to the model. Activation energy was determined from Arrhenius equation which related kinetic constant and temperature. The results reveal that pyrolysis of waste biomass, teak sawdust, could be the effective thermochemical route for bioenergy.

Keywords

Activation energy · Kinetic constants · Pyrolysis · Teak sawdust

12.1 Introduction

Thermochemical biomass conversion processes involve the production of high energy products from biomass by the application of heat and chemicals. They do not produce direct useful energy products. They produce energy carriers, such as producer gas, oils or methanol, under controlled temperature and oxygen conditions.

S. Aswin · S. P. Ranjithkumar
Department of Biotechnology, Kumaraguru College of Technology, Coimbatore, India

S. Sivamani (✉)
Chemical Engineering Section, Engineering Department, Salalah College of Technology, Salalah, Sultanate of Oman

N. Srivastava et al. (eds.), *Biofuel Production Technologies: Critical Analysis for Sustainability*, Clean Energy Production Technologies,
https://doi.org/10.1007/978-981-13-8637-4_12

The carriers reduce transport costs and good combustion characteristics allowing them to be utilised in engines and turbines because of their high density (Yaman 2004). Thermochemical biomass conversion processes include carbonisation, gasification, liquefaction and pyrolysis. Carbonisation, gasification and liquefaction produce activated charcoal, syngas and bio-oil, respectively. But, pyrolysis produces bio-char, bio-oil and syngas simultaneously (Bridgwater 2012). Hence, pyrolysis received attention in bio-energy research.

Pyrolysis is the process of heating at high temperature and pressure in the absence of or limited oxygen. Temperature and residence time decide the quality and proportion of product. Longer residence time and lower process temperatures are optimum for charcoal production. Short residence time and moderate temperature favour the production of bio-oil. Longer residence time and higher temperature enhance syngas synthesis. But, all the three products are always produced with varied composition and process parameters (French and Czernik 2010). A more comprehensive understanding of the physical and chemical properties of thermal reactions allowed the improvement of nuclear reactors (Carlson et al. 2008).

Teak (*Tectona grandis*) is native to South and Southeast Asian countries such as India, Sri Lanka, Indonesia, Malaysia, Thailand, Myanmar and Bangladesh. It is also cultivated in some parts of Africa. It is one of the crucial lignocellulosic biomass valued for its water resistance and durability (Ismadji et al. 2005). It contains predominantly hemicellulose, cellulose and lignin. Being a species of hardwood, it contains more cellulose and lignin and less hemicellulose. The mechanism of biomass pyrolysis is explained as follows: Moisture evaporates completely at around 120 °C and then decomposition of hemicellulose starts at around 250 °C followed by cellulose and lignin at around 350 and 500 °C, respectively. When temperature reaches 500 °C during heating, the pyrolysis reactions are nearly completed (Miura et al. 2004).

Hence, the present work aims at developing a mathematical model for pyrolysis of teak sawdust. The objectives of the present work are to (1) perform experimental studies to collect data on weight loss of teak sawdust as a function of time at different temperatures, (2) develop a mathematical model for pyrolysis of teak sawdust and (3) evaluate activation energy for teak sawdust pyrolysis.

12.2 Materials and Methods

12.2.1 Materials

Waste teak biomass was collected from Forest College and Research Institute, Mettupalayam, Coimbatore district, Tamil Nadu, India. It was dried to constant weight in hot air oven (Narang scientific works Private Limited, New Delhi) at 60 °C for 48 h. It was crushed into small pieces in jaw crusher (Almech Enterprises, Coimbatore) and then sieved to obtain (−52 + 60) mesh particles in gyratory sieve shaker (Lawrence and Mayo (India) Private Limited, Mumbai). Finally, particles are transferred to airtight cover for further experimental studies.

12.2.2 Experimental Studies

Fast pyrolysis reaction was carried out in chemical vapour deposition (CVD) tubular furnace (VB ceramic consultants, Chennai) with an inert nitrogen gas atmosphere maintained at a heating rate of 10 °C/min. 1 g of teak sawdust was pyrolysed, and the loss in weight of biomass was noted down as function of time at different temperatures. The experiment was conducted till biomass weight reaches constant value.

12.2.3 Modelling of Pyrolysis of Steak Sawdust

Biomass pyrolysis products are a complex combination of the products from the individual pyrolysis of cellulose, hemicellulose and extractives, each of which has its own kinetic characteristics. In addition, secondary reaction products result from cross-reactions of primary pyrolysis products and between pyrolysis products and original feedstock molecules. Pyrolysis of each constituent is itself a complex process that is dependent on many factors (Collard and Blin 2014). Pyrolysis modelling is classified into three different types: (1) one-step global models, (2) one-step multi-reaction models and (3) two-stage semi-global models (oochit et al. 2017; Bridgwater 2015).

The first category of models considers pyrolysis as a single-step first-order reaction:

Primary interaction:

Virgin biomass → gases + volatiles
Virgin biomass → char

Secondary interaction:

Gases + volatiles + char → char + gases

The second category of models discusses those mechanisms, which consider simultaneous and competing first-order reactions in which virgin wood decomposes into different constituents of pyrolysis products, namely, tar, gases and char.

The third-class models consider pyrolysis to be a two-stage reaction, in which the products of the first reaction (tar and gases) further react with the char produced by the second reaction to produce tar and gases and char of different compositions. Thus, the primary pyrolysis products participate in secondary interactions causing a modified final product distribution. As particle size increases, the residence time of the volatiles inside the solid increases and the effect of secondary reactions also increases (Venderbosch 2015).

During pyrolysis, biomass (B) produces tar and gases (G) and char (C), which is represented as follows:

Biomass (B) → Volatiles + Gases (G) (R1)

Biomass (B) → Char (C) (R2)

For reaction (R1),

$$-\left(\frac{dC_B}{dt}\right) = k_1.C_B^{n1} \tag{12.1}$$

where n_1 and k_1 are the order and rate constant of reaction (R1), respectively.

For reaction (R2),

$$-\left(\frac{dC_B}{dt}\right) = k_2.C_B^{n2} \tag{12.2}$$

where n_2 and k_2 are the order and rate constant of reaction (R2), respectively.

Combining Eqs. (12.1) and (12.2)

$$-\left(\frac{dC_B}{dt}\right) = k_1.C_B^{n1} + k_2.C_B^{n2} \tag{12.3}$$

According to Shafizadeh (1981), $n_1 = n_2 = 1$.

Substituting the above condition in Eq. (12.3),

$$\begin{aligned} -\left(\frac{dC_B}{dt}\right) &= k_1.C_B + k_2.C_B \\ &= C_B(k_1 + k_2) \\ -(dC_B/C_B) &= dt(k_1 + k_2) \end{aligned} \tag{12.4}$$

$$\begin{aligned} [\ln C_B]_1^{C_B} &= [(k_1 + k_2)t]_0^t \\ -(\ln 1 - \ln C_B) &= (k_1 + k_2)(0 - t) \\ \ln C_B &= -(k_1 + k_2)t \\ C_B &= e^{-(k_1+k_2)t} \end{aligned} \tag{12.5}$$

Equation 12.5 is used to predict the concentration of biomass as a function of time:

$$\begin{aligned} \frac{dc_c}{dt} &= k_2.C_B{}^{n_2} \\ &= k_2.e^{-(k_1+k_2)t} \\ dC_c &= k_2 e^{-(k_1+k_2)t}.dt \\ [C_C] &= -\left[k_2.\left(e^{-(k_1+k_2)t}\right)/(k_1 + k_2)\right] \end{aligned} \tag{12.6}$$

Equation 12.6 is used to calculate the concentration of char as a function of time.

When pyrolysis occurs, teak sawdust loses its weight as a result of escaping of all the gases from CVD chamber. So, only unreacted biomass and produced char can be measured from the residue. Let W be the mass of biomass and char together:

$$\begin{aligned}
W &= C_B + C_C \\
W &= e^{-(k_1+k_2)t} + k_2\left[\frac{1 - e^{-(k_1+k_2).t}}{(k_1+k_2)}\right] \\
&= e^{-(k_1+k_2).t} + \frac{k_2 - k_2.e^{-(k_1+k_2).t}}{(k_1+k_2)} \\
&= e^{-(k_1+k_2).t} + \frac{k_2}{k_1+k_2} - \frac{k_2.e^{-(k_1+k_2).t}}{(k_1+k_2)} \\
&= \frac{k_2}{k_1+k_2} + e^{-(k_1+k_2).t}.\left[1 - \frac{k_2}{(k_1+k_2)}\right] \\
&= \frac{k_2}{k_1+k_2} + e^{-(k_1+k_2).t}.\left[\frac{k_1}{(k_1+k_2)}\right] \\
W &= \frac{1}{(k_1+k_2)}\left[k_1.e^{-(k_1+k_2).t} + k_2\right] \\
\frac{dW}{dt} &= \frac{k_1}{(k_1+k_2)}.e^{-(k_1+k_2).t}. - (k_1+k_2) + C \\
-\frac{dW}{dt} &= k_1.e^{-(k_1+k_2)t} \\
\ln\left(-\frac{dW}{dt}\right) &= \ln k_1 - (k_1+k_2)t
\end{aligned} \tag{12.7}$$

Equation 12.7 is used to predict the concentration of biomass and char together as a function of time.

Rate constants k_1 and k_2 can be evaluated by plotting ln ($-dW/dt$) versus time from slope and intercept. Then, activation energy can be calculated by plotting natural logarithm of rate constants versus 1/T as per Arrhenius equation as given by.

$$k = A.e^{-E/RT}$$

$$\ln k = \ln A - (E/RT)$$

Slope of curve gives ($-$E/R) and intercept gives ln A from which activation energy and pre-exponential factor can be calculated and interpreted.

12.3 Results and Discussion

12.3.1 Pyrolysis of Teak Sawdust at 300 °C

Figure 12.1 shows the weight of residue as a function of time at 300 °C. At time $t = 0$, weight of teak sawdust is 1 g. As pyrolysis progresses, weight of biomass reduces gradually as a result of releasing of pyrolytic gases, leaving behind char and

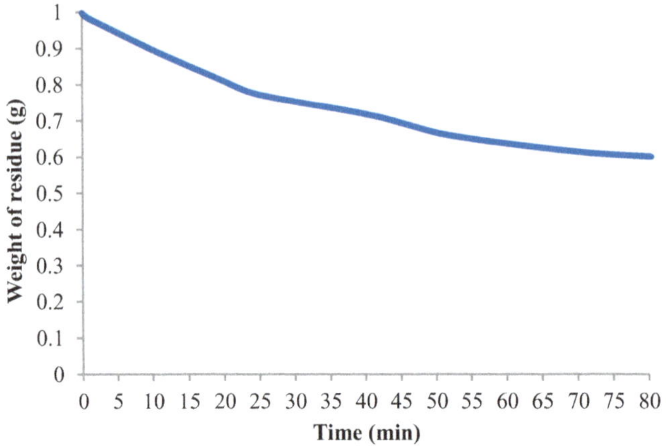

Fig. 12.1 Weight of residue of teak sawdust as a function of time at 300 °C

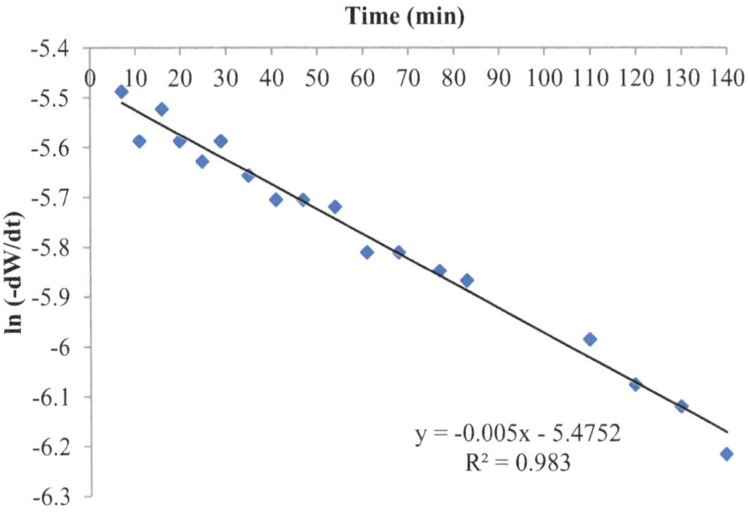

Fig. 12.2 A graph showing the calculation of rate constants from experimental data at 300 °C

unconverted biomass as residue. After 140 min, pyrolysis reaction is stopped which is indicated from the constant weight of residue. Bio-char produced from pyrolysis at 300 °C for 140 min was 0.489 g. Finally, negligible quantity of unreacted teak sawdust (<0. mg) was observed because of their high surface area.

By using Eq. (12.7), ln $(-dW/dt)$ was calculated and plotted against time as shown in Fig. 12.2. Slope gives $-(k_1 + k_2)$, and intercept gives ln k_1. From Fig. 12.2, the values of k_1 and k_2 were 0.0114 min^{-1} and 0.0093 min^{-1}. It means that conversion of biomass to volatiles and gases is faster than conversion of biomass

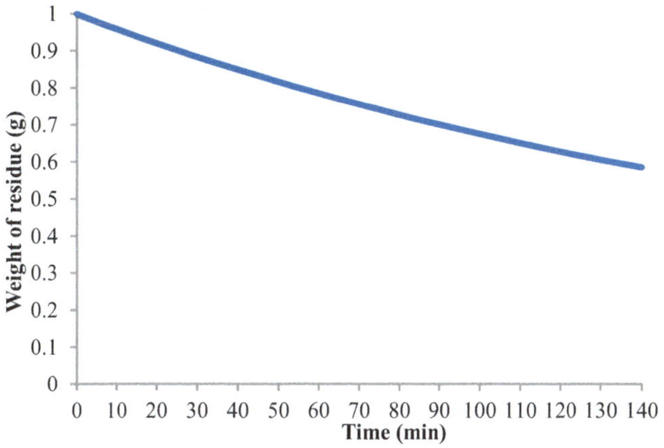

Fig. 12.3 Weight of residue of teak sawdust as a function of time at 400 °C

to char (Robinson et al. 2015; Demirbas and Arin 2002). These values will be used to calculate activation energy of pyrolysis of biomass at 300 °C.

12.3.2 Pyrolysis of Teak Sawdust at 400 °C

Figure 12.3 shows the weight of residue as a function of time at 400 °C. Initial weight of teak sawdust is 1 g. As pyrolysis proceeds, reduction in weight of biomass is observed due to formation of gases and volatiles, leaving behind unconverted biomass and char as residue. Pyrolysis reaction is completed after 85 min which is indicated from the constant weight of residue. Bio-char produced from pyrolysis at 400 °C for 140 min was 0.199 g. Finally, negligible quantity of unreacted teak sawdust (1–2 mg) was observed because of their high surface area.

By using Eq. (12.7), ln $(-dW/dt)$ was calculated and plotted against time as shown in Fig. 12.4. Slope gives $-(k_1 + k_2)$, and intercept gives ln k_1. From Fig. 12.2, the values of k_1 and k_2 were 0.0042 min^{-1} and 0.0008 min^{-1}. It means that conversion of biomass to volatiles and gases is faster than conversion of biomass to char. These values will be used to calculate activation energy of pyrolysis of biomass at 400 °C.

12.3.3 Pyrolysis of Teak Sawdust at 500 °C

Figure 12.5 shows the weight of residue as a function of time at 500 °C. At time $t = 0$, weight of teak sawdust is 1 g. As pyrolysis progresses, weight of biomass reduces gradually as a result of releasing of pyrolytic gases, leaving behind char and unconverted biomass as residue. After 80 min, pyrolysis reaction is stopped which is indicated from the constant weight of residue. Bio-char produced from pyrolysis at

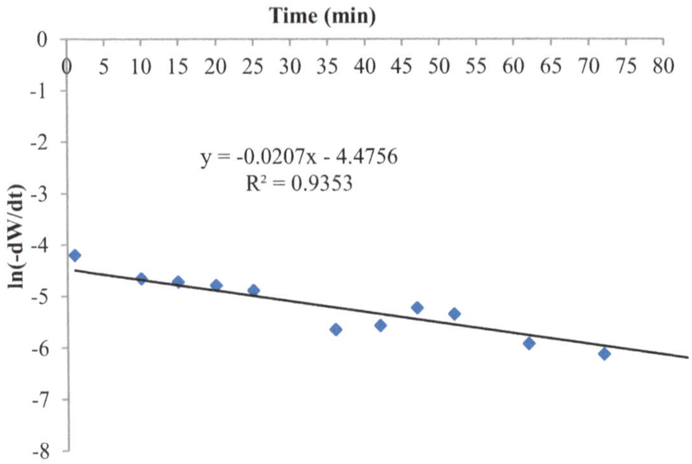

Fig. 12.4 A graph showing the calculation of rate constants from experimental data at 400 °C

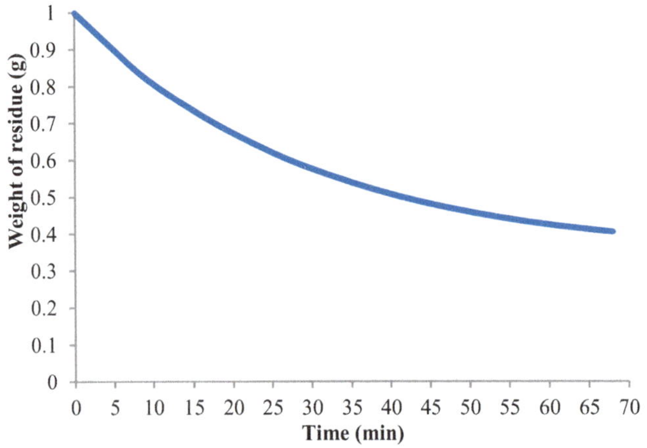

Fig. 12.5 Weight of residue of teak sawdust as a function of time at 500 °C

500 °C for 80 min was 0.6 g. Finally, negligible quantity of unreacted teak sawdust (1–2 mg) was observed because of their high surface area.

By using Eq. (12.7), ln $(-dW/dt)$ was calculated and plotted against time as shown in Fig. 12.6. Slope gives $-(k_1 + k_2)$, and intercept gives ln k_1. From Fig. 12.2, the values of k_1 and k_2 were 0.0244 min^{-1} and 0.0094 min^{-1}. It means that conversion of biomass to volatiles and gases is faster than conversion of biomass to char. These values will be used to calculate activation energy of pyrolysis of biomass at 500 °C.

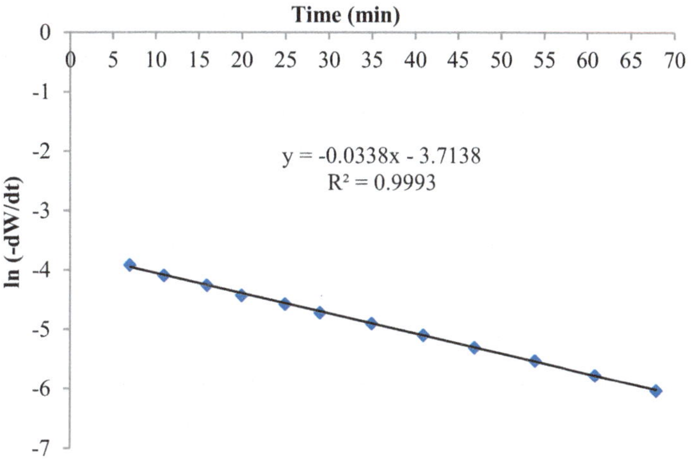

Fig. 12.6 A graph showing the calculation of rate constants from experimental data at 500 °C

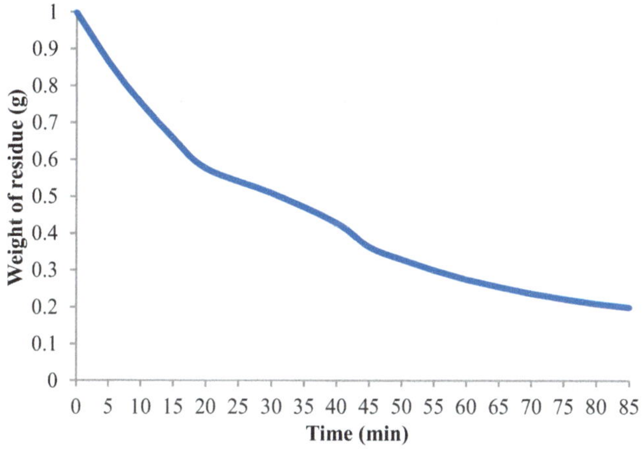

Fig. 12.7 Weight of residue of teak sawdust as a function of time at 600 °C

12.3.4 Pyrolysis of Teak Sawdust at 600 °C

Figure 12.7 shows the weight of residue as a function of time at 600 °C. Initial weight of teak sawdust is 1 g. As pyrolysis proceeds, reduction in weight of biomass is observed due to formation of gases and volatiles leaving behind unconverted biomass and char as residue. Pyrolysis reaction is completed after 70 min which is indicated from the constant weight of residue. Bio-char produced from pyrolysis at 600 °C for 70 min was 0.407 g. Finally, negligible quantity of unreacted teak sawdust (1–2 mg) was observed because of their high surface area (Fig. 12.8).

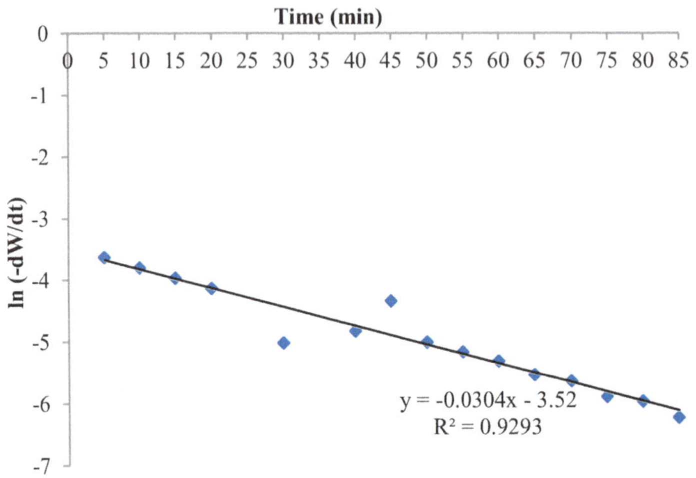

Fig. 12.8 A graph showing the calculation of rate constants from experimental data at 600 °C

Table 12.1 Rate constants of pyrolysis reactions at different temperatures

Temperature (°C)	Quantity of bio-char (g)	Slope (min^{-1})	Intercept	k_1 (min^{-1})	k_2 (min^{-1})
300	0.600	−0.0050	−5.4752	0.0042	0.0008
400	0.586	−0.0207	−4.4756	0.0114	0.0093
500	0.407	−0.0338	−3.7138	0.0244	0.0094
600	0.199	−0.0304	−3.5200	0.0296	0.0008

By using Eq. (12.7), ln (−dW/dt) was calculated and plotted against time as shown in Fig. 12.4. Slope gives $-(k_1 + k_2)$, and intercept gives ln k_1. From Fig. 12.2, the values of k_1 and k_2 were 0.0296 min^{-1} and 0.0008 min^{-1}. It means that conversion of biomass to volatiles and gases is faster than conversion of biomass to char. These values will be used to calculate activation energy of pyrolysis of biomass at 600 °C (Table 12.1).

12.3.5 Calculation of Activation Energy

The pre-exponential factor is a measure of the probability that two (or more) molecules involved in a reaction collide. It is worth reviewing the kinetic theory of gases to get a better understanding of what it is. As for the activation energy, it can be seen as the barrier of energy that has to be overcome so the reaction can occur. Both parameters are very important and can certainly be used in reaction engineering, process modelling and optimisation process.

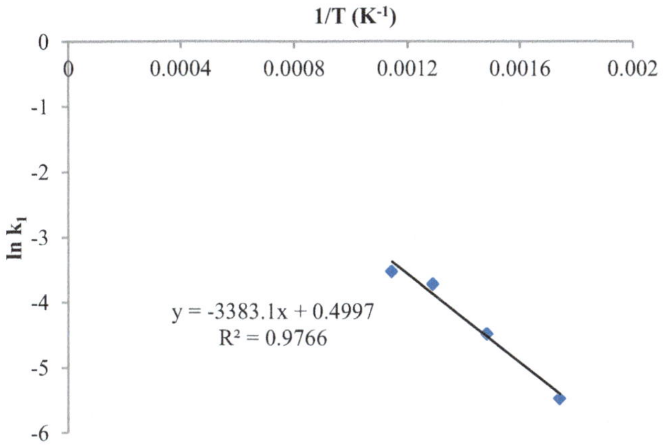

Fig. 12.9 Arrhenius plot for conversion of teak sawdust biomass to volatiles and gases

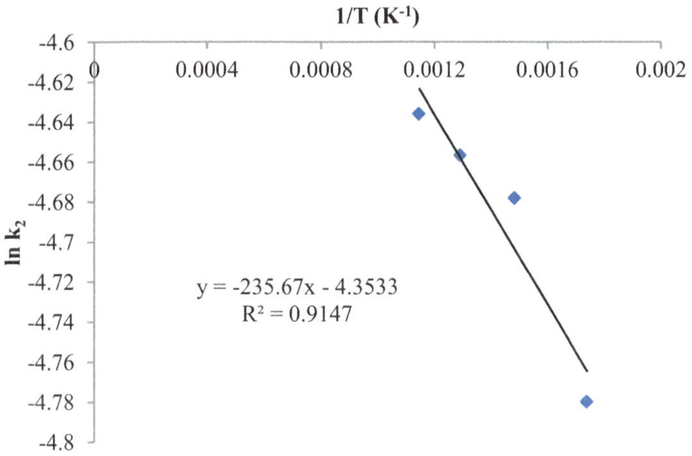

Fig. 12.10 Arrhenius plot for conversion of teak sawdust biomass to char

Arrhenius plot for conversion of teak sawdust biomass to volatiles and gases reveals the activation energy of 28.12 kJ/mol and pre-exponential factor of 1.648 min^{-1} (Fig. 12.9). Arrhenius plot for conversion of teak sawdust biomass to char reveals the activation energy of 1.959 kJ/mol and Arrhenius factor of 0.013 min^{-1} (Fig. 12.10). Reactions with high activation energies are very temperature sensitive; reactions with low activation energies are relatively temperature insensitive (Babu 2008; Prakash and Karunanithi 2008; Kumar et al. 2010). Hence, reaction from teak sawdust biomass to volatiles and gases is more temperature sensitive than to char (Koufopanos et al. 1989; Jalan and Srivastava 1999; Levenspiel 1999).

12.4 Review of Literature

French and Czernik (2010) assessed a series of commercial and laboratory-synthesised catalysts via the pyrolysis/catalytic cracking route for their hydrocarbon production performance. For the production of stationary energy using boilers or turbines, rapid pyrolysis bio-oils currently produced in demonstration and semi-commercial facilities can be used as fuel, but they must be appreciably modified to become sustainable transport fuel. To remove oxygen from organic compounds and transform into hydrocarbons by catalytic upgrading of pyrolysis vapours with zeolite is a favourable method. Cellulose, lignin and wood were the types of raw materials used. Batch experiments were conducted in which raw materials, quartz boats and the catalysts pyrolysed at a temperature between 400 and 600 °C. The ratio of catalyst to biomass ratio of 5:10 (by weight) was used. Molecular beam mass spectrometry (MBMS) was used to analyse the product for vapour and gas composition. The total highest hydrocarbon yield of 16 wt%, of which 3.5 wt% is toluene, was attained with ZSM-5 substituted with nickel, cobalt, iron and gallium. Tests conducted in a semi-continuous flow reactor reported changes in the composition of the volatile products produced by pyrolysis/steam cracking with respect to time. The deoxygenation activity decreases over time due to the coke deposits formed on the catalyst.

Zhang et al. (2005) pyrolysed a biomass in a fluidised-bed unit (5 kg/h) to maximise the liquid yield. The liquid product formed during pyrolysis was separated into aqueous and oil phases, respectively. The oil phase reinforced by a sulphide catalyst of Co-Mo-P in an autoclave. The optimum conditions were determined by studying the effects of reaction conditions on product distribution. The comparison was made by analysis between the crude oil phase and the improved liquid fuel. The former was soluble in methanol, while the latter was soluble in oil.

Czernik et al. (2007) proposed a two-step process, rapid pyrolysis of biomass, which produces high yields of a liquid product, the bio-oil, followed by catalytic steam reforming of the bio-oil to produce biomass, hydrogen. Thermoconversion of biomass is one of the most important short-term options for the production of renewable hydrogen and can supply an important part of the transport fuel needed in the future. A big advantage of this concept is that bio-oil is much easier and cheaper to transport than biomass or hydrogen. As a result, biomass processing and hydrogen production can be carried out at different locations, optimised for the supply of raw materials and the infrastructure for the distribution of hydrogen. This approach makes the process very suitable for centralised and distributed hydrogen production. This work demonstrates the reforming of bio-oil in a fluidised-bed system at scale and provides the hydrogen yield obtained using various commercial and tailor-made catalysts.

Aho et al. (2008) carried out catalytic pyrolysis in a fluid bed reactor at 450 °C using pine biomass and zeolite acid catalyst structures which are used as the bed material in the reactor. In non-catalytic pyrolysis, quartz sand was used as reference material, while in pine pyrolysis, the proton forms of Beta-, Y-, ZSM-5 and modernite were tested as catalysts. The pyrolysis product phase yield is slightly affected by structures of zeloite acid catalysts, but there is dependency of chemical composition of the bio-oil on the structures. Ketones and phenols were the dominant groups of compounds in the bio-oil. The formation of ketones was greater than that of ZSM-5, and the amount of acids and alcohols was lower than that of the other bed materials tested. Mordenite and quartz sand produced lower amounts of polyaromatic hydrocarbons compared to other catalysts tested. Finished zeolites can be regenerated successfully without deforming the structures of zeolite.

Wan et al. (2009) aimed at assessing the effects of the catalysts on the selectivity of the pyrolysis product assisted by microwave cornstalks and aspen. Oxides, salts and metal, including $K_2Cr_2O_7$, Al_2O_3, KAc, H_3BO_3, Na_2HPO_4, $MgCl_2$, $AlCl_3$, $CoCl_2$ and $ZnCl_2$, were premixed with corn stalks or wood pellets especially for pyrolysis using microwave heating. The thermal process produced three product fractions, namely, bio-oil, gas and charcoal. The effects of catalysts on fractional yields have been studied. It was found that the yield of the bio-oil efficiency of coal or gas increased with KAc, Al_2O_3, $MgCl_2$, H_3BO_3 and Na_2HPO_4. These catalysts can accelerate as absorbents for microwave heating or take part in a recovery in situ pyrolytic vapour during the pyrolysis of biomass assisted by microwaves. GC-MS analysis of the bio-oils revealed that the chloride salts favoured some reactions while suppressing most of the other reactions observed for the control samples. In a biomass of 8 g MgC12/100, the total ion chromatograms GC-MS bio-oil from the corn stalk or aspen treated shows one main peak of furfural covering about 80% of the acreage spectrum. It was concluded that some catalysts enhance bio-oil yields, and in particular chloride salts simplify the chemical composition of the resulting bio-oil and thereby improve product selectivity of the pyrolysis process.

Fahmi et al. (2008) focused on the pyrolysis of four reference fuels and three low-lignin Lolium-Festuca grasses to produce pyrolysis oils. The oils were analysed to determine their quality and stability, which made it possible to identify the properties of the raw materials that influence the stability of the oil. Two washed raw materials were also subjected to pyrolysis to determine if a wax could improve the quality of the pyrolysis oil. Minerals appeared to have a dominant effect on pyrolysis compared to the lignin content, in terms of pyrolysis yields for organic matter, coal and gases. However, the higher molecular weight compounds present in the pyrolysis oil are due to the lignin-derived compounds as determined by the results of GC and liquid GC/MS. The yield of the light organic fraction also increased, but its water content was lowered as the metals increased at the expense of the lignin content. It has been found that the fresh oil and the aged oil have different intensities/concentrations of compounds, which is the result of a large number of reactions that occur during daily oil aging. These results are consistent with previous reports suggesting that a large amount of repolymerisation occurred, since the levoglucose yields increased during aging, while hydroxyacetaldehyde decreased. In summary, the article describes a window for producing a more stable pyrolysis oil using energy crops and also shows that washing with biomass can improve the quality and stability of the oil for high-quality raw materials, in the ash, but less for energy crops.

Demirbas (2002) reported that at the desired temperature, three different biomass samples were subjected to direct and catalytic pyrolysis to obtain gaseous products rich in hydrogen. The pyrolysis products are obtained into a volatile fraction consisting of gaseous vapour and tar components and a carbon-rich solid residue. The pyrolysis process consists of a very complex set of reactions in which radicals are formed. Biomass gasification is a heat treatment that results in an increased production of gaseous products and small amounts of coal and ash. Hydrogen is produced from solid waste by pyrolysis. The untreated and catalyst-impregnated samples were pyrolysed at temperatures of 775 K, 925 K, 975 K and 1025 K. As temperature increases, there is increase in total volume and the gas yield.

Sonobe and Worasuwannarak (2008) investigated the behaviour of pyrolysis of various agricultural residues using thermogravimetric analysis. The evolution rates of gaseous products during pyrolysis, such as H_2, CH_4, H_2O, CO and CO_2, were also measured using TG-MS techniques. Distribution activation energy model (DAEM) proposed by Miura and Maki (1998) used to obtain kinetic parameter activation energy f(E) and frequency factor $k_o(E)$ of the pyrolysis. Increases in values of f (E) peaks of the f(E) curve for rice straw, rice waste, corn and cellulose were found to be 170, 174, 183 and 185 kJ/mol, respectively, which increased by an order of 10^{11} to an order of 10^{18} s^{-1}, while E increased from 120 to 250 kJ/mol. The variation of f (E) curve of different types of biomass is due to the alkali and alkaline earth metal catalytic during pyrolysis.

Adam et al. (2005) included four Al-MCM-41 type catalysts with an Si/Al ratio of 20. Mesoporous Al-MCM-41 catalysts were used to convert the pyrolysis vapours of spruce wood into better bio-oil properties. The catalytic properties of the Al-MCM-41 catalyst have been modified by the enlargement of the pores, allowing the treatment of larger molecules and the introduction of Cu cations into the structure. The pyrolysis of the spruce wood at 500 °C was carried out, and the products were analysed by means of online pyrolysis gas/mass spectrometry (Py-GC/MS). In addition, thermogravimetry/mass spectrometry (TG/MS) experiments were used to monitor the evolution of the product under slow heating conditions (20 °C/min) of 50 to 800 °C. Levoglucosan is completely eliminated, while acetic acid, furfural and furans become important components of the cellulose pyrolysis products compared to the unmodified Al-MCM-41 catalyst. The dominance of high molecular weight phenolic compounds is greatly reduced among lignin products. The increase in the yield of acetic acid and furan and the decrease of large methoxyphenols are suppressed to some extent in comparison with dilated pore catalysts. The Cu-modified catalyst exhibited a performance comparable to that of the expanded pore size catalyst in converting pyrolysis vapours into wood, although its pore size corresponded to that of unmodified Al-MCM-41.

Stefanidis et al. (2014) carried out thermogravimetric (TG) analyses as well as rapid thermal and catalytic pyrolysis experiments of cellulose, hemicellulose, lignin and their mixtures to study and determine their pyrolysis products if the pyrolysis behaviour of a given lignocellulose-containing biomass is possible when the content is known in these three components. The limited heat transfer had no significant effect on the TG curves but affected the product distribution in the fast pyrolysis experiments, resulting in an inaccurate calculation of product yields with the use of a simple law on additives. In addition, the pyrolysis products of each component of the biomass have been characterised to study their contribution to the yield and composition of products from complete biomass pyrolysis. A study of the pyrolysis reaction paths of each component was also carried out, using the characterisation data of the bio-oil of this study and those found in the literature.

Zhang et al. (2007) carried out the co-pyrolysis of biomass and coal in a free-fall reactor under atmospheric pressure with nitrogen as equilibrium gas. The chosen coal sample was Dayan brown coal, while the biomass used was leguminous straw. The working temperature was between 500 and 700 °C, and the mixing ratio of the biomass in the mixtures varied between 0% and 100% by weight. The results showed that there were synergistic effects in the co-pyrolysis of biomass and coal. Under the conditions of the higher mixing ratio, the carbonisation yields are lower than the theoretical values calculated for the pyrolysis of each individual fuel, and therefore the liquid yields are higher. Moreover, the experimental results showed that the compositions of the gaseous products from the mixed samples do not all conform to those of the parental fuels. CO_2 reactivities of carbonates from co-pyrolysis under higher mixing ratio conditions (about 70% by weight) are about twice as high as those of the carbonator alone or even higher than those of the biomass alone.

Chen et al. (2003) focused on the use of catalysts for the production of gaseous hydrogen from biomass. The use of cheap biomass as a source of thermochemical conversion is a good way to produce hydrogen. Hydrogen is a clean and efficient energy source and should play an important role in future energy demand. Various types of catalysts have been studied on our test bench at wide operating temperature ranges. The results show that the catalyst has a positive influence on the hydrogen-rich gas yield. The hydrogen concentration of the pyrolytic gas is significantly improved by some types of catalysts. The results obtained here can be very useful for large-scale hydrogen production based on the biomass source.

Shen et al. (2009) investigated the effect of the particle size of the biomass (0.18–5.6 mm) on the yield and the composition of the result bio-oil from pyrolysis of woody biomass *Dutch mallee* in a reactor fluidised bed at 500 °C. The yield of bio-oil decreased when the average particle size of the biomass increased from 0.3 to about 1.5 mm. Subsequent increase in the particle size of the biomass did not result in a further reduction in the yield of bio-oil. These results are mainly due to the impact of particle size in the production of lignin-derived compounds. The possible interactions between vapour particles of bio-oil and coal particles or vapours of homogeneous reactions are not responsible for the decrease in the yield of bio-oil. Samples of bio-oil were characterised by thermogravimetric analysis, UV fluorescence spectroscopy titration Karl Fischer and precipitation in cold water. It was found that the yield of light bio-oil fractions increased and those heavy bio-oil fractions decreased with increasing the size of the biomass particles. The pyrolytic formation of water at low temperature (<500 °C) is little influenced by the temperature or the particle size. It is believed that the decrease in the heating rate of coarse particles is an important factor responsible for the low yields of large particle bio-oil and changes in the overall composition of the oils obtained. Changes in the cell structure of biomass during grinding can also affect the yield and composition of the bio-oil.

Güllü and Demirbas (2001) focused on the pyrolysis of lignocellulose-containing biomass to produce methanol. Methanol can be used to replace conventional petrol and diesel. Thermal depolymerisation and decomposition of biomass include cellulose, hemicellulose and lignin, liquid and gaseous products and a solid residue of charcoal. A promising route for the treatment of biomass is pyrolytic conversion, which has been carried out under different experimental conditions, in which coal, tarry materials, an aqueous fraction and gaseous products have been produced. Pyroligneous acid consists of about 50% methanol, acetone, phenols and water. Methanol can be produced by pyrolysis of biomass. The methanol mainly comes from the methoxyl groups of uronic acid and the decomposition of methyl esters and/or ethers from the decomposition of pectin-like plant material. Acetic acid mainly comes from acetyl groups of hemicelluloses.

Foster et al. (2012) concentrated on the conversion of glucose, furan and maple wood using various types of ZSM-5 catalysts in semi-batch and fixed-bed reactors. The aromatic yield of glucose conversion is maximised by the ratio of silica to alumina (SAR) of ZSM-5 with an optimum at SAR = 30. This suggests that the concentration of acidic sites in the zeolite is critical for maximising the aromatic yield. The formation of hierarchical mesopores in the zeolite slightly increased the formation of coke and reduced the formation of monocyclic aromatics. It has also been observed that mesoporous ZSM-5 favours the production of larger alkylated monoaromatics. The selective removal of the external acid sites of the ZSM-5 catalysts not only slightly increases the activity of the catalyst but also reduces the selectivity for the desired aromatics.

12.5 Conclusion

The present study aimed at developing a mathematical model for pyrolysis of teak sawdust. Rate constants and activation energies of reactions are calculated from experimental studies on pyrolysis of teak sawdust. 1 g of teak sawdust was pyrolysed in CVD chamber at four different temperatures from 300 to 600 °C. The mathematical model was developed for pyrolysis of teak sawdust. Rate constants reveal that conversion of biomass to volatiles and gases is faster than conversion of biomass to char. Activation energies disclose that conversion of biomass to volatiles and gases is more temperature sensitive than conversion of biomass to char. The results reveal that pyrolysis of waste biomass, teak sawdust, could be the effective thermochemical route for bioenergy.

References

Adam J, Blazso M, Meszaros E, Stöcker M, Nilsen MH, Bouzga A et al (2005) Pyrolysis of biomass in the presence of Al-MCM-41 type catalysts. Fuel 84(12–13):1494–1502
Aho A, Kumar N, Eränen K, Salmi T, Hupa M, Murzin DY (2008) Catalytic pyrolysis of woody biomass in a fluidized bed reactor: influence of the zeolite structure. Fuel 87(12):2493–2501
Babu BV (2008) Biomass pyrolysis: a state-of-the-art review. Biofuels Bioprod Biorefin 2 (5):393–414
Bridgwater AV (2012) Review of fast pyrolysis of biomass and product upgrading. Biomass Bioenergy 38:68–94
Bridgwater AV (2015) Pyrolysis of biomass. In: van Swaaij W, Kersten S, Palz W (eds) *Transformations to effective use: biomass power for the world*, vol 6. Pan Stanford Publishing, Boca Raton, pp 473–514
Carlson TR, Vispute TP, Huber GW (2008) Green gasoline by catalytic fast pyrolysis of solid biomass derived compounds. Chem Sus Chem 1(5):397–400
Chen G, Andries J, Spliethoff H (2003) Catalytic pyrolysis of biomass for hydrogen rich fuel gas production. Energy Convers Manag 44(14):2289–2296
Collard FX, Blin J (2014) A review on pyrolysis of biomass constituents: Mechanisms and composition of the products obtained from the conversion of cellulose, hemicelluloses and lignin. Renew Sust Energ Rev 38:594–608
Czernik S, Evans R, French R (2007) Hydrogen from biomass-production by steam reforming of biomass pyrolysis oil. Catal Today 129(3–4):265–268
Demirbaş A (2002) Gaseous products from biomass by pyrolysis and gasification: effects of catalyst on hydrogen yield. Energy Convers Manag 43(7):897–909
Demirbas A, Arin G (2002) An overview of biomass pyrolysis. Energy Sources 24(5):471–482
Fahmi R, Bridgwater AV, Donnison I, Yates N, Jones JM (2008) The effect of lignin and inorganic species in biomass on pyrolysis oil yields, quality and stability. Fuel 87(7):1230–1240
Foster AJ, Jae J, Cheng YT, Huber GW, Lobo RF (2012) Optimizing the aromatic yield and distribution from catalytic fast pyrolysis of biomass over ZSM-5. Appl Catal A Gen 423:154–161
French R, Czernik S (2010) Catalytic pyrolysis of biomass for biofuels production. Fuel Process Technol 91(1):25–32
Güllü D, Demirbaş A (2001) Biomass to methanol via pyrolysis process. Energy Convers Manag 42 (11):1349–1356
Ismadji S, Sudaryanto Y, Hartono SB, Setiawan LEK, Ayucitra A (2005) Activated carbon from char obtained from vacuum pyrolysis of teak sawdust: pore structure development and characterization. Bioresour Technol 96(12):1364–1369
Jalan RK, Srivastava VK (1999) Studies on pyrolysis of a single biomass cylindrical pellet—kinetic and heat transfer effects. Energy Convers Manag 40(5):467–494
Koufopanos CA, Lucchesi A, Maschio G (1989) Kinetic modelling of the pyrolysis of biomass and biomass components. Can J Chem Eng 67(1):75–84
Kumar R, Pandey KK, Chandrashekar N, Mohan S (2010) Effect of tree-age on calorific value and other fuel properties of Eucalyptus hybrid. J For Res 21(4):514–516
Levenspiel O (1999) Chemical reaction engineering. Ind Eng Chem Res 38(11):4140–4143
Miura K, Maki T (1998) A simple method for estimating f(E) and $k_o(E)$ in the distributed activation energy model. Energy Fuel 12(5):864–869
Miura M, Kaga H, Sakurai A, Kakuchi T, Takahashi K (2004) Rapid pyrolysis of wood block by microwave heating. J Anal Appl Pyrolysis 71(1):187–199
Oochit D, Selvarajoo A, Arumugasamy SK (2017) Pyrolysis of biomass. In: Waste biomass management–a holistic approach. Springer, Cham, pp 215–229
Prakash N, Karunanithi T (2008) Kinetic modeling in biomass pyrolysis–a review. J Appl Sci Res 4 (12):1627–1636

Robinson J, Dodds C, Stavrinides A, Kingman S, Katrib J, Wu Z et al (2015) Microwave pyrolysis of biomass: control of process parameters for high pyrolysis oil yields and enhanced oil quality. Energy Fuel 29(3):1701–1709

Shafizadeh F (1981) Basic principles of direct combustion. In: Biomass conversion processes for energy and fuels. Springer US, pp 103–124

Shen J, Wang XS, Garcia-Perez M, Mourant D, Rhodes MJ, Li CZ (2009) Effects of particle size on the fast pyrolysis of oil mallee woody biomass. Fuel 88(10):1810–1817

Sonobe T, Worasuwannarak N (2008) Kinetic analyses of biomass pyrolysis using the distributed activation energy model. Fuel 87(3):414–421

Stefanidis SD, Kalogiannis KG, Iliopoulou EF, Michailof CM, Pilavachi PA, Lappas AA (2014) A study of lignocellulosic biomass pyrolysis via the pyrolysis of cellulose, hemicellulose and lignin. J Anal Appl Pyrolysis 105:143–150

Venderbosch RH (2015) A critical view on catalytic pyrolysis of biomass. Chem Sus Chem 8 (8):1306–1316

Wan Y, Chen P, Zhang B, Yang C, Liu Y, Lin X, Ruan R (2009) Microwave-assisted pyrolysis of biomass: catalysts to improve product selectivity. J Anal Appl Pyrolysis 86(1):161–167

Yaman S (2004) Pyrolysis of biomass to produce fuels and chemical feedstocks. Energy Convers Manag 45(5):651–671

Zhang S, Yan Y, Li T, Ren Z (2005) Upgrading of liquid fuel from the pyrolysis of biomass. Bioresour Technol 96(5):545–550

Zhang L, Xu S, Zhao W, Liu S (2007) Co-pyrolysis of biomass and coal in a free fall reactor. Fuel 86(3):353–359

The manufacturer's authorised representative in the EU is Springer Nature Customer Service Centre GmbH, Europaplatz 3, 69115 Heidelberg, Germany. If you have any concerns regarding our products, please contact ProductSafety@springernature.com

Printed and bound by CPI Group (UK) Ltd, Croydon, CR0 4YY
30/09/2026
02225449-0010